UNDERWATER ACOUSTICS

UNDERWATER ACOUSTICS

ANALYSIS, DESIGN AND PERFORMANCE OF SONAR

Richard P. Hodges
Sonalysts, Inc.

WILEY
A John Wiley and Sons, Ltd., Publication

This edition first published 2010

Registered office
John Wiley & Sons Ltd, The Atrium, Southern Gate, Chichester, West Sussex, PO19 8SQ, United Kingdom

For details of our global editorial offices, for customer services and for information about how to apply for permission to reuse the copyright material in this book please see our website at www.wiley.com.

Library of Congress Cataloging-in-Publication Data

Hodges, Richard P.
Underwater acoustics : analysis, design, and performance of sonar / by Richard P. Hodges.
p. cm.
Includes index.
ISBN 978-0-470-68875-5 (cloth)
1. Underwater acoustics. 2. Sonar–Mathematical models. 3. Elastic wave propagation. I. Title.
QC242.2.H63 2010
621.389'5–dc22

2010003321

A catalogue record for this book is available from the British Library.

ISBN 978-0-470-68875-5

Typeset in 10/12pt Times by Aptara Inc., New Delhi, India

Contents

About the Author

Richard P. Hodges has an SB degree in physics from the Massachusetts Institute of Technology (MIT), three years of graduate studies in physics at Boston University, and forty years' experience in sonar, operations analysis, modeling, and the simulation of military systems. He is currently working for Sonalysts, Inc. as a principal analyst and is a core member of the Sensor Optimization Working Group (SOWG), which makes recommendations for models to support tactical decision aids for the U.S. Navy. He is a member of the Acoustical Society of America.

He has taught courses at the Naval Undersea Warfare Center (NUWC) and elsewhere in naval analysis of sonar, acoustics, TMA, tactics, weapons, damage and kill mechanisms, C4I, nonacoustic sensors, platform dynamics weapons, tactics, and on the use of NUWC's SIM II Naval Engagement Simulation. He has been one of SIM II's principal architects, engineers, analysts, and a user. He was responsible for the development of the expert systems approach to tactics and for the physical algorithmic modeling of sensors, weapons, environments, fire control systems, and platform characteristics. Mr. Hodges has also developed ASW, ASUW, Land Strike, AAW, and Weapon Models, as well as investigations of the tactical employment of existing and proposed sea-based military systems using SIM II.

Preface

I started writing this book in 2004 when I realized that all of the books I used to teach new sonar analysts were out of print since then several have been made available from other publishers.

In writing this book, I have tried to tread the middle ground between theoretical highly mathematical texts and popular nonmathematical texts. I have included not just graphical results but the accompanying equations that can be used in model development. To present what I consider useful and important results, both theoretical and experimental, I have generally left the details to the references. Over the years, I have spent countless hours reading points from historical graphs and developing curve fits in an effort to save others the time.

I primarily view this book as a source for sonar analysts and those interested in naval operations.

Acknowledgements

I would like to express my thanks to all the giants of the field, to my colleagues at Sonalysts, Inc. and the Naval Undersea Warfare Center, from whom I have tried to learn, to Micheley Angelina for her patience and particularly to Stephanie Schuller, without whom this book might have been unreadable.

1

Introduction to Sonar

SONAR (SOund NAvigation and Ranging) systems have many similarities to radar and electro-optical systems. The operation of sonar is based on the propagation of waves between a target and a receiver. The two most common types of sonar systems are passive and active. In a passive sonar system, energy originates at a target and propagates to a receiver, analogous to passive infrared detection. In an active sonar system, waves propagate from a transmitter to a target and back to a receiver, analogous to pulse-echo radar. In addition to these two types, there is also daylight or ambient sonar, where the environment is the source of the sound, which bounces off or is blocked by the target, and the effects of which are observed by the receiver. This latter type of sonar is analogous to human sight.

Sonar differs fundamentally from radar and electro-optical systems because the energy observed by sonar is transferred by mechanical vibrations propagating in water, solids, gases, or plasma, as opposed to electromagnetic waves. Today, sonar refers not only to systems that detect and/or transmit sound, but to the science of sound technology as well.

In military applications, sonar systems are used for detection, classification, localization, and tracking of submarines, mines, or surface contacts, as well as for communication, navigation, and identification of obstructions or hazards (e.g., polar ice). In commercial applications, sonar is used in fish finders, medical imaging, material inspection, and seismic exploration.

Figures 1.1, 1.2, and 1.3 illustrate the basic passive, active, and daylight/ambient sonar systems.

1.1 Acoustic Waves

The term "acoustic" refers to sound waves in any medium. Acoustic waves come in two types: longitudinal or compression and transverse or shear. In fluids, only longitudinal or compression waves are supported because fluids lack shear strength. The easiest way to visualize these two types of waves is to consider a Slinky (see Figure 1.4). If the end or middle portion of a Slinky is moved side to side or up and down, a transverse or shear wave will move along it. This method displaces the material of the Slinky in a direction perpendicular to the direction of travel. As the material is moved off the axis, the spring force exerts a restoring force that pulls it back on axis. If several of the Slinky coils are compressed or stretched, then releasing them

Underwater Acoustics: Analysis, Design and Performance of Sonar Richard P. Hodges

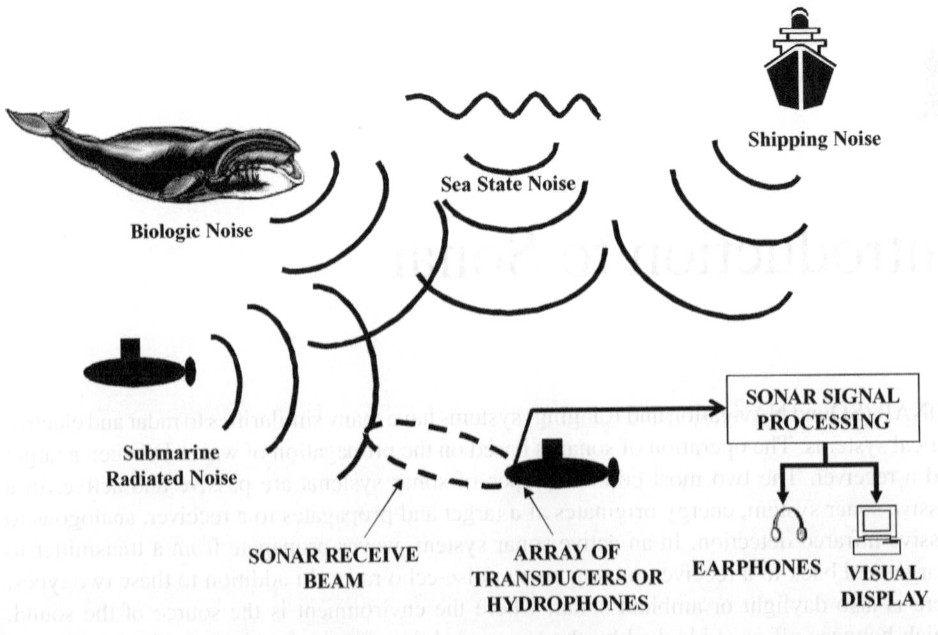

Figure 1.1 Passive sonar system

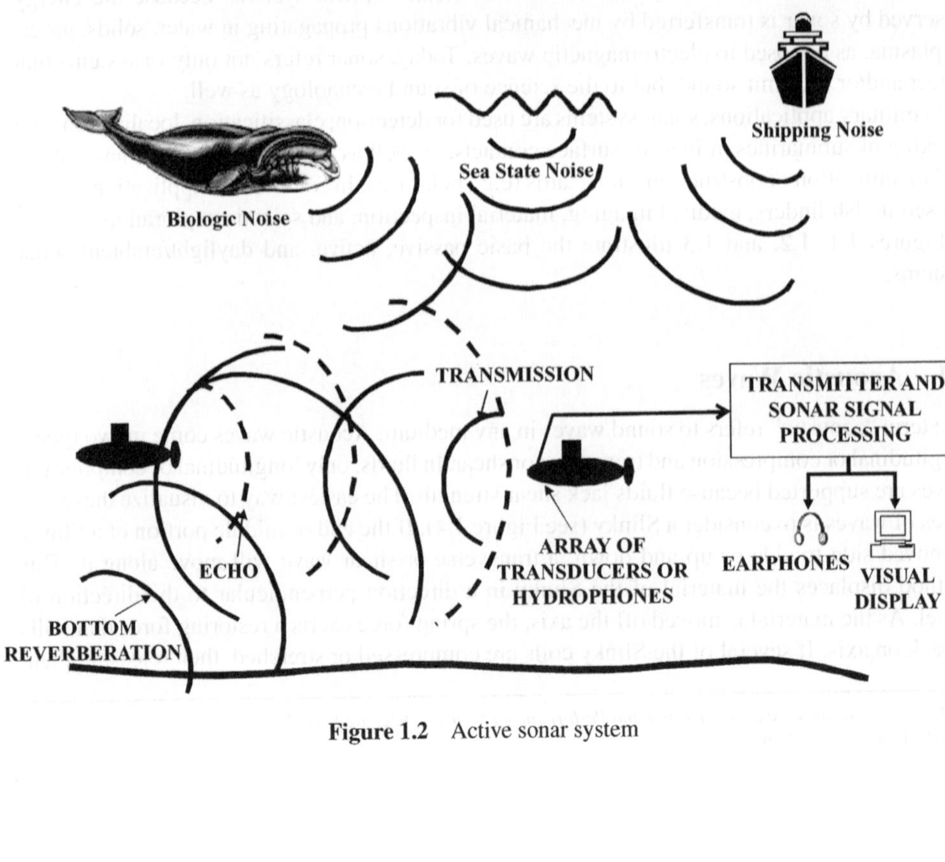

Figure 1.2 Active sonar system

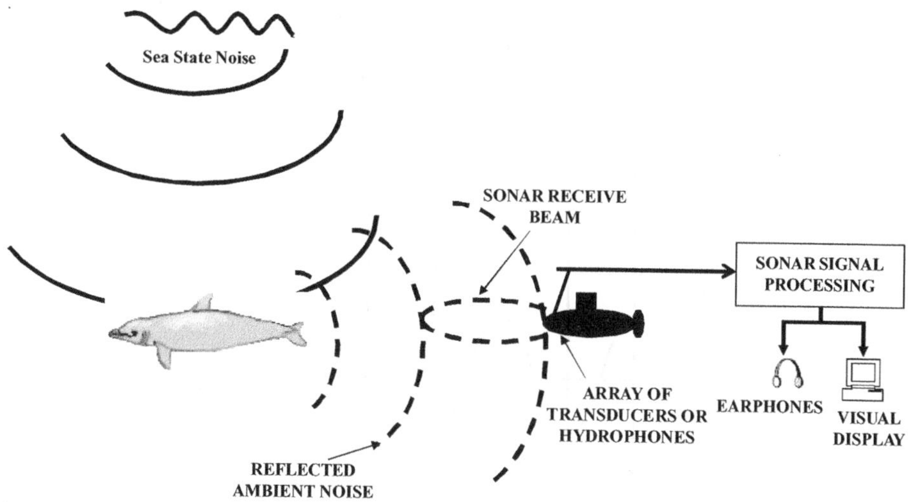

Figure 1.3 Daylight/ambient sonar system

will propagate a longitudinal or compression wave along the Slinky. This method displaces the material of the Slinky along the direction of travel. Again, the restoring force will tend to push the material back into place. In this book, we will deal with transverse or shear waves only occasionally, so unless specifically stated, longitudinal or compression waves are assumed.

1.1.1 Compressions and Rarefactions

Longitudinal waves are composed of compressions, where the parts of the medium (coils of the Slinky) are closer together than normal, and rarefactions, where the parts of the medium are farther apart than normal.

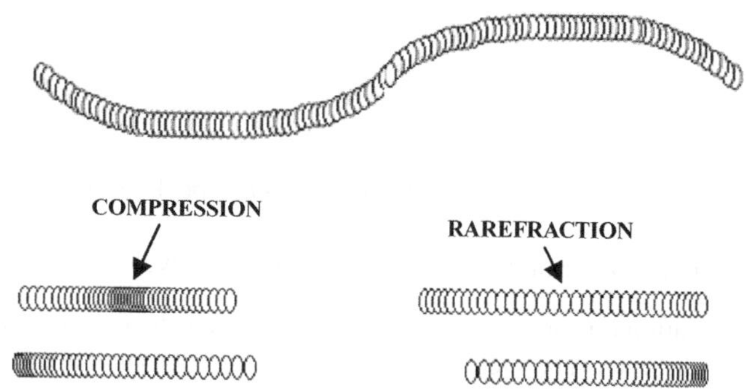

Figure 1.4 Transverse (top) and longitudinal (bottom) waves on a Slinky

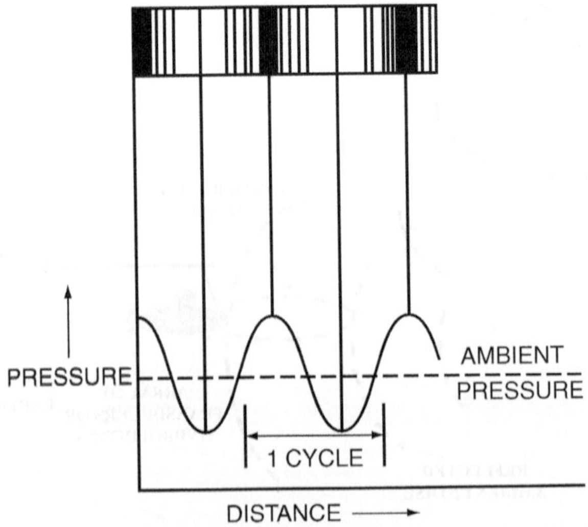

Figure 1.5 Pressure wave

The fundamental parameter of an acoustic wave is pressure. When water or air molecules are pushed or pulled apart, they exert a restoring force that resists the motion. The force will be felt locally as pressure or force per unit area. The amplitude of the wave will be the peak pressure reached in one cycle. The disturbance of the medium that propagates is the distance between molecules. Figure 1.5 illustrates the fundamentals of a pressure wave.

1.2 Speed of Propagation

For a nondispersive medium, one in which different wavelengths propagate at the same phase velocities (e.g., water), we would expect the same type of relationship between wavelength and frequency as with electromagnetic waves:

$$c = \lambda f \tag{1.1}$$

where

λ = wavelength, the distance between corresponding points (peak to peak or valley to valley) on a wave

f = frequency, the number per unit time the wave performs a cycle

The speed of propagation for sound waves is much slower than electromagnetic radiation, on the order of 1500 m/s in water. The speed of propagation is a function of ambient temperature (T), pressure (p) and salinity (S) of the water [1]. Therefore, we can write:

$$c = F(T, p, S) \tag{1.2}$$

Given the complexity of this function, the following rules of thumb apply:

+1 °C change in temperature	= +4.6 m/s at 0.0 °C increase in speed +2.5 m/s at 21.1 °C increase in speed
+100 m of depth increase	= +1.7 m/s increase in speed
+1 ppt (part per thousand) increase in salinity	= +1.4 m/s increase in speed
+1 °F change in temperature	= +8.4 ft/s at 32 °F increase in speed +4.6 ft/s at 70 °F increase in speed
+100 ft of depth increase	= +1.7 ft/s increase in speed
+1 ppt (part per thousand) increase in salinity	= +4 ft/s increase in speed

As these rules show, the greatest variation in speed occurs with changes in temperature. Fluctuations in temperature, by as much as 30 °C, are possible in submarine operational areas. The change in depth required to change the speed of propagation by the same amount is more than 5000 m or 16 000 ft. (Note that large variations in salinity are limited to regions where fresh and salt water mix, e.g., river outflows or under melting sea ice, which are frequently beyond the regions where antisubmarine warfare (ASW) operations take place.)

A simple empirical equation for the speed of sound in sea water, with reasonable accuracy for the world's oceans, is due to Mackenzie [2]:

$$\begin{aligned} c(T, S, D) = {} & A_1 + A_2T + A_3T^2 + A_4T^3 + A_5(S - 35) \\ & + A_6D + A_7D^2 + A_8T(S - 35) + A_9TD^3 \end{aligned} \tag{1.3}$$

where

T = temperature in degrees Celsius
S = salinity in parts per thousand
D = depth in meters

The constants used are

$A_1 = 1448.96$	$A_4 = 2.374 \times 10^{-4}$	$A_7 = 1.675 \times 10^{-7}$
$A_2 = 4.591$	$A_5 = 1.340$	$A_8 = -1.025 \times 10^{-2}$
$A_3 = -5.304 \times 10^{-2}$	$A_6 = 1.630 \times 10^{-2}$	$A_9 = -7.139 \times 10^{-13}$

Solving this equation gives 1550.74 m/s for $T = 25$ °C, $S = 35$ ‰, $D = 1000$ m. The standard error for salinities between 25 and 40 ppt is 0.070 m/s. Other, far more complicated, equations for sound speed in sea water are accurate over a wider range of conditions (e.g., Del Grosso [3] and Chen and Millero [4]).

Table 1.1 shows sound speeds for select liquids at 1 atmosphere and 25 °C, unless otherwise noted.

1.3 Acoustic Wave Parameters

The two fundamental parameters of an acoustic wave are frequency and amplitude, a pressure measured in units of force per area. The System International (SI) unit of pressure is the Pascal

Table 1.1 Sound speeds for select liquids

Liquid	Sound velocity (m/s)	Liquid	Sound velocity (m/s)	Liquid	Sound velocity (m/s)
Acetic acid	1584	Chloroform	995	Phenol	1274
Acetone	1174	Ether	985	Toluene	1275
Alcohol, ethyl (ethanol)	1144	Ethylene Glycol	1644	Turpentine	1240
Alcohol, methyl	1205	Glycerol (Glycerine)	1904	Water	1493
Alcohol, propyl	1205	Heptane	1138	Water, 0 °C	1402
Benzene	1298	Hexane	1203	Water, 20 °C	1482
Carbon disulfide	1149	Kerosene	1324	Water, sea	1533
Carbon tetrachloride	926	Mercury	1450	Water, sea, 20 °C	1522
Castor oil	1474	Octane	1171		

(Pa), where 1 Pa = 1 N/m^2. For those readers not comfortable with this unit, Table 1.2 gives the most commonly encountered units of pressure and their equivalents.

As anyone who has flown in an airplane knows, changes of a fraction of an atmosphere can be quite painful. The human ear is capable of hearing sound with pressure changes as small as 2×10^{-10} atmospheres. This value was used in older literature as a reference pressure for measurements, 0.0002 μbar.

In general, longitudinal waves are the most important type of waves in acoustics, particularly when discussing the underwater environment. Consequently, the analysis in this book will be specifically for this type of wave. When a simple harmonic wave propagates, the magnitude of the acoustic disturbance varies sinusoidally in time at every place in the medium. The spatial distribution of the disturbance at any fixed time is also sinusoidal.

The surface joining regions within the medium undergoing the same amount of perturbation during the same compression or rarefaction cycle is known as a wave front. The shape of the wave front enables the classification of acoustic waves to be subdivided further. Waves generated in a homogeneous medium, from a point source that is very small compared to the wavelength, propagate with spherical symmetry. These wave fronts are spherical in shape and are known as spherical waves. If instead the medium is bounded by two parallel planes (as in the case of the sea), waves generated by a point source will eventually spread with circular symmetry only in the horizontal plane. These wave fronts will be cylinders and are known as

Table 1.2 Common units of pressure

Normal atmospheric pressure atmospheres (atm) = 1.01325 bar
= 1.01325×10^5 Pascal (Pa)
= 1.01325×10^{10} dynes/m^2
= 1.01325×10^6 dynes/cm^2
= 14.6960 pounds/in^2 (psi)
= 29.9213 inches of Hg
= 760 mm of Hg (torr)
= 406.8 inches of water

cylindrical waves. If the source is an infinite plane surface, the resulting wave fronts are also a plane. No spreading occurs and the waves are known as plane waves. Although plane waves cannot be generated in practice, both spherical and cylindrical waves approximate plane waves when they are sufficiently far from their source.

An acoustic pressure wave applies a stress to successive elements of the medium through which it propagates. The resulting particle motion in each element is determined by the mechanical properties of the medium, i.e., its elastic modulus describing the difficulty with which it is compressed and its density (ρ).

In a solid, the elastic modulus is frequently dependent upon the orientation of the medium relative to the acoustic wave. In a completely anisotropic solid, 21 constants are required to specify completely the stress–strain relationship. Being isotropic, fluids and gases require only one elastic constant, compressibility, s (m^2/N), defined as the volumetric strain produced per unit of applied stress:

$$s = \frac{\Delta v / v_0}{p} \tag{1.4}$$

where Δv is the change in the original volume, v_0, caused by the application of a pressure, p. The reciprocal of compressibility is known as the volume elasticity or bulk modulus, κ, and is usually used in acoustic expressions instead of compressibility:

$$\kappa = \frac{1}{s} = \frac{p}{\Delta v / v_0} \tag{1.5}$$

For the following discussion, we will assume that the bulk modulus is constant. This is essentially true for low-amplitude acoustic waves. As an acoustic wave moves though a medium it can be characterized by certain parameters that vary periodically with both time and space.

Particle displacement, ξ, is the amount of displacement of a particle from its mean position within the medium under the action of the acoustic pressure.

Particle velocity, u, is the velocity of a particle in the medium, given by the time derivative of the particle displacement:

$$u = \frac{\mathrm{d}\xi}{\mathrm{d}t} \tag{1.6}$$

Particle acceleration, a, is the time derivative of the particle velocity:

$$a = \frac{\mathrm{d}u}{\mathrm{d}t} = \frac{\mathrm{d}^2\xi}{\mathrm{d}t^2} \tag{1.7}$$

Acoustic or excess pressure, p, is the change in pressure from the mean value. It is the difference between the instantaneous pressure, P, and the ambient pressure, P_0:

$$p = P - P_0 \tag{1.8}$$

Condensation, S, is the fractional change in density resulting from the acoustic pressure:

$$S = \frac{\rho - \rho_0}{\rho_0} \tag{1.9}$$

where ρ and ρ_0 are the instantaneous and mean densities respectively. Since mass must be conserved, this can also be written in terms of volume as

$$S = \frac{\Delta \nu}{\nu_0} \tag{1.10}$$

Combining these last two equations gives an equation relating bulk modulus and excess pressure:

$$p = \kappa S \tag{1.11}$$

Propagation speed, c, is the speed with which the acoustic wave passes through a fluid medium. It is determined by the mechanical properties of the medium, by

$$c = \sqrt{\frac{\kappa}{\rho_0}} \tag{1.12}$$

The derivation of this equation assumes that the propagation takes place at a constant temperature, i.e., isothermal. In reality, the temperature rises during compression and drops during rarefaction. In general, the actual temperature gradient formed is small if the pressure changes are small (this would not be true for a shock wave). This, combined with the short duration between compressions and rarefactions is usually insufficient for any significant heat flow to occur. As a result, a better assumption is that the propagation is an adiabatic process. A more general form for this is

$$c = \sqrt{\frac{\mathrm{d}p}{\mathrm{d}\rho}} \tag{1.13}$$

For isothermal and adiabatic conditions, the excess pressure and density are related by:

$$p = A\,(\rho - \rho_0) \tag{1.14}$$

and

$$P = A'\,(\rho - \rho_0)^{\gamma} \tag{1.15}$$

respectively, where γ is the ratio of specific heats, C_p and C_v, of the medium, measured under constant pressure and constant volume respectively. The isothermal and adiabatic propagation speeds are

$$
\begin{aligned}
c_{\text{is}}^2 &= A = \frac{p}{(\rho - \rho_0)} \\
c_{\text{ad}}^2 &= A'\gamma\,(\rho - \rho_0)^{\gamma-1} = \gamma\frac{p}{(\rho - \rho_0)}
\end{aligned}
\tag{1.16}
$$

or

$$
c_{\text{ad}}^2 = \gamma c_{\text{is}}^2 = \gamma\frac{\kappa}{\rho} \tag{1.17}
$$

For sea water at 13 °C, $\gamma = 1.01$, $\kappa = 2.28 \times 10^9$ Pa (N/m^2), $\rho = 1026$ kg/m^3, and the speed of propagation is 1498 m/s. This value assumes one atmosphere (at the surface) and a salinity of 35 ppt. Note that the calculation assuming isothermal speeds would have been 1491 m/s or an error of about 0.5 %.

1.4 Doppler Shift

If a sound source is moving relative to the medium with a component of v_{s} towards a fixed receiver, the observed frequency must change. At the receiver, the observed frequency (f_{r}) is given by

$$
f_{\text{r}} = f_{\text{s}}\frac{c}{c - v_{\text{s}}} \tag{1.18}
$$

where

c = speed of sound in the medium
f_{s} = source frequency
v_{s} = source relative speed (positive if toward the receiver)

Similarly, if the receiver is moving with a component, v_{r}, relative to the source, the frequency is altered:

$$
f_{\text{r}} = f_{\text{s}}\frac{c - v_{\text{r}}}{c} \tag{1.19}
$$

In general, if both the source and medium are moving, the received frequency is given by

$$
f_{\text{r}} = f_{\text{s}}\frac{c - v_{\text{r}}}{c - v_{\text{s}}} \tag{1.20}
$$

The frequency change resulting from relative motion of the source and receiver is known as the Doppler shift. The change in frequency is defined as

$$
\Delta f = f_{\text{s}}\frac{v_{\text{s}} - v_{\text{r}}}{c - v_{\text{s}}} \tag{1.21}
$$

Frequently, the speeds are very small relative to the speed of sound, therefore

$$\Delta f = f_s \frac{\Delta v}{c} \tag{1.22}$$

where Δv is the combined relative speed component (positive for closing). If frequency is in kHz, Δv is in knots, and Δf is in Hz, then the change in frequency in the ocean is defined as

$$\Delta f(\text{ocean}) \cong 0.35 f_s \Delta v \tag{1.23}$$

The change in frequency in air is about five times larger because of the lower speed of sound:

$$\Delta f(\text{air}) \cong 1.7 f_s \Delta v \tag{1.24}$$

A special case of interest is the Doppler shift as observed by a monostatic active sonar (Figure 1.2), where the signal has twice the shift (out and return):

$$\Delta f(\text{monostatic active sonar in ocean}) \cong 0.7 f_s \Delta v \tag{1.25}$$

1.5 Intensity, SPL, and Decibels

The energy flow in an acoustic wave is similar to that of radar and electro-optics. The power per unit area in an acoustic wave, referred to as intensity, I, varies as the square of the pressure. This relationship is written as

$$I \propto p^2$$

For the purpose of this discussion, we will use a term called sound pressure level, or SPL, which is defined as

$$\text{SPL} = 20 \log \left(\frac{p}{p_0} \right) \tag{1.26}$$

where p_0 is the reference pressure, frequently identified as 1 μPa, 1 μbar, and 0.0002 μbar (see Table 1.2). Here, the units are decibels (dB). Therefore, the exact coefficients of the proportionality are irrelevant, since SPL is the log of a ratio.

It should be noted that a factor of 20 is used because intensity is proportional to the square of the pressure and 10 log of power is the decibels. Decibels express a ratio of powers, which in this case is proportional to the square of pressure. Using the properties of logarithms, the exponent is brought down in front and multiplies the normal factor of 10. For example, 10 log (x^2) = 20 log (x). This is a very important lesson to understand. Too often decibels are manipulated without truly being understood.

If a reference pressure is not stated, then the SPL is **absolutely meaningless**. For example, if an SPL is assumed to have a reference pressure of 1 μPa and it is actually 1 μbar, the numerical value will be off by 120 dB or one has missed the power by a multiplicative factor of 1 trillion. What may have been thought of as 1 milliwatts could actually be 10 megawatts! The problem could be even worse if the focus was on not only a single frequency but power over a band of frequencies or because sometimes measurements are made at a distance (1 yd, 1 m, 100 yd, etc.).

A true description of SPL must include: (1) the reference pressure (or sometimes power density, e.g., W/cm^2), (2) the frequency range over which the power is measured (e.g., 1 Hz or one-third of an octave), and (3) in the case of signals radiated from a source, the reference range being used (e.g., 1 yd). Surprisingly, most people do not know that an octave is a factor of 2 in frequency. In music, middle C (the center of a piano keyboard) is about 261.6 Hz, with each note above it increasing by a twelfth of an octave, a factor of $2^{1/12}$. Seen as a geometric progression, C sharp, which is also D flat, is about 277.2 Hz and C flat is 246.9 Hz, and so forth. Middle A is actually the reference at 440.0 Hz. The third of an octave mentioned above is a factor of $2^{1/3}$ or 1.2599. In other words, the frequency band from 1000 to 1260 Hz is roughly a one-third of the octave band, and would be from 100 to 126 Hz. Another common bandwidth is a tenth decade of $10^{1/10}$, which is a factor of 1.2589 and is so close to the one-third octave that many analysts use the two interchangeably.

Decibels come in two forms: the first form, just discussed, which represents a power density, and the second form, which represents a gain or loss. This latter form is dimensionless. For example, a linear amplifier might provide 20 dB gain; i.e., the signal provided within the linear range of the amplifier will be raised by 20 dB (in power, a factor of 100, or in voltage, a factor of 10). As a result, the value is not dependent on what units are being used for expressing the input level.

Acoustic analysis is done in decibels because it is a convenient way to handle the wide range of values and because the addition of decibels is equivalent to the multiplication of the underlying quantities. For example, the human ear is capable of hearing sound with pressure changes as small as 2×10^{-10} atmospheres. The level at which sound becomes painful is about 120 dB higher. Thus, the human ear has an enormous dynamic range of about 10^{12} in power.

1.6 Combining Acoustic Waves

Consider two acoustic waves arriving at some point in a medium. To compute the combined intensity we must superimpose the signals. At each instant in time, the pressures will add. Therefore, in order to compute the intensity we must sum the pressures and average over time. There are two ways that these signals can combine: coherently and incoherently, the latter sometimes referred to as power addition. In reality, signals always combine coherently in a medium in the sense that the pressure signatures sum. However, averaged over time, the two signals might or might not maintain coherence that is in phase for a sine wave.

Let us start with two sine waves with the same frequency, amplitude and phase, as stated below:

$$S_1 = A \sin(\omega t)$$
$$S_2 = A \sin(\omega t)$$

The average power per unit area, at the reference distance (R) of each individual signal is

$$\text{Power}_1 = \text{Power}_2 = \frac{A^2}{2\rho c R^2} \tag{1.27}$$

$$\text{Power(surface)} = \frac{4\pi R^2 A^2}{2\rho c R^2} = \frac{2\pi A^2}{\rho c} \tag{1.28}$$

When these are added:

$$S_c = S_1 + S_2 = 2A\sin(\omega t) \tag{1.29}$$

the average power per unit area (W) at a distance R is

$$W = \frac{\int_0^{2\pi/\omega} \frac{[2A\sin(\omega t)]^2}{\rho c R^2}\,dt}{\int_0^{2\pi/\omega} 1\,dt} = \frac{2A^2}{\rho c R^2} \tag{1.30}$$

Note that this is four times the power per unit area of each signal individually. If we were to assume that this happens over the surface of a sphere of radius R then we get

$$W(\text{over sphere?}) = 4\pi R^2 \frac{2A^2}{\rho c R^2} = \frac{8\pi A^2}{\rho c} \tag{1.31}$$

Clearly this cannot be the case because energy must be conserved, so what went wrong?

Consider two point sources, one radiating S_1 above and the other S_2, separated by a distance d. The pressure arriving from each (P_1 and P_2) at a point (x, y, z) is given by

$$P_1 = \frac{A\sin(\omega t + \Delta_1)}{\sqrt{x^2 + y^2 + (z - d/2)^2}}$$

$$\Delta_1 = \frac{\sqrt{x^2 + y^2 + (z - d/2)^2}}{c}$$

$$P_2 = \frac{A\sin(\omega t + \Delta_2)}{\sqrt{x^2 + y^2 + (z + d/2)^2}}$$

$$\Delta_2 = \frac{\sqrt{x^2 + y^2 + (z + d/2)^2}}{c} \tag{1.32}$$

Assuming the distances to the spherical surface from each source is very large compared to d, the distances can be replaced by a single value R. Summing the pressure at that point yields

$$P_1 + P_2 = \frac{A}{R}\{\sin[\omega(t + \Delta_1)] + \sin[\omega(t + \Delta_2)]\} \tag{1.33}$$

The average power of the sum is then given by

$$\text{Power(average)} = \frac{\displaystyle\int_0^{2\pi/\omega} \frac{A^2}{\rho c R^2} \{\sin[\omega(t+\Delta_1)] + \sin[\omega(t+\Delta_2)]\}^2 \, dt}{\int_0^{2\pi/\omega} 1 \, dt} \tag{1.34}$$

Converting to polar coordinates:

$$x = R\sin(\theta)\sin(\phi)$$
$$y = R\cos(\theta)\sin(\phi)$$
$$z = R\cos(\phi)$$

gives

$$\Delta_1 c = \sqrt{R^2[\sin(\phi)]^2 + R^2[\cos(\phi)]^2 - Rd\cos(\phi) + d^2/4}$$
$$\Delta_2 c = \sqrt{R^2[\sin(\phi)]^2 + R^2[\cos(\phi)]^2 + Rd\cos(\phi) + d^2/4} \tag{1.35}$$

Since $R \gg d$:

$$\Delta_1 \cong \frac{R}{c}\left[1 - \frac{d\cos(\phi)}{2R}\right]$$
$$\Delta_2 \cong \frac{R}{c}\left[1 + \frac{d\cos(\phi)}{2R}\right] \tag{1.36}$$

Substituting and simplifying yields

$$\text{Power (average)} = \frac{\int_0^{2\pi/\omega} \frac{A^2}{\rho c R^2}\left[4\sin(\omega t)^2 \cos\left(\frac{\omega d\cos(\phi)}{2c}\right)^2\right] dt}{\int_0^{2\pi/\omega} 1 \, dt}$$
$$= \frac{2A^2 \cos\left(\frac{\omega d\cos(\phi)}{2c}\right)^2}{\rho c R^2} \tag{1.37}$$

Integrating over the whole surface yields

$$\text{Power(surface)} = \int_0^{\pi}\int_0^{2\pi} \frac{2A^2\left[\cos\left(\frac{\omega d\cos(\phi)}{2c}\right)\right]^2}{\rho c R^2} R^2 \sin(\phi)\, d\theta \, d\phi \tag{1.38}$$

$$\text{Power(surface)} = \frac{4\pi A^2}{\rho c}\left[1 + \frac{\sin\left(\frac{2\pi d}{\lambda}\right)}{\left(\frac{2\pi}{\lambda}\right)}\right] \tag{1.39}$$

Since $\omega = 2\pi f$ and $\lambda = c/f$.

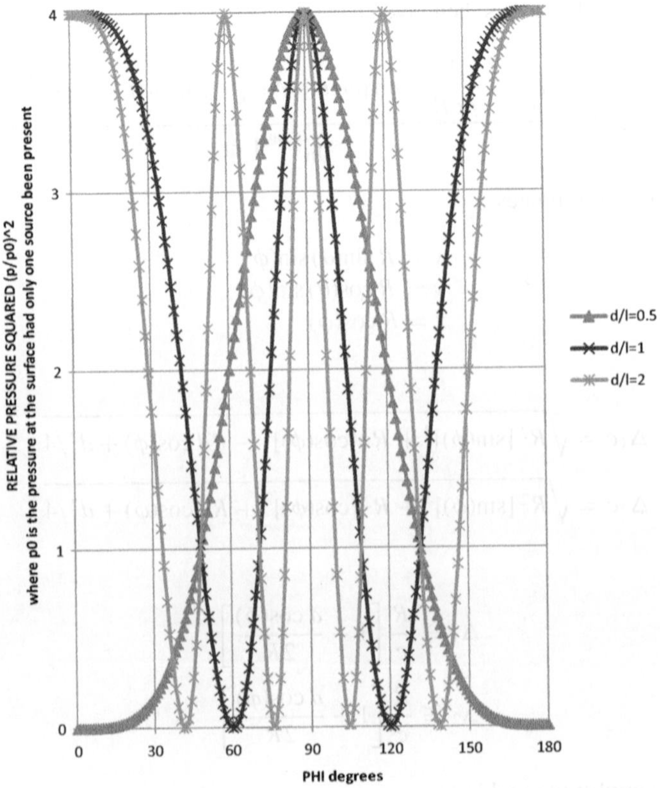

Figure 1.6 Power versus conical angle for two separated equal intensity in-phase sources

The first term is the sum of the powers of the two sources individually, as expected. The resolution to the possible violation of the conservation of energy is simply when two sources combine; the power per unit area will vary from zero (out of phase) to four times the individual power density (in phase).

The second term is the near field effect. If the two sources are very close together, more power is required to make each source put out the desired level because of the pressure field generated by the other source. This term disappears if the separation, d, is an integer multiple of half the wavelength. This term becomes very small if $d \gg \lambda$, such as when the interaction between the sources becomes very small. Figure 1.6 shows the power per unit area for a north–south oriented dipole located at the center of a sphere as a function of the angle from the pole.

1.7 Comparative Parameter for Sound in Water and Air

Table 1.3 shows select sound levels of interest for water. Table 1.4 shows similar levels and corresponding parameters of air.

Table 1.3 Select sound pressure levels and associated parameters for water (impedance of sea water = $\rho c = 1.5 \times 10^6$ kg/m^2 s)

Sound level	Intensity I(W/m^2)	Pressure rms p(N/m^2)	Particle velocity u(m/s)	Particle displacement u/ω (m) at 3 kHz
Sea state 3 (50 dB re: 1 μPa)	6.7×10^{-14}	3.2×10^{-4}	2.1×10^{-10}	1.1×10^{-14}
Active source (220 dB re: 1 μPa)	6.7×10^{3}	1.0×10^{5}	6.7×10^{-2}	3.5×10^{-6}

Note. The pressure for active source is 1 atmosphere.

Table 1.4 Select sound pressure levels and associated parameters for air (impedance of sea water = $\rho c = 415$ kg/m^2 s)

Sound level	Intensity I(W/m^2)	Pressure rms p(N/m^2)	Particle velocity u(m/sec)	Particle displacement u/ω (m) at 440 Hz
Threshold of human hearing (26 dB re: 1 μPa)	9.6×10^{-13}	2.0×10^{-5}	4.8×10^{-5}	1.7×10^{-11}
Normal conversation (106 dB re: 1 μPa)	9.6×10^{-7}	2.0×10^{-2}	4.8×10^{-5}	1.7×10^{-8}
Rock band / threshold of pain (146 dB re: 1 μPa)	1	2.0×10^{1}	4.8×10^{-2}	1.7×10^{-3}

Note. Human hearing works over a vast power range, 10^{12}. The particle displacement at the threshold of human hearing at middle C is about 6 % of the diameter of a hydrogen molecule, so the ear is able to detect tiny movements.

References

[1] Wong, G. S. K., and Zhu, S.-m., "Speed of Sound in Seawater as a Function of Salinity, Temperature and Pressure," *Journal of the Acoustical Society of America*, **97**, March 1995, 1732.

[2] Mackenzie, K. V., "Discussion of Sea-Water Sound-Speed Determinations," *Journal of the Acoustical Society of America*, **70**, 1981, 801–806.

[3] Del Grosso, V. A. "New Equation for Speed of Sound in Natural Waters (with Comparisons to Other Equations)," *Journal of the Acoustical Society of America*, **56**, 1974, 1084–1091.

[4] Chen, C. T., and Millero, F. J., "Speed of Sound in Seawater at High Pressures," *Journal of the Acoustical Society of America*, **62**, 1977, 1129.

2

The Sonar Equations

The sonar equations are a method of combining all parts of the sonar process. These equations are used to estimate performance and to assist in the design of sonar systems. These equations allow one to calculate the performance of a system–target–environment combination.

2.1 Signal-to-Noise Ratio

First, we start with the passive sonar equation because it is simpler and will allow us to define some terms and notation. Of fundamental importance to sonar is the signal-to-noise power ratio (SNR) achieved:

$$S/N = \frac{\text{Signal}}{\text{Prop Loss}} \times \frac{\text{Array Gain}}{\text{Total Noise}} \tag{2.1}$$

In dB, this would be

$$\text{SNR} = \text{Ls} - \text{Nw} - \text{Ln} + \text{NDI (Horton [1])}$$

or

$$\text{SNR} = \text{SL} - \text{TL} - \text{NL} + \text{DI} \quad \text{(Urick [2])} \tag{2.2}$$

where

SNR = 10 log (S/N) = signal-to-noise ratio (dB)
Ls = SL = 10 log (Signal) = radiated signal (dB)
Nw = TL = 10 log (Propagation loss) = transmission or propagation loss (dB)
Ln = NL = 10 log (Total noise) = total noise (dB)
NDI = DI = 10 log (Array gain) = directivity index (dB)
= gain in signal-to-noise ratio due to the antenna array

Each of these terms will be addressed in further detail in subsequent chapters.

Equation (2.1) simply stated is a radiated signal originating at a target is diminished by transmission or propagation loss as it travels to the receiver. The noise level at the receiver

Underwater Acoustics: Analysis, Design and Performance of Sonar Richard P. Hodges

may degrade the signal received, while the array gain at the receiver may improve the signal-to-noise ratio. The first variation of this equation uses the notation developed by Horton in 1957 [1]. The second variation uses the notation developed by Urick in 1967 [2]. Throughout this text, both sets of notation will be used to limit confusion and allow for comparison with other sources. In Horton's notation, any variable starting with the letter "N" is a gain or loss and is therefore dimensionless, whereas any variable starting with the letter "L" is a level and therefore has dimensions (see Chapter 1, Section 1.5, Intensity, SPL, and Decibels).

A signal (Ls or SL) is what we are looking for; it could be a submarine or a whale, depending on our objective. The radiated signal is dependent on several factors: the type of target, its operating mode, speed, depth, acceleration, frequency, aspect (0° for the bow of a target, 90° for the beam of a target), and the up/down or depression/elevation (D/E) angle. A signal comes in many forms; it may be continuous, intermittent, or transient in duration. For example, a broadband signal, as the name suggests, has acoustic energy over a wide range of frequencies, whereas a narrowband signal presents as tones. Most targets emit several types of signals. Many sonar systems are designed to detect all of these types over a wide range of frequencies (see Chapter 10, Radiated Noise, for more details).

Transmission or propagation loss (Nw or TL) includes all the effects of energy spreading out, being absorbed, or scattered (other contributors to this term will be discussed in Chapter 5, Transmission Loss, and Chapter 6, Transmission Loss: Interaction with Boundries). This loss term is highly dependent on the parameters of the environment: sound speed versus depth (the sound speed profile (SSP) is often inappropriately called the SVP, or sound velocity profile), volume absorption/scattering, bottom loss, surface loss, frequency, and depths (transmitter, target, and receiver). It is this term, in particular, that makes sonar much more difficult to predict and operate as compared to radar and has necessitated the development of very complex programs and databases to allow for its estimation.

Total noise (Ln or NL) is simply everything we are not looking for. Fish are a source of noise if we are looking for submarines and submarines are a source of noise if we are looking for fish. As will be discussed in Chapter 11, Self Noise, and Chapter 7, Ambient Noise, total noise is made up of many components, making it a function of the specific sensor self noise (flow, electrical, structure borne), sea state, shipping level, relative bearing, D/E angle, and frequency.

The traditional NDI and DI terms are used here for noise gain, but strictly speaking these terms are defined as gain against omnidirectional far-field noise. A more general term, array gain (Nag or AG), takes into account directional and correlation characteristics of the various components of the noise field. The noise gains observed here are a function of array dimensions, frequency, D/E angle, and the horizontal angle relative to the array. In general, we need to combine each source of noise with its appropriate gain and then combine the results using power addition (see Chapter 1, Section 1.5, Intensity, SPL, and Decibels). As we will see later, in Chapter 3, Transducers, Directionality, and Arrays, it is necessary to define different gains for different types of noise. For example, arrays perform differently toward near-field flow-related noise than towards far-field ambient noise.

2.2 Active Sonar Equation

The active sonar equation is the same as the passive sonar equation in the sense that the signal-to-noise ratio is calculated. When calculating the active sonar equation, the signal starts at the transmitter, propagates to the target, is reflected, and then propagates back to the receiver. This transmitted signal, normally very powerful, is also reflected by other elements in the environment, like the bottom or surface of the ocean, generating additional "noise," called reverberation. As we will explore in later chapters, noise is any signal that is not the signal of interest. For the purposes of this initial discussion, we will assume a monostatic active sonar, namely one in which the transmitter and receiver are co-located. Thus,

$$S/N = \frac{\text{Signal}}{\text{Prop Loss}} \times \frac{\text{Array Gain}}{\text{Total Noise}} \tag{2.3}$$

In dB, this would be

$$\text{SNR} = \text{Lp} - 2\text{Nw} + \text{Nts} - \text{Ln} + \text{Nag} \quad \text{(Horton [1])}$$

or

$$\text{SNR} = \text{SL} - 2\text{TL} + \text{TS} - \text{NL} + \text{AG} \quad \text{(Urick [2])} \tag{2.4}$$

where

SNR = signal-to-noise ratio (dB)
Lp = SL = source level (dB)
Nw = TL = transmission or propagation loss (dB)
Nts = TS = target strength (dB)
Ln = NL = total noise (dB)
Nag = AG = 10 log (Array gain) = array gain (dB)
= gain in the signal-to-noise ratio due to the antenna array

Array gain (Nag or AG) has been used here instead of directivity because the gain against ambient noise and reverberation will almost certainly be different. Given this consideration, the active sonar equation is actually more complex in that the terms (–Ln + Nag) or (−NL + AG) must be expanded as

$$-\text{Ln} + \text{Nag} = -\text{NL} + \text{AG} = 10 \log \left[10^{(\text{Self}-\text{AG}_{\text{self}})/10} + 10^{(\text{Ambient}-\text{AG}_{\text{amb}})/10} + 10^{(\text{Reverberation}-\text{AG}_{\text{rev}})/10}\right]$$

where each of the terms has been converted from decibels to power, adjusted by the appropriate array gain, summed, and then converted back to decibels. Even this is an oversimplification of the equation because the array gain against reverberation is also a function of its source (surface or bottom or volume scattering), as will be discussed in Chapter 8, Reverberation.

Active sonars come in three forms: monostatic, bistatic, and multistatic. Monostatic active sonars use the same array for both transmitting and receiving. Bistatic active sonars use different arrays for transmitting and receiving; an example would be the Sonar 2087 on the Royal Navy Type 23 frigates. The bistatic sonar equation would then replace "two times the propagation loss" with the sum of the propagation loss from the transmitter to the target and the propagation loss from the target to the receiver. Multistatic active sonar is most often a single transmitter with many receivers distributed in space. In this case, multiple bistatic sonar equations are used, one for each combination of the source and receiver.

2.3 Signal Excess

In order to perform a function such as detection or tracking, there is a required SNR to achieve a specified level of performance. Signal excess is simply the amount one has relative to the requirement. For detection, the required amount is referred to as the detection threshold (DT), recognition differential (Nrd or RD), or signal differential (Nsd or SD). The most common definition of this term is the required signal-to-noise ratio for a 50 % probability of detection in a specified time period, at a specified probability of false alarm, for a specific system/signal and bandwidth. We will find in Chapter 12, Statistical Detection Theory, and Chapter 13, Methodology for Calculation of the Recognition Differential, that Nrd is a function of bandwidth, time, signal processing, and false alarm rate. Consequently, the passive sonar equation is usually written as

$$\mathrm{XS} = \mathrm{Ls} - \mathrm{Nw} - \mathrm{Ln} + \mathrm{NDI} - \mathrm{Nrd} \quad \text{(Horton[1])}$$

or

$$\mathrm{SE} = \mathrm{SL} - \mathrm{TL} - \mathrm{NL} + \mathrm{DI} - \mathrm{DT} \quad \text{(Urick [2])} \tag{2.5}$$

where

XS = SE = signal excess (dB)
= the amount of extra SNR above the required amount (Nrd or DT)

By this definition, if the signal excess is greater than zero, the probability of detection is greater than 50 %. In Chapter 16, Modeling Detection and Tactical Decision Aids, and Chapter 17, Cumulative Probability of Detection, we will discuss how to estimate the probability of detection for nonzero values of signal excess.

The active sonar equation is usually written as

$$\mathrm{XS} = \mathrm{Lp} - 2\mathrm{Nw} - \mathrm{Ln} + \mathrm{Nag} - \mathrm{Nrd}$$

or

$$\mathrm{SE} = \mathrm{SL} - 2\mathrm{TL} - \mathrm{NL} + \mathrm{AG} - \mathrm{DT} \tag{2.6}$$

Table 2.1 shows the definitions, parameters and reference locations for each variable in the sonar equation. An additional complication is that reverberation changes with time (range) and is proportional to the source level. Therefore, if a target is reverberation limited, meaning the reverberation experienced is much louder than other noise, then increasing the source level (Lp or SL) will not change the signal excess. Instead, both the signal and reverberation will increase by exactly the same amount.

2.4 Figure of Merit

The term “figure of merit” (FOM) is frequently encountered when discussing the sonar equations. Simply stated, FOM is the amount of transmission loss that will result in a signal excess

Table 2.1 Sonar variables. (Reference intensity is that of a plane wave with the rms pressure level of the reference pressure, usually 1 μPa)

Variable	Definition	Reference distance
Source level (Ls or Lp or SL)	10 log(intensity at source/reference intensity)	Distance from source, typically 1 yd or 1 m
Transmission loss (Nw or TL)	10 log(intensity of signal at reference distance/intensity of signal at receiver or target)	Distance from source, typically 1 yd or 1 m
Noise level (Ln or NL)	10 log(intensity of noise/ reference intensity)	At hydrophone
Target strength (Nts or TS)	10 log(intensity of the echo at the reference distance from target/incident intensity)	Distance from acoustic center of target, typically 1 yd or 1 m
Reverberation (Lr or RL)	10 log(intensity of reverberation at receiver/reference intensity)	At hydrophone
Detection threshold (Nrd or DT)	10 log(signal-to-noise power at output of array)	At array output
Array gain (Nag or AG)	10 log(gain in signal to noise relative to a single omnidirectional hydrophone)	At array output
Directivity index (NDI or DI)	Same as AG but noise is assumed to be uniform in all directions (omnidirectional)	At array output

of zero dB. By knowing the FOM, we can look at the propagation loss versus range and determine where we can expect to detect the target. FOM is calculated by setting the signal excess to zero and solving the sonar equations for transmission loss:

$$\text{FOM} = \text{SL} - \text{NL} + \text{DI} - \text{NRD} = \text{Ls} - \text{Ln} + \text{NDI} - \text{Nrd Passive}$$

or

$$\text{FOM} = \text{SL} - \text{NL} + \text{AG} - \text{DT} = \text{Lp} - 2\text{Nw} - \text{Ln} + \text{Nag} - \text{Nrd Active} \tag{2.7}$$

The figure of merit is most useful for passive sonar calculations because the terms on the right-hand side of the equation are fairly independent of the target range. For an active sonar, this is not the case, as reverberation is a function of target range. (Note that although the name implies that FOM is a measure of the sonar, the FOM equation includes target and environmental characteristics, so is only a good comparative metric for that target and environment.)

References

[1] Horton, J. W., *Fundamentals of Sonar*, Annapolis, MD: United States Naval Institute, 1957.
[2] Urick, R. J., *Principles of Underwater Sound for Engineers*, New York: McGraw-Hill, 1967.

3

Transducers, Directionality, and Arrays

A sonar could be as simple as a single omnidirectional hydrophone, but most applications benefit from, or require, directional functionality. Directionality allows for the determination of the bearing of the target and reduces noise by not responding equally in all directions. Directionality can be obtained from a single hydrophone if the hydrophone is large enough, or mechanically using reflectors or lenses. The most common sonars today are comprised of an array of hydrophones combined with appropriate time delays to form beams in the desired direction or multiple directions. This chapter will focus on this type of sonar. Due to the requirement of directionality, such arrays must have processing to enable searching on all bearings of interest. In older systems, this was accomplished by electronically or mechanically steering the beam through the angles of interest. With the advent of inexpensive computer chips, modern systems are able to electronically form beams in all directions simultaneously. This is referred to as pre-formed beam processing or DIMUS (DIgital MUltibeam Sonar).

The word "transducer" describes devices that convert energy from one form to another. For our use, a transducer converts acoustic energy into electrical energy or vice versa. Transducers used exclusively as receivers are called hydrophones and those used exclusively for transmitting are called transmitters or projectors. The use of these terms is not consistent and hydrophone or transducer is likely to be used for either of these versions. However, it is common for the same transducers to serve both functions in an array or group of transducers. This type of transducer is said to have the characteristic of reversibility. Figure 3.1 shows the bow arrays for the U.S. AN/BSY-2 sonar on the Seawolf Class. The large sphere is the passive element of the system. Notice the large number of transducers on the sphere; these are beamformed to look at many horizontal and vertical angles. The framework surrounding the sphere is for the conformal array, which has fewer more widely space transducers, which, as we shall see, suggests a lower frequency of operation. On the submarine these would both be inside an enclosing sonar dome. Figure 3.2 shows a U.S. Navy SH-60 helicopter with a compact dipping sonar, whose transducers are not visible because of the covering sonar dome.

Underwater Acoustics: Analysis, Design and Performance of Sonar Richard P. Hodges

Figure 3.1 U.S. Navy spherical and conformal sonar arrays for the *Seawolf* Class SSN. (U.S. Navy photo)

Figure 3.2 U.S. Navy SH-60 helicopter with dipping sonar. (U.S. Navy photo)

3.1 Transducer Response

Another desirable characteristic for a transducer is linearity; i.e., the output is directly proportional to the input as opposed to some power of the input. For linear transducers, the sensitivity or response is the proportionality constant expressed in decibels. The receive response is expressed as the number of decibels relative to 1 volt produced by an acoustic pressure of a standard level (usually either 1 microPascal (μPa) (10^{-6} N/m^2) or 1 dyne/cm^2). It is typically written as –100 dB relative to 1 V/μPa. This means that the open circuit *rms* voltage out of the transducer is 10^{-5} V when placed in a plane wave acoustic field having an *rms* pressure of 1 microPascal. For projectors, the response is referenced to a standard distance from the transducer, normally 1 meter or 1 yard (a difference of 0.89 dB). This measurement requires measuring the pressure in the far field and adjusting it back to the reference distance using the propagation loss. Projector sensitivity or response is expressed as the number of decibels relative to a standard pressure, usually 1 μPa or 1 dyne/cm^2 at the reference distance (1 yd or 1 m) produced by a 1 ampere current into the transducer or by 1 volt at the input. Transmitter or projector responses are written as: 100 dB relative to 1 μPa/ampere at 1 m, or 100 dB relative to 1 μPa/volt at 1 m, which means a 1 ampere current or 1 volt will generate 10^5 μPa at a distance of 1 meter.

3.2 Beam Pattern Response

Individual transducers and arrays of transducers may, and in general do, have outputs that are a function of direction. This property is desirable in order to improve the received signal-to-noise ratio, to allow the determination of the direction of signals, or to limit the directional distribution of transmitted energy in the case of projectors. If $R(\theta, \phi)$ is the output of a transducer or an array in polar coordinates, it is normal to write

$$R(\theta, \phi) = R(0, 0)\, r(\theta, \phi) \tag{3.1}$$

where $R(0, 0)$ is the response in the direction 0, 0 and $r(\theta, \phi)$ is the normalized response in the direction θ, ϕ, that is, $r(0, 0) = 1$. It is conventional to define the coordinate system such that 0, 0 is the direction of maximum output, known as the maximum response axis (MRA). The power beam pattern response or pattern function is then r squared:

$$B(\theta, \phi) = [r(\theta, \phi)]^2 \tag{3.2}$$

$B(\theta, \phi)$ is the mean square voltage out for a receiver or the mean square pressure at the reference distance for a projector. It is common to express this in decibels as a beam pattern correction (BPC) relative to the maximum response axis:

$$\mathrm{BPC}(\theta, \phi) = 10 \log [B(\theta, \phi)] \tag{3.3}$$

Directionality arises from the beamforming process because of the difference in properties between noise and signals. Noise from random flow near the array, or from ambient noise arising from the sea surface, arrives at the transducers with little correlation. Signals from spatially discrete sources arrive with well-defined or high correlations at the hydrophones of an array. For example, a plane wave sinusoidal signal arriving normal to a planar array of hydrophones will have the same phase and amplitude at each hydrophone. Noise arriving at these same hydrophones will have random phase and amplitude. Consider two hydrophones, separated by a distance, *d*. If the signal arriving at each hydrophone has the same phase (ϕ) then the output of this array for the signal will be

$$S_{\text{array}} = 2S\sin(\omega t + \phi) \tag{3.4}$$

If we square this value and average it over time, the mean square output is given by

$$\frac{\int_0^{2\pi/\omega}\int_0^{2\pi}[2S\sin(\omega t + \phi)]^2\,d\phi\,dt}{\int_0^{2\pi/\omega}\int_0^{2\pi}d\phi\,dt} \tag{3.5}$$

$$\text{Power(signal)} = 2S^2 \tag{3.6}$$

A single hydrophone would not have the 2 in front of the *S* in the integral and therefore would have an average power out of $S^2/2$. Consequently, the signal power out has been multiplied by a factor of four.

For noise, the average power is given by

$$\frac{\int_0^{2\pi/\omega}\int_0^{2\pi}[S\sin(\omega t) + S\sin(\omega t + \phi)]^2\,d\phi\,dt}{\int_0^{2\pi/\omega}\int_0^{2\pi}l\,d\phi\,dt} \tag{3.7}$$

$$\text{Power(noise)} = S^2 \tag{3.8}$$

The noise output has only doubled, so the signal to noise has been improved by a factor of two. This gain is referred to as the array gain. If the noise is truly uncorrelated and the signal has a correlation of one, the improvement is equal to the number of hydrophones summed with unit gain.

Earlier, we assumed that the signal phase was the same at the two hydrophones. If this is not true because the signal is not arriving at the two hydrophones at the same time, we can adjust it by adding the correct time delay (or fixed phase shift in the case of a sine wave) to the first hydrophone prior to summing (see Figure 3.3).

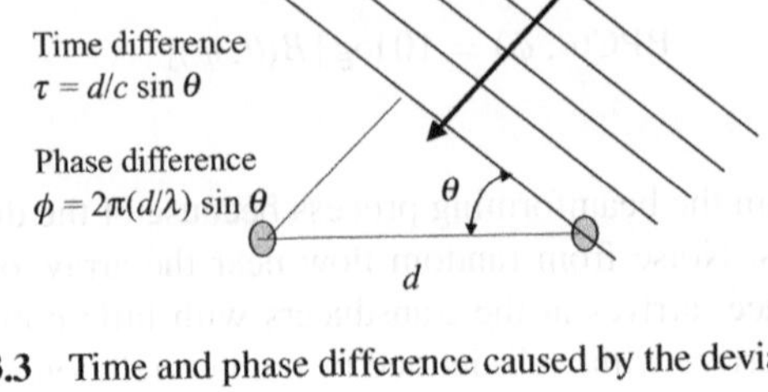

Figure 3.3 Time and phase difference caused by the deviation angle

3.3 Linear Arrays

The beam pattern for an array composed of m evenly spaced hydrophones in a line can be calculated as follows. The signal arriving at each hydrophone is given by

$$S_1 = S\cos\left[\omega\left(t + \tau_j\right)\right] \tag{3.9}$$

where

$\tau_j = j\frac{d}{c}\sin(\theta)$
$\theta =$ conical angle or polar angle measured from the line

The arrival time is independent of which side the signal comes from. A line array therefore suffers from a left–right ambiguity, as well as a depression/elevation angle (D/E angle) ambiguity. Line arrays measure a conical angle only. If the hydrophones are added with unity weighting, the output will be given by

$$\begin{aligned} R(\theta) &= \sum S\cos\left[\omega\left(t + \tau_j\right)\right] = S\sum\cos(\omega t)\cos(\omega\tau_j) - S\sum\sin(\omega t)\sin(\omega\tau_j) \\ R(\theta) &= S\cos(\omega t)\sum\cos(\omega\tau_j) - S\sin(\omega t)\sum\cos(\omega\tau_j) \end{aligned} \tag{3.10}$$

where

$\omega\tau_j = \omega j\, d/c\, \sin(\theta)$

Two useful trigonometric relationships for dealing with the sum of sines and cosines are

$$\begin{aligned} \sum_{j=0}^{N-1}\cos(jx) &= \frac{\sin\left(\frac{Nx}{2}\right)\cos\left[(N-1)\frac{x}{2}\right]}{\sin\left(\frac{x}{2}\right)} \\ \sum_{j=0}^{N-1}\sin(jx) &= \frac{\sin\left(\frac{Nx}{2}\right)\sin\left[(N-1)\frac{x}{2}\right]}{\sin\left(\frac{x}{2}\right)} \end{aligned} \tag{3.11}$$

Setting $jx = \omega j\, d/c\, \sin(\theta)$ yields

$$R(\theta) = S\cos(\omega t)\frac{\sin\left(\frac{Nx}{2}\right)\cos\left[(N-1)\frac{x}{2}\right]}{\sin\left(\frac{x}{2}\right)} - S\sin(\omega t)\frac{\sin\left(\frac{Nx}{2}\right)\sin\left[(N-1)\frac{x}{2}\right]}{\sin\left(\frac{x}{2}\right)} \tag{3.12}$$

$$R(\theta) = \frac{\sin\left(\frac{Nx}{2}\right)}{\sin\left(\frac{x}{2}\right)}\cos\left[\omega t + \frac{(N-1)x}{2}\right] \tag{3.13}$$

Given that the second term is simply the cosine wave, the first term is the angular dependence:

$$R(\theta) = \frac{\sin\left[N\pi d\dfrac{\sin(\theta)}{\lambda}\right]}{\sin\left[\pi d\dfrac{\sin(\theta)}{\lambda}\right]} \tag{3.14}$$

since

$$\lim_{x\to 0}\frac{\sin\left(\dfrac{Nx}{2}\right)}{\sin\left(\dfrac{x}{2}\right)} = N \tag{3.15}$$

The normalized response is

$$r(\theta) = \frac{\sin\left[N\pi d\dfrac{\sin(\theta)}{\lambda}\right]}{N\sin\left[\pi d\dfrac{\sin(\theta)}{\lambda}\right]} \tag{3.16}$$

The form sin(x)/x will occur in many equations in this book. Important values of x are 1.391 559, 1.895 494, and 2.318 578, which yield –3 dB, −6 dB and –10 dB respectively.

The power response is

$$B(\theta) = [r(\theta)]^2 = \left\{\frac{\sin\left[N\pi d\dfrac{\sin(\theta)}{\lambda}\right]}{N\sin\left[\pi d\dfrac{\sin(\theta)}{\lambda}\right]}\right\}^2 \tag{3.17}$$

$B(\theta)$ is the mean square voltage out for a receiver or the mean square pressure at the reference distance for a projector. It is common to express this value in decibels as a beam pattern correction (BPC) relative to the maximum response axis:

$$\mathrm{BPC}(\theta) = 10\log\left\{\left[\frac{\sin\left(N\pi d\dfrac{\sin(\theta)}{\lambda}\right)}{N\sin\left(\pi d\dfrac{\sin(\theta)}{\lambda}\right)}\right]^2\right\} \tag{3.18}$$

Figure 3.4 shows the beam pattern correction versus angle for several choices of N, with an even spacing of $d = \lambda/2$ with equal weighting (no shading) on each hydrophone in a linear multispot array. No shading means that all hydrophones are summed equally. As we will explore later, there are reasons not to do this. Note that the first side lobes have about the same level, −13.46 dB if N is large. As expected, the main lobe becomes narrower and narrower as the number of hydrophones increases.

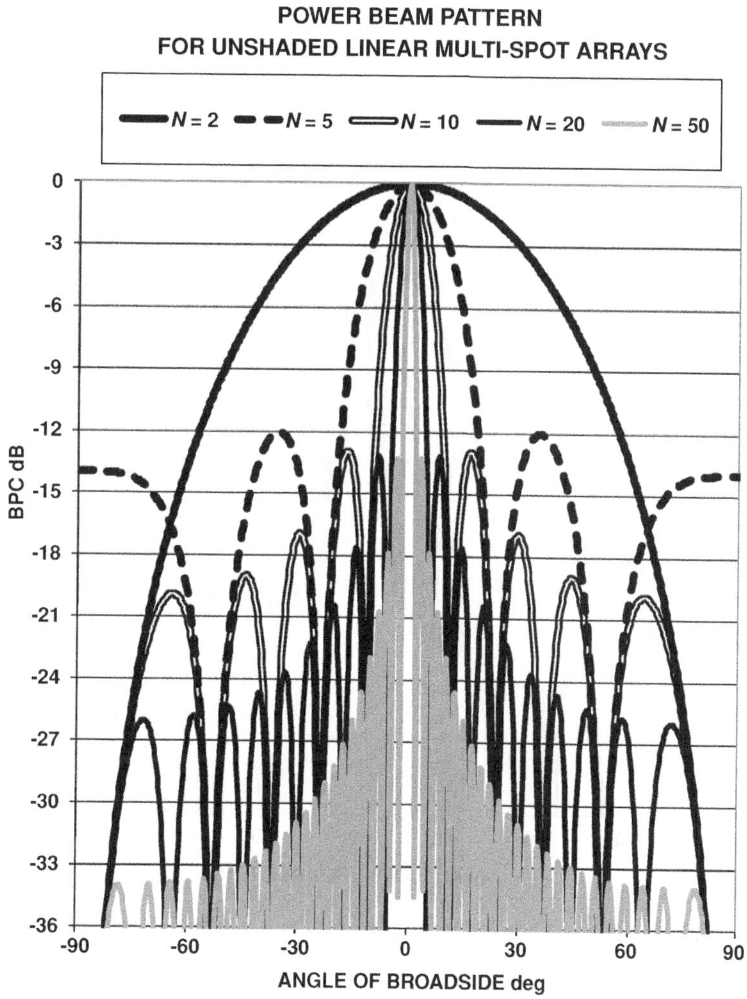

Figure 3.4 Power beam pattern for unshaded, linear Multispot arrays

Figure 3.5 shows $N = 20$ equal weightings (no shading) for various hydrophone spacings. A spacing of $\lambda/2$ is normally referred to as the "design frequency" of a linear array. If the frequencies above the design frequency are processed, large side lobes appear; particularly if the array is steered off broadside. The figure shows side lobes that are equal to the main lobe appear for twice the design frequency or one wavelength spacing. When the array is operated below the design frequency, the beams become wider in inverse proportion to the frequency.

Steering or pointing the maximum response axis is accomplished by applying a time delay to each hydrophone such that a signal from the steered direction will be added in phase. For a linear multispot array with uniform spacing (d), the delays (τd) need to be

$$td\ (j\text{th hydrophone}) = j\ d/c\ \sin(\varphi) \tag{3.19}$$

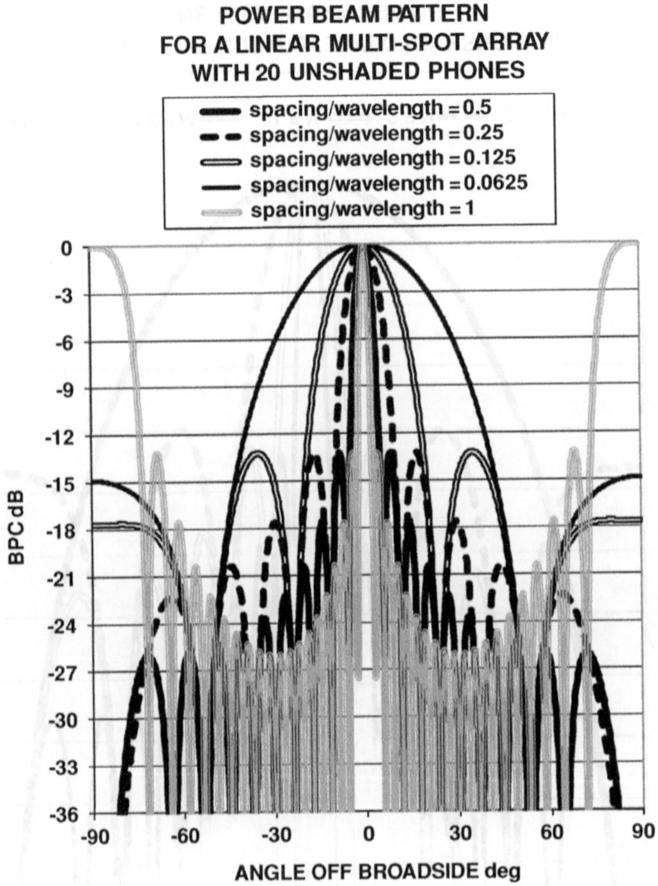

Figure 3.5 Power beam pattern for linear multispot array with 20 unshaded hydrophones

where

φ = steered direction
d = distance between hydrophones
$(N-1)d$ = length of the array

$$R(\theta) = \sum S\cos\left[\omega\left(t+\tau_j-\tau_d(i)\right)\right]$$
$$R(\theta) = S\sum\cos(\omega t)\cos\left[\omega(\tau_j-\tau_d(i)\right] - S\sum\sin(\omega t)\cos\left[\omega(\tau_j-\tau_d(i)\right] \quad (3.20)$$
$$R(\theta) = s\cos(\omega t)\sum\cos(\omega T_j) - S\cos(\omega t)\sum\sin(\omega T_j)$$

where:

$$\omega\tau_j = \omega j\frac{d}{c}\left[\sin(\theta)-\sin(\phi)\right]$$

So letting

$$jx = \varpi j\frac{d}{c}\left[\sin(\theta)-\sin(\phi)\right]$$

yields

$$R(\theta, \phi) = S\cos(\omega t) - \frac{\sin\left(N\frac{x}{2}\right)\cos\left[(N-1)\frac{x}{2}\right]}{\sin\left(\frac{x}{2}\right)} - S\sin(\omega t)\frac{\sin\left(N\frac{x}{2}\right)\sin\left[(N-1)\frac{x}{2}\right]}{\sin\left(\frac{x}{2}\right)}$$

$$R(\theta, \phi) = \frac{\sin\left(N\frac{x}{2}\right)}{\sin\left(\frac{x}{2}\right)}\cos\left[\omega t + (N-1)\frac{x}{2}\right] \tag{3.21}$$

Given that the second term is simply the cosine wave, the first term is the angular dependence:

$$R(\theta) = \frac{\sin\left\{N\pi d\frac{[\sin(\theta) - \sin(\phi)]}{\lambda}\right\}}{\sin\left\{\pi d\frac{[\sin(\theta) - \sin(\phi)]}{\lambda}\right\}} \tag{3.22}$$

Since

$$N = \lim_{x\to 0}\frac{\sin\left(N\frac{x}{2}\right)}{\sin\left(\frac{x}{2}\right)} \tag{3.23}$$

The normalized response is

$$r(\theta) = \frac{\sin\left\{N\pi d\frac{[\sin(\theta) - \sin(\phi)]}{\lambda}\right\}}{N\sin\left\{\pi d\frac{[\sin(\theta) - \sin(\phi)]}{\lambda}\right\}} \tag{3.24}$$

The power response is

$$B(\theta) = [r(\theta)]^2 = \left[\frac{\sin\left\{N\pi d\frac{[\sin(\theta) - \sin(\phi)]}{\lambda}\right\}}{N\sin\left\{\pi d\frac{[\sin(\theta) - \sin(\phi)]}{\lambda}\right\}}\right]^2 \tag{3.25}$$

$B(\theta)$ is the mean square voltage out for a receiver or the mean square pressure at the reference distance for a projector. It is common to express this value in decibels as a BPC relative to the maximum response axis

$$\mathrm{BPC}(\theta) = 10\log\left\{\left[\frac{\sin\left\{N\pi d\frac{[\sin(\theta) - \sin(\phi)]}{\lambda}\right\}}{N\sin\left\{\pi d\frac{[\sin(\theta) - \sin(\phi)]}{\lambda}\right\}}\right]^2\right\} \tag{3.26}$$

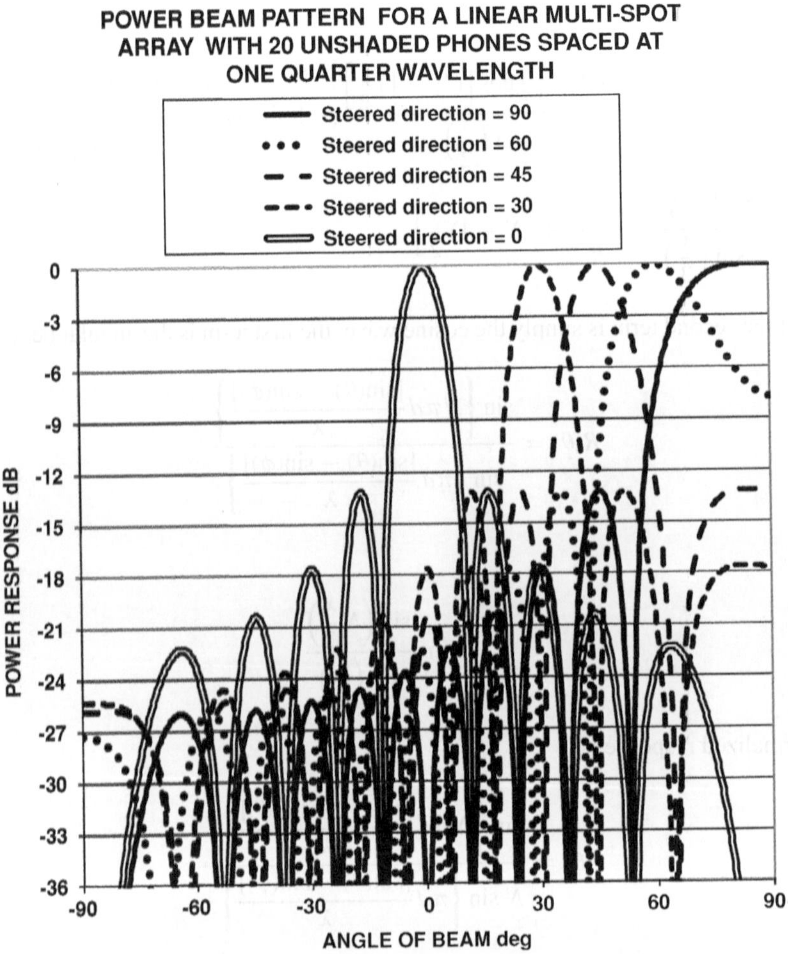

Figure 3.6 Power beam pattern for a linear multispot array with 20 unshaded hydrophones for different steering directions

Figure 3.6 shows an example of steering to various angles for a linear multispot array with 20 hydrophones with equal weightings (no shading).

A single-line hydrophone with length, L, has the same response pattern as a multispot array of identical length with hydrophones spaced at one half wavelength. The pressure or voltage response is

$$r(\theta) = \frac{\sin\left[\dfrac{kL}{2}\sin(\theta)\right]}{\dfrac{kL}{2}\sin(\theta)} \tag{3.27}$$

where

$k = 2\pi/\lambda$

The power response in decibels is

$$\mathrm{BPC}(\theta) = 10\log\left\{\left[\frac{\sin\left[\dfrac{kL}{2}\sin(\theta)\right]}{\dfrac{kL}{2}\sin(\theta)}\right]^2\right\} \tag{3.28}$$

3.3.1 *Triplet Towed Array*

Line arrays cannot determine on which side, left or right, a contact is because this type of array only measures conical angles. Triplet element processing, and potentially other combinations of hydrophones, is a means for resolving the bearing ambiguity of a contact. As will be discussed in Section 3.15, if two closely spaced hydrophones, one of which has been time delayed by the horizontal spacing, are subtracted, the result is a cardioid beam pattern response. A triplet towed array is constructed by using three hydrophones within the towed array hose, where a single hydrophone would normally be on a linear multispot array. These hydrophones normally form an equilateral triangle oriented perpendicular to the line axis. The center and right phones are processed to form a cardioid to the right and the center and left are processed to form a cardioid to the left. Figure 3.7 shows an example of the beam pattern response for a triplet array. Note that the response on the ambiguous bearing (180°–360°) is greatly reduced, instead of being a mirror image of the 0–180° region.

3.3.2 *Multiline Towed Arrays*

Multiline towed arrays use multiple linear multispot arrays towed from a single cable or towed body. The arrays may employ lifting bodies or separate tow points on a towed body to provide separation between arrays either horizontally, vertically, or both. These arrays may be as simple as two arrays or as complex as several arrays. The objective is to acquire additional gain in a fixed length and to achieve good left–right discrimination. The beamforming used may be a nominal time delay weight and add scheme, employ directional groups as in the triplet array, or use adaptive beamforming to gain against interfering sources (Section 3.18). Figure 3.8 shows an example of the beam pattern response of a four-line square multiline towed array. Note that the ambiguous bearing response is reduced.

3.4 Rectangular Planar Array

The beam pattern for a flat, rectangular array of hydrophones evenly spaced parallel to the edges can be calculated as follows. This is a different case from the line array in that the diagonal spacing of hydrophones is not the same as the spacing parallel to the edge. If we have $m \times n$ hydrophones (Figure 3.9) the signal arriving at each hydrophone is given by

$$\begin{aligned} S_{ij} &= S\sin\left[\omega\left(t+\tau_{ij}\right)\right] \\ \tau_{ij} &= i\frac{d}{c}\cos(\theta)\sin(\phi) + j\frac{d}{c}\sin(\theta)\cos(\phi) \end{aligned} \tag{3.29}$$

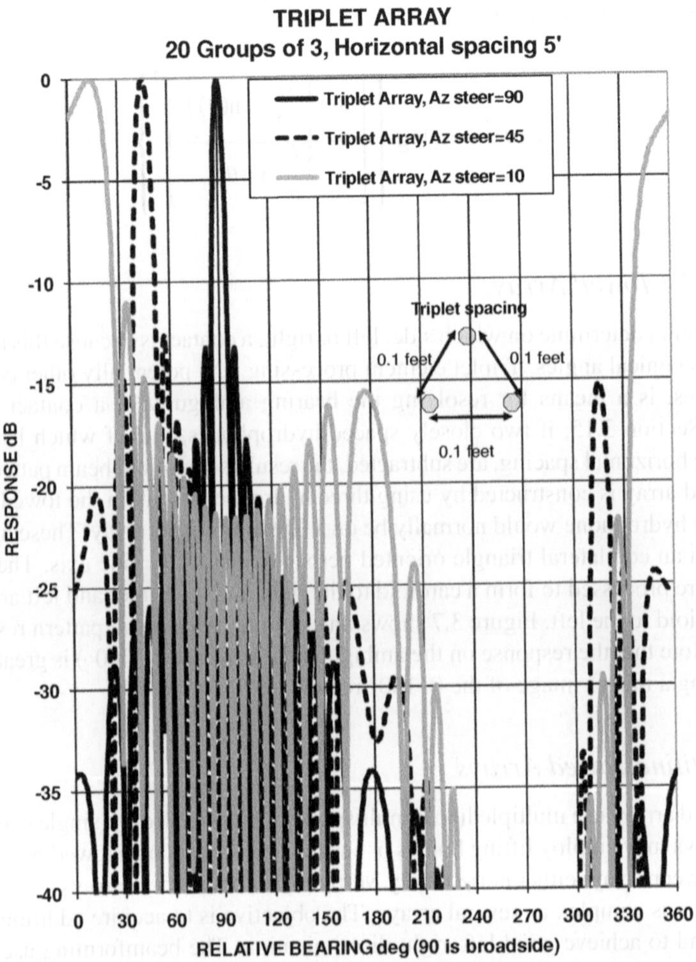

Figure 3.7 Beam pattern response of a triplet towed array at 480 Hz

where φ is the polar angle in a spherical coordinate system and $\theta = 0$ is lined up with the m hydrophones, and i runs from $-(m-1)/2$ to $(m-1)/2$ and j is from $-(n-1)/2$ to $(n-1)/2$. If the hydrophones are added with unity weighting, then the output is given by

$$R(\theta,\phi) = \sum_{i=-(m-1)/2}^{(m-1)/2} \sum_{j=-(n-1)/2}^{n-1/2} S\cos\left\{\omega\left[t + \frac{id}{c}\cos(\theta)\sin(\phi) + \frac{jd}{c}\sin(\theta)\sin(\phi)\right]\right\} \tag{3.30}$$

Looking at the beam pattern parallel to the m hydrophones ($\theta = 0$),

$$R(\theta,\phi) = \sum_{i=-(m-1)/2}^{m-1/2} \sum_{-(n-1)/2}^{n-1/2} S\cos\left\{\omega\left[t + i\frac{d}{c}\sin(\phi)\right]\right\} \tag{3.31}$$

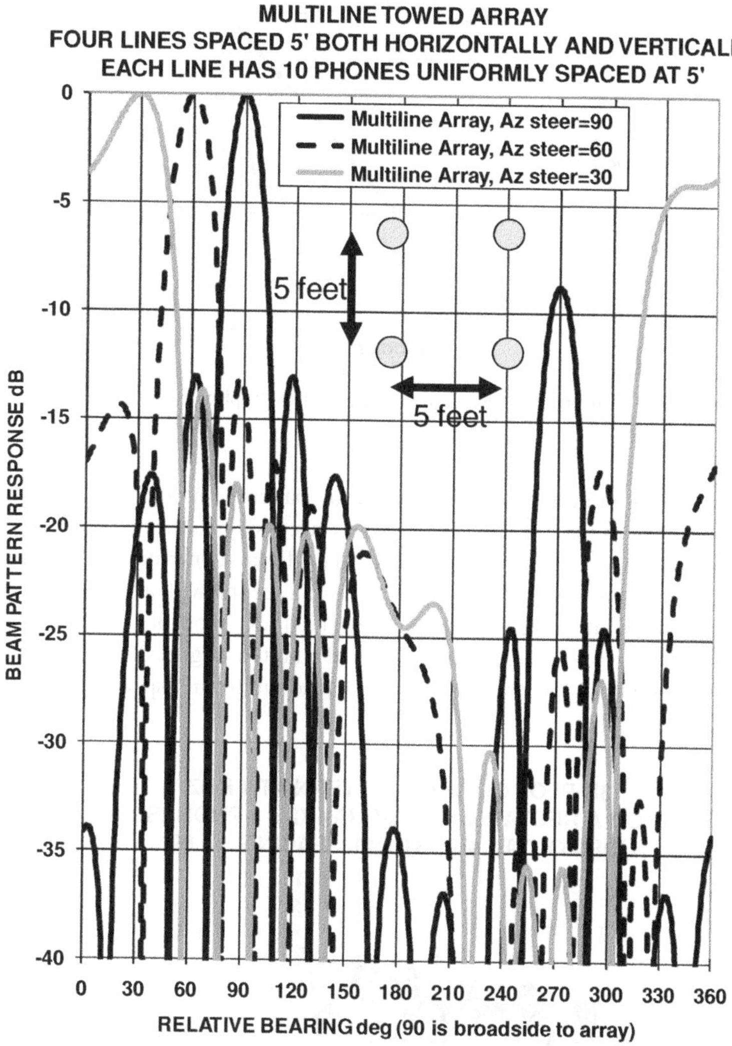

Figure 3.8 Beam pattern response of a multiline towed array at 200 Hz

or

$$R(0, \phi) = nS \sum_{-(m-1)/2}^{m-1/2} \cos\left\{\omega\left[t + i\frac{d}{c}\sin(\phi)\right]\right\} \tag{3.32}$$

which, except for the leading *n,* is identical to the equation for a line array with *m* equally spaced elements. Therefore, the beam pattern is identical to the line array parallel to this side and by symmetry a similar result will be true parallel to the perpendicular direction (the same pattern as a line array with *n* equally spaced elements).

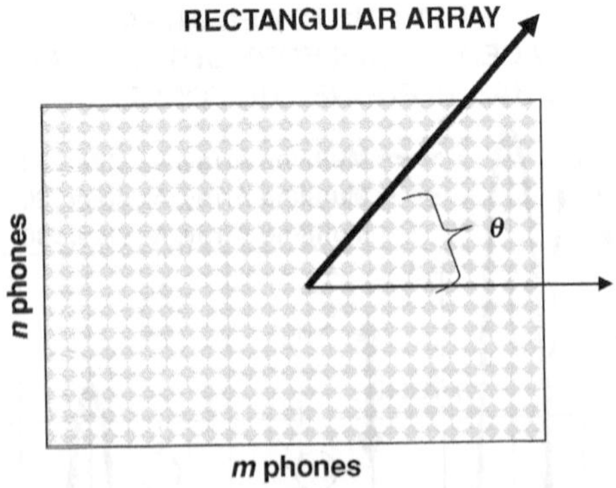

Figure 3.9 Rectangular array of hydrophones

It is simple to write a program to compute the beam patterns for any array of hydrophones. When I first needed to do this, around 1972, it was necessary to meet with the head of the computer department to arrange to take over the main-frame computer for the entire night. Today, the same calculations can be done in fractions of a second with an Excel© spreadsheet on a personal computer. Unlike the labor-intensive hand drawn plots of 1972, the results today are graphed in beautiful plots taking the mere push of a button. Figure 3.10 shows an example of the beam pattern response for a rectangular, unshaded, uniformly spaced array.

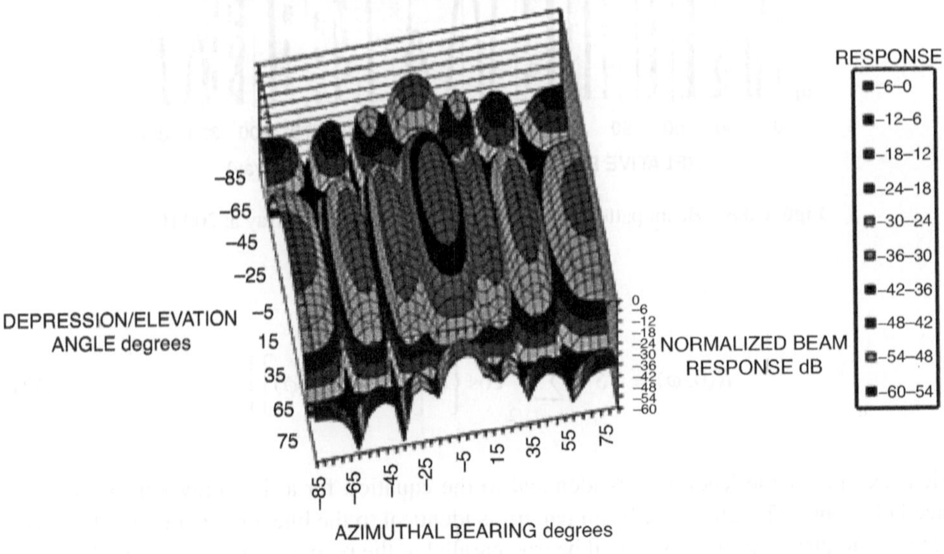

Figure 3.10 Beam pattern response for an unshaded rectangular array

3.5 Amplitude Shading

All of the previous examples assume equal weighting for each hydrophone. It is easy to apply a weighting or shading coefficient to each hydrophone and evaluate the beam pattern using modern programs. However, it is difficult to get closed-form solutions for all but the simplest of cases.

Shading is most commonly used to suppress side lobes (responses away from the main response lobe) or to suppress responses in noisy directions (known as adaptive beam forming (ABF)). Many methodologies exist for calculating shading coefficients depending on what the objective is. An inspection of equation 3.47 for array gain shows that having all coefficients equal offers the maximum array gain in an isotropic noise field. It is a general rule that shading increases the width of the main lobe and decreases side lobes and reduces array gain (in a uniform noise field). Figures 3.11 through 3.13 show examples of beam patterns for linear multispot arrays with different sets of shading coefficients. Table 3.1 shows the effects of applying various weighing schemes to a 50 hydrophone linear multispot array of length, L, operated at its design frequency at broadside. Note that with shading, we are trading off the main lobe width and side lobe level.

3.6 Continuous Arrays

Consider a plane intersecting a planar or line array. The response is given by

$$R(k) = \int_{-L/2}^{L/2} T(r)\cos(\omega t - kr)\,\mathrm{d}r \tag{3.33}$$

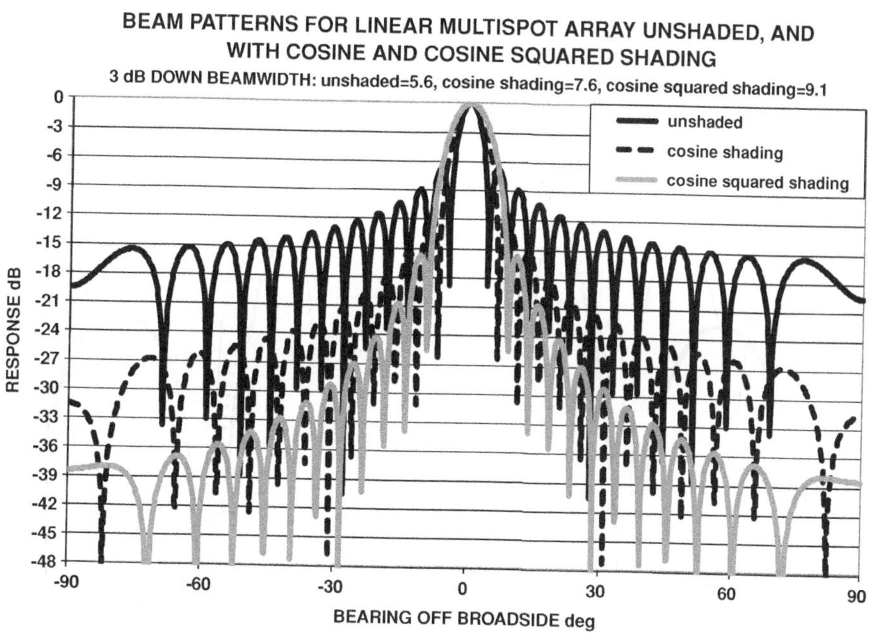

Figure 3.11 Beam pattern response for a linear multispot array unshaded, with cosine and cosine squared shading

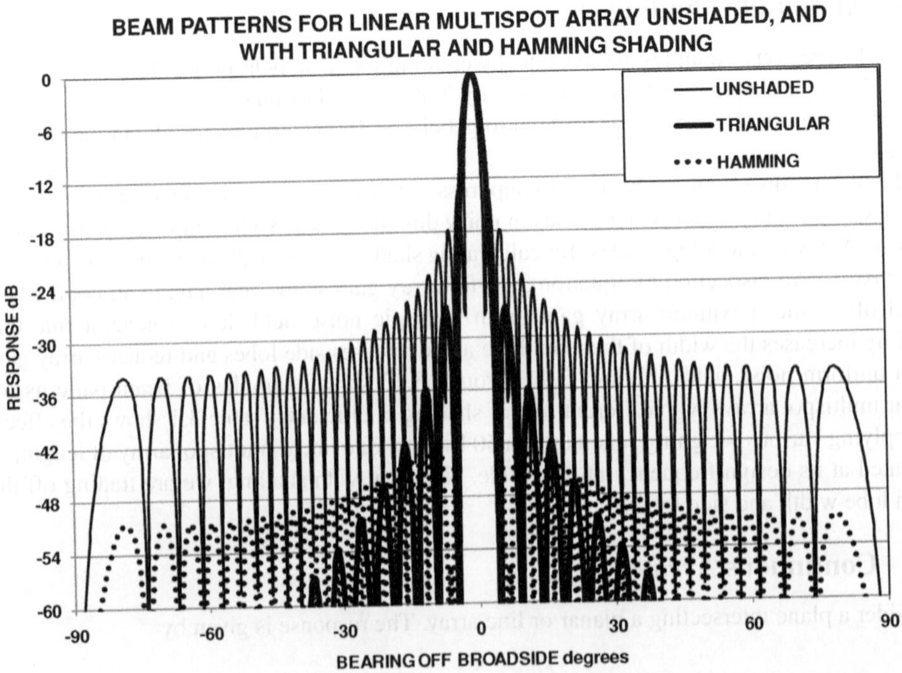

Figure 3.12 Beam pattern response for a linear multispot array unshaded, with triangular and Hamming shading

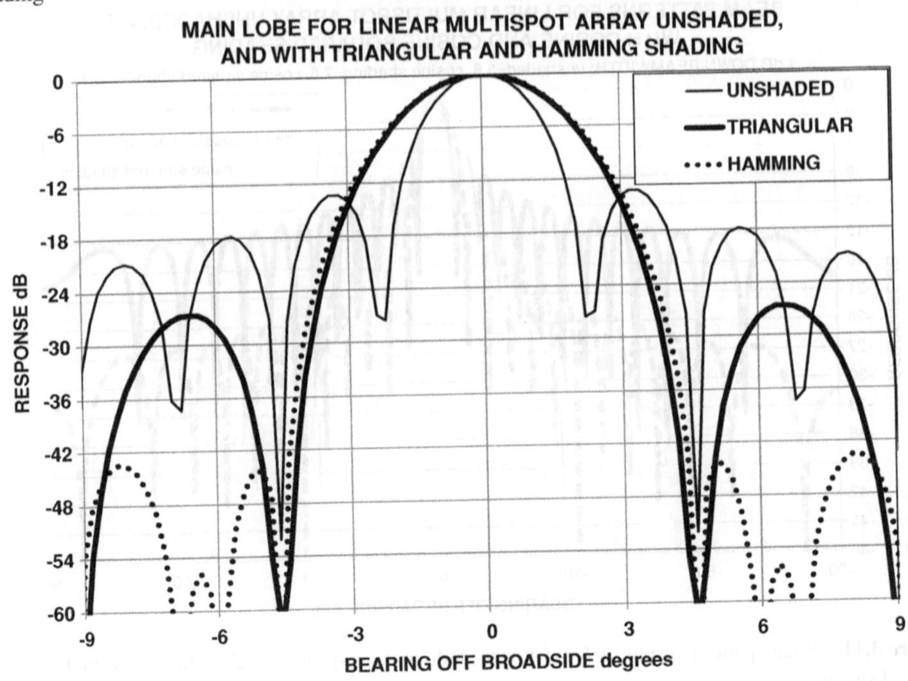

Figure 3.13 Beam pattern response of the main lobe for a linear multispot array unshaded, with triangular and Hamming shading

where $T(r)$ is the transducer sensitivity along the line defined by r, $\omega = 2\pi f$, $k = 2\pi/\lambda$, and L is the length along the array. Thus,

$$R(k) = \cos(\omega t)\int_{-L/2}^{L/2} T(r)\cos(kr)\,\mathrm{d}r + \sin(\omega t)\int_{-L/2}^{L/2} T(r)\sin(kr)\,\mathrm{d}r \tag{3.34}$$

Note that the first integral is the cosine Fourier transform (see Appendix A) and the second is the sine Fourier transform. Typically, we do not care about the phase of the response, but only the amplitude. This can be obtained by

$$R(k) = \sqrt{\left[\int_{-L/2}^{L/2} T(r)\cos(kr)\,\mathrm{d}r\right]^2 + \left[\int_{-L/2}^{L/2} T(r)\sin(kr)\,\mathrm{d}r\right]^2} \tag{3.35}$$

Frequently, arrays are symmetric about the center, in which case $T(r)$ is an even function and the second integral is zero.

Now consider a line array with linear sensitivity along its length, e.g., the diagonal of a square hydrophone. Then

$$R(k) = \int_{-L/2}^{L/2} \frac{\cos(kr)}{L}\,\mathrm{d}r$$

$$R(k) = \frac{\sin\left(\frac{kL}{2}\right)}{\frac{kl}{2}} \tag{3.36}$$

Now consider an array with a cosine taper sensitivity:

$$T(r) = \frac{2}{L}\cos\left(\pi\frac{r}{L}\right) \tag{3.37}$$

Table 3.1 Comparison of array gain and beamwidths for a 50 hydrophone linear multispot array of length, L, operated at its design frequency at broadside

Shading	AG isotropic noise	Half beamwidth (−3 dB)	Weight equation $-L/2 < x < L/2$	AG loss	Ratio of beamwidths
Unshaded	17.0	1.02	1	0	1
Tri	11.0	1.46	If $(x < 0, (L/2 + x)/(L/2), (L/2 - x)/(L/2))$	−6.0	1.43
cos	13.1	1.36	$\cos(\pi x/L)$	−3.9	1.33
$\cos^2$ Hanning	11.0	1.65	$\cos(\pi x/L)\cos(\pi x/L)$	−6.0	1.62
Hamming	11.6	1.49	$0.54 + 0.46\cos(2\pi x/L)$	−5.4	1.46

This array has a unit response at the center, tapering to zero at each end:

$$R(k) = \int_{-L/2}^{L/2} \frac{2}{L} \cos\left(\frac{\pi r}{L}\right) \cos(kr)\, \mathrm{d}r \tag{3.38}$$

$$R(k) = \frac{\pi}{(-\pi + kL)(\pi + kL)} \tag{3.39}$$

The normalized amplitude response is shown in Figure 3.14 with the uniform line array shown for comparison. Note that the first side lobes have amplitudes of –0.07 and –0.21

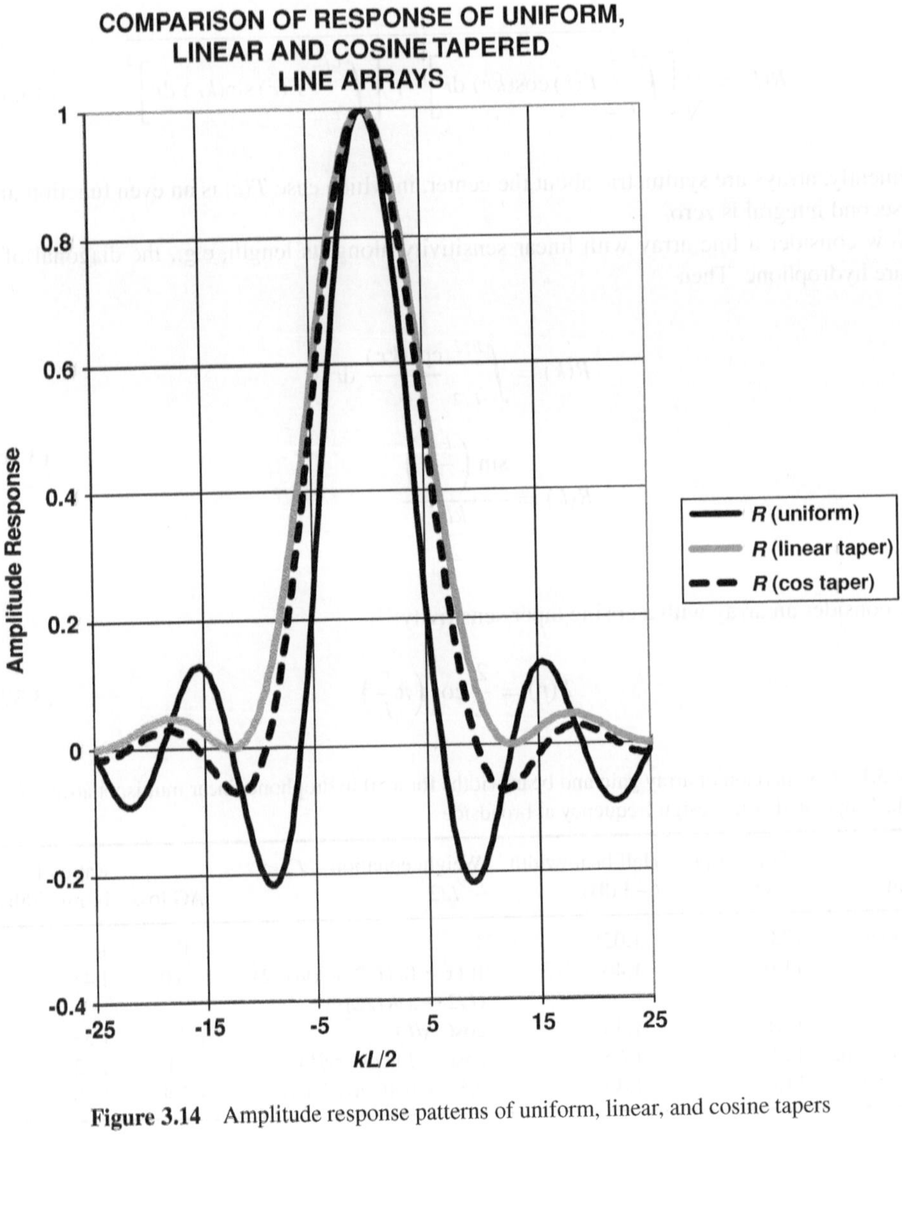

Figure 3.14 Amplitude response patterns of uniform, linear, and cosine tapers

respectively, or in decibels –23.1 dB and –13.6 dB, or an improvement of 9.5 dB, but at a cost of a 28 % wider main lobe at the half power (−3 dB) points.

A square array evaluated along the diagonal has a linear taper response:

$$R(k) = \int_{-L/2}^{0} \frac{\left(1 + \frac{2r}{L}\right)}{L} \cos(kr)\, \mathrm{d}r + \int_{0}^{L/2} \frac{\left(1 - \frac{2r}{L}\right)}{L} \cos(kr)\, \mathrm{d}r \tag{3.40}$$

$$\begin{aligned} R(k) &= \frac{4}{k^2L^2} - \frac{4\cos\left(\frac{kL}{2}\right)}{k^2L^2} \\ R(k) &= \frac{8}{k^2L^2}\left[\sin\left(\frac{kL}{4}\right)\right]^2 \end{aligned} \tag{3.41}$$

Normalizing yields

$$R(k) = \frac{\left[\sin\left(\frac{kL}{4}\right)\right]^2}{\left(\frac{kL}{4}\right)^2} \tag{3.42}$$

This is plotted in Figure 3.15. Note that all side lobes are positive. If this were truly a square, then L would be $\sqrt{2}l$ <add vinc over 2l>, where l is one side of the square. Figure 3.15 shows plots of the two amplitude responses.

A circular array has a response given by

$$\begin{aligned} R(k) &= \int_{-L/2}^{L/2} \sqrt{\frac{L^2}{4} - r^2} \cos(kr)\, \mathrm{d}r \\ R(k) &= \frac{J_1\left(\frac{kL}{2}\right)}{\frac{kL}{2}} \end{aligned} \tag{3.43}$$

where J_1 is the Bessel function of the first kind and first order. Table 3.2 shows the nonshaded beam widths for various arrays.

3.7 Volumetric Arrays

Thus far only linear and planar array have been considered. There is the possibility of having hydrophones distributed in all three dimensions. The first problem that arises is how physically to support the hydrophones. Usually, this is accomplished by imbedding them in a "transparent" material. Going from a line array to a square array, we gained a beam pattern response

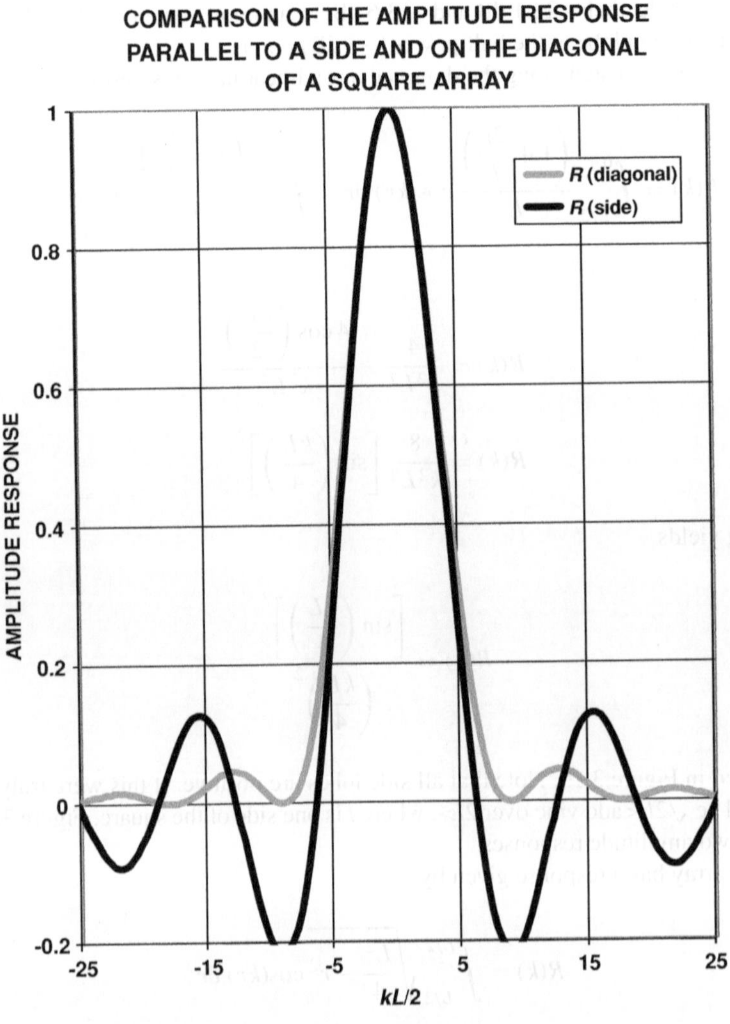

Figure 3.15 Amplitude response of a square array

Table 3.2 Beam widths for line and circular arrays between −3, −6 and −10 dB power response points, unshaded

	Beam width	
	Line	Circular
−3 dB	$2a\sin[1.39c/(\pi Lf)]$	$2a\sin[0.54c/(Df)]$
−6 dB	$2a\sin[1.90c/(\pi Lf)]$	$2a\sin[0.74c/(Df)]$
−10 dB	$2a\sin[2.32c/(\pi Lf)]$	$2a\sin[0.90c/(Df)]$

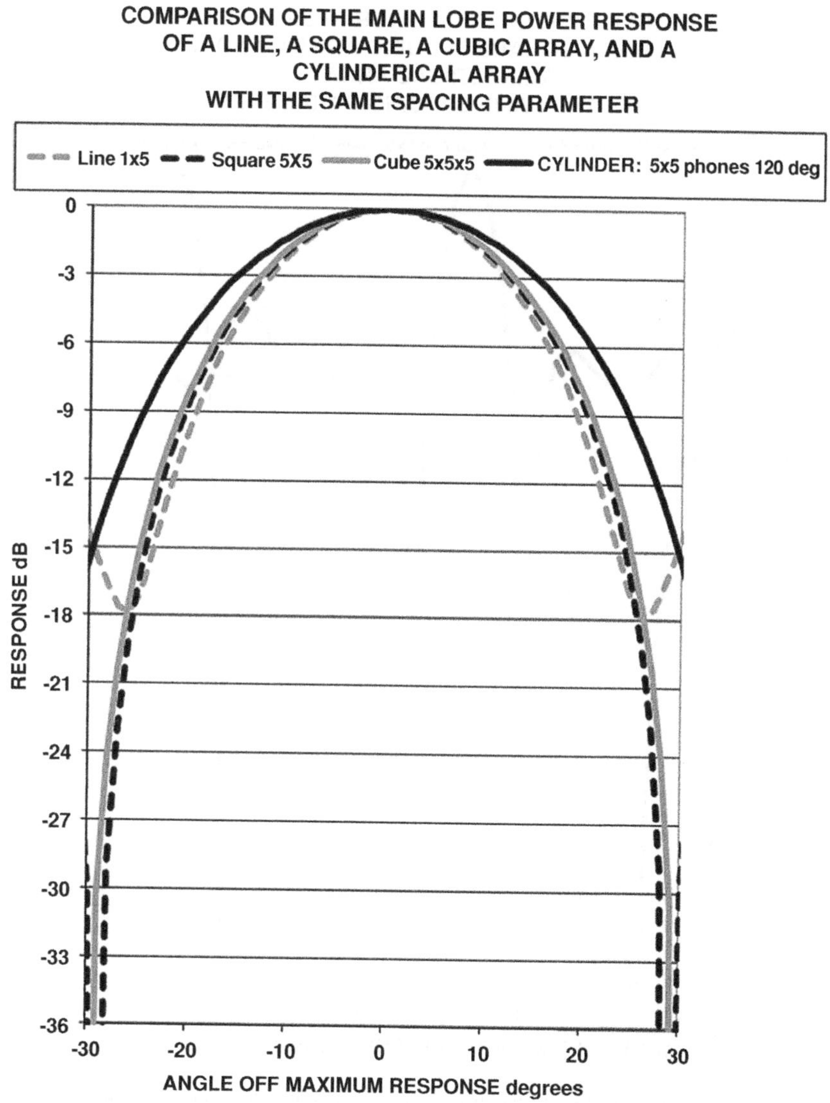

Figure 3.16 Comparison of the main lobe power response of a line, a square, a cubic array, and a cylindrical array with the same spacing parameter

with a vertical response, as well as a horizontal response. For a response on the normal to the line array at the design frequency, the beam pattern response covered about $1/n$ of the solid angle and for the square, about $1/n^2$ (with n^2 hydrophones). The expectation might be $1/n^3$ for a cubical array, but this is not the case (see Figures 3.16 and 3.17). If the $n \times n \times n$ hydrophones is uniformly distributed in a cube and beamformed normal to one side, the result is a main lobe that is virtually identical to an $n \times n$ square array with the same spacing parameter. The gain here is a reduction in the back lobe response. This is also the case for three-dimensional surface areas like a cylinder, where the curvature reduces the back lobe response. In

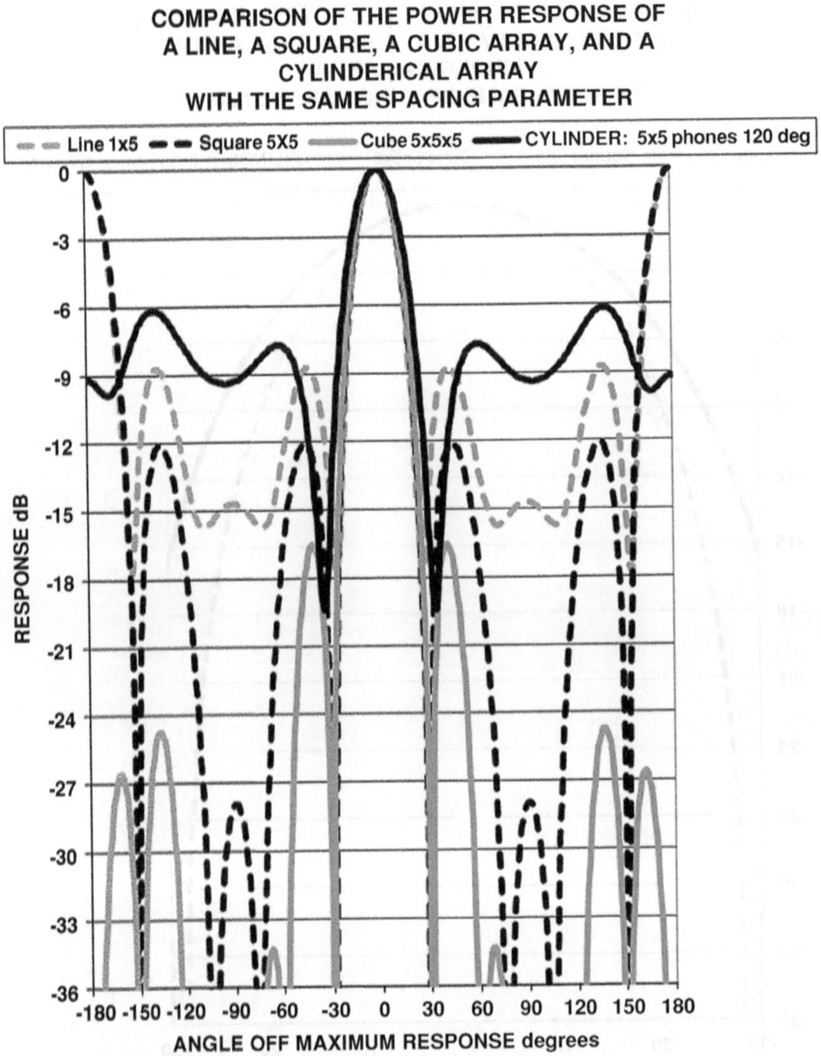

Figure 3.17 Comparison of the power response of a line, a square, a cubic array, and a cylindrical array with the same spacing parameter

Section 3.14, we will learn that there are additional complications in positioning the hydrophones in the third dimension.

3.8 Product Theorem

If the individual hydrophones are themselves directional, the resulting response can be shown to be the product of the two responses:

$$R(\theta, \phi) = R_{\text{array}}(\theta, \phi) R_{\text{phone}}(\theta, \phi) \tag{3.44}$$

This theorem only applies to additive arrays with the hydrophones aligned in the same direction.

3.9 Broadband Beam Patterns

All of the beam patterns discussed above assume a single frequency. If a broad frequency band is processed, the effect generally smoothes the beam pattern. Figures 3.18 and 3.19 show the azimuthal beam response for an unshaded 50 hydrophone linear multispot array with a spacing of 0.8 feet and a design frequency of 3030 Hz. Figure 3.18 shows the response for a flat signal 200 Hz wide. For the center frequency of that band, the main lobes are very similar, but the peaks and valleys in the side lobes are smoothed for the broadband. Figure 3.19 shows the response for a 1 octave flat input compared to the center frequency. Here, the main lobes are very similar, but the side lobes are almost smooth.

3.10 Directivity and Array Gain

The parameters directivity (NDI or DI) and array gain (AG) are both the decibel gain in the signal-to-noise ratio of an array compared to what would be achieved by a single

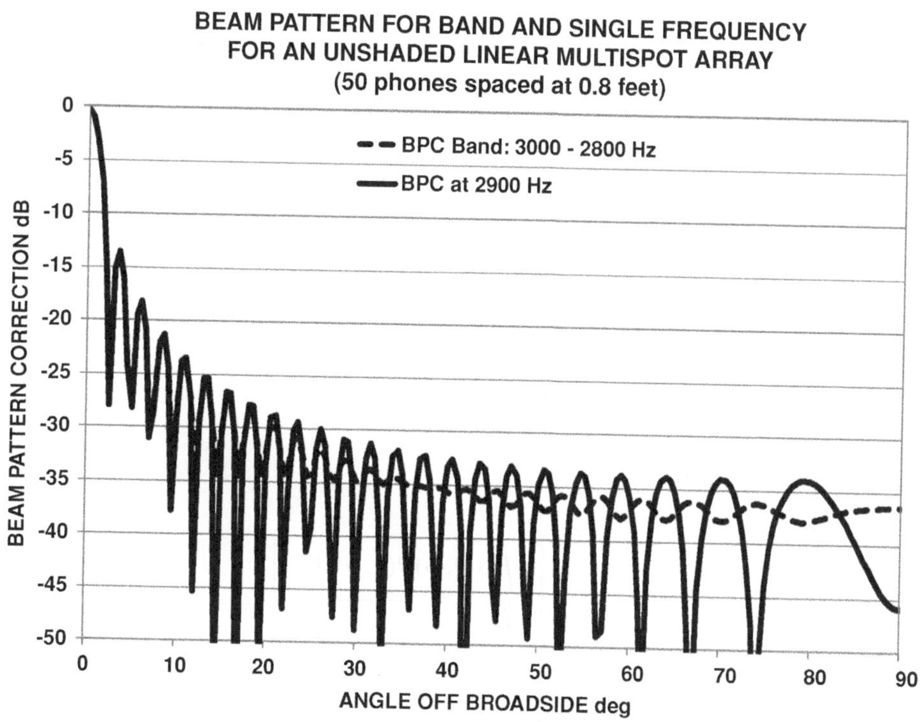

Figure 3.18 Comparison of the beam pattern response for a linear array for a 200 Hz wide flat input with the single frequency response

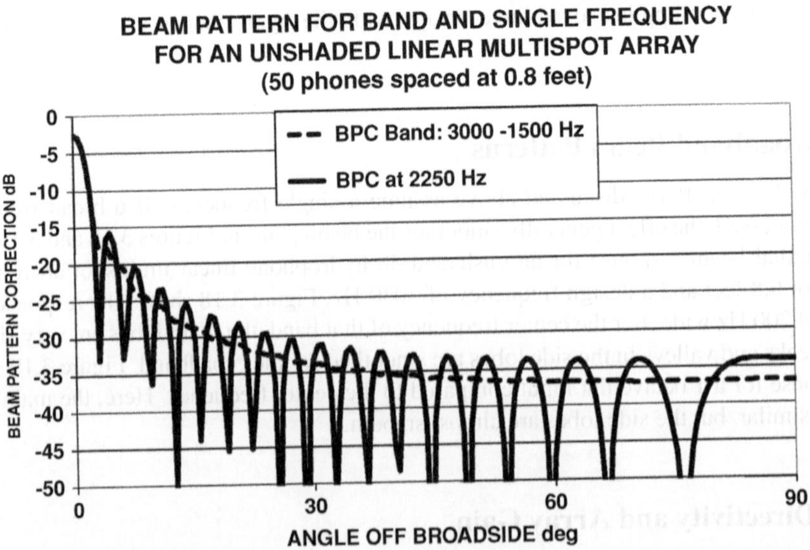

Figure 3.19 Comparison of the beam pattern response for a linear array for a 1 octave wide flat input with the single frequency response

omnidirectional hydrophone. The difference in definition has to do with the assumed noise field and signal. Directivity is the gain achieved for a plane wave signal in an isotropic, constant in all directions, noise field. Array gain does not make these assumptions. If the signal is a plane wave and the noise field is truly isotropic, directivity and array gain are identical. If we assume that the signal will remain fully correlated ($\rho = 1$) across the array then the directivity is

$$\mathrm{NDI} = \mathrm{DI} = 10\log\left[\frac{4\pi}{\displaystyle\int_0^{\pi}\int_0^{2\pi} R(\theta,\phi)\sin(\phi)\,\mathrm{d}\theta\,\mathrm{d}\phi}\right] \tag{3.45}$$

The array gain is given by

$$\mathrm{AG} = 10\log\left[\frac{\dfrac{\displaystyle\int S(\theta,\phi)R(\theta,\phi)\,\mathrm{d}\Omega}{\displaystyle\int N(\theta,\phi)R(\theta,\phi)\,\mathrm{d}\Omega}}{\dfrac{\displaystyle\int S(\theta,\phi)\,1\,\mathrm{d}\Omega}{\displaystyle\int N(\theta,\phi)\,1\,\mathrm{d}\Omega}}\right] \tag{3.46}$$

where the numerator is the signal to noise out of the array and the denominator is the signal to noise from an omnidirectional transducer. For a linear additive array of discrete hydrophones this can be written in terms of the cross-correlation coefficients:

$$\mathrm{AG} = 10\log\left[\frac{\sum_i\sum_j a_i a_j \rho_{ij}(s)}{\sum_i\sum_j a_i a_j \rho_{ij}(n)}\right] \tag{3.47}$$

where

$i, j =$ indices for the hydrophones
$\rho(s)_{ij} =$ cross-correlation coefficient for the signal between the ith and jth hydrophones
$\rho(n)_{ij} =$ cross-correlation coefficient for noise between the ith and jth hydrophones
$a_i =$ weighting (shading coefficient) applied the ith hydrophone

Using the cross-correlation coefficients from Table 3.3, consider the array gain for a linear, unshaded, uniformly spaced array of n hydrophones. Note that the noise cross-correlation is zero whenever $d = m\lambda/2$. If the hydrophones are spaced at $\lambda/2$, then all noise correlations are zero except when $i = j$ and the sum of these noise correlations will be n. Similarly, if the signal has a cross-correlation of 1 for all cases, the sum in the numerator will be n^2 and the array gain will be 10 log (n).

3.11 Noise Cross-Correlation between Hydrophones

The correlation coefficient for two variables (x, y) is given by the expected value of the product of the variables minus the mean of each variable, divided by the product of the root mean squares of each variable, or

$$\rho(x, y) = \frac{E(xy) - \mu(x)\mu(y)}{\sigma(x)\sigma(y)} \tag{3.48}$$

The correlation coefficient for two sine waves with one wave delayed by a time, τ, is given by

$$\rho(\tau) = \frac{\int_0^{2\pi/\omega} \sin(\omega t)\sin[\omega(t+\tau)]\,\mathrm{d}t}{\int_0^{2\pi/\omega} [\sin(\omega t)]^2\,\mathrm{d}t}$$

$$\rho(\tau) = \frac{\left[\frac{-1}{4}\frac{[-4\cos(\omega\tau)\pi + \sin(\omega\tau)]}{\omega} + \frac{1}{4}\frac{\sin(\omega\tau)}{\omega}\right]}{\pi} \tag{3.49}$$

$$\rho(\tau) = \cos(\omega\tau) \tag{3.50}$$

Consider two hydrophones separated by a distance, d, on the z axis of a spherical coordinate system. For a patch of surface ($R^2 \sin(\varphi)\,\mathrm{d}\theta\,\mathrm{d}\varphi$) the difference in arrival time, Δt, if R is very

Table 3.3 Cross-correlation coefficients for noise and signal

	Point source signal	Isotropic noise
Zero time delay Single frequency	$\cos(\omega\tau)$	$\dfrac{\sin\left(\dfrac{\omega d}{c}\right)}{\dfrac{\omega d}{c}}$
Time delayed Single frequency	$\cos\left[\omega\left(\tau+\tau_e\right)\right]$	$\dfrac{\sin\left(\dfrac{\omega\tau}{c}\right)}{\dfrac{\omega\tau}{c}}\cos(\omega\tau_e)$
Zero time delay Flat band	$\dfrac{\sin\left[\dfrac{(\omega_2-\omega_1)\tau}{2}\right]\cos\left[\dfrac{(\omega_2+\omega_1)\tau}{2}\right]}{\dfrac{(\omega_2-\omega_1)\tau}{2}}$	$\dfrac{S_i\left(\dfrac{\omega_2 d}{c}\right)-S_i\left(\dfrac{\omega_1 d}{c}\right)}{\dfrac{(\omega_2-\omega_1)d}{c}}$
Time delayed Flat band	$\dfrac{\sin\left[\dfrac{(\omega_2-\omega_1)(\tau+\tau_e)}{2}\right]\cos\left[\dfrac{(\omega_2+\omega_1)(\tau+\tau_e)}{2}\right]}{\dfrac{(\omega_2-\omega_1)d}{2}}$	$\dfrac{S_i\left[\omega_2\left(\dfrac{d}{c}+\tau_e\right)\right]-S_i\left[\omega_1\left(\dfrac{d}{c}+\tau_e\right)\right]+S_i\left[\omega_2\left(\dfrac{d}{c}-\tau_e\right)\right]-S_i\left[\omega_1\left(\dfrac{d}{c}-\tau_e\right)\right]}{2(\omega_2-\omega_1)\dfrac{d}{c}}$

large is given by

$$\Delta t = \frac{d\cos(\phi)}{c} \tag{3.51}$$

The correlation coefficient for that patch is given by

$$\rho(\phi, \theta) = \cos(\omega\,\Delta t) \tag{3.52}$$

If it is assumed that the noise is coming uniformly from the entire surface of the sphere, then the correlation coefficient for two hydrophones in omnidirectional noise is calculated as

$$\rho(d) = \frac{\displaystyle\int_0^{\pi}\int_0^{2\pi} \rho(\phi, \theta)\sin(\phi)\,\mathrm{d}\theta\,\mathrm{d}\phi}{\displaystyle\int_0^{\pi}\int_0^{2\pi} \sin(\phi)\,\mathrm{d}\theta\,\mathrm{d}\phi}$$

$$\rho(d) = \frac{\sin\left(\dfrac{\omega d}{c}\right)}{\left(\dfrac{\omega d}{c}\right)}$$

$$\omega = 2\pi f = \frac{2\pi c}{\lambda} \tag{3.53}$$

$$\rho(d) = \frac{\sin(kd)}{kd} \tag{3.54}$$

where k is the wave number ($k = 2\pi/\lambda$).

$\omega = 2\pi \times$ frequency
$\tau =$ travel time between hydrophones of signal $= d\cos(\theta)/c$
$d =$ distance between hydrophones
$c =$ speed of sound
$\theta =$ angle of line between hydrophones to signal direction
$\tau_e =$ difference in beamforming delay between hydrophones
$\omega_2 = 2\pi \times$ (highest frequency of band)
$\omega_1 = 2\pi \times$ (lowest frequency of band)

and

$$S_i(x) = \int_0^x \frac{\sin(y)}{y}\,\mathrm{d}y \tag{3.55}$$

3.12 Directivity of Line Arrays

Figure 3.20 shows the directivity of a linear, uniformly spaced, unshaded array versus frequency and steered direction. The design frequency is the frequency at which the spacing is 1 half wavelength. Note particularly that the end-fire beam, which points parallel to the line of the

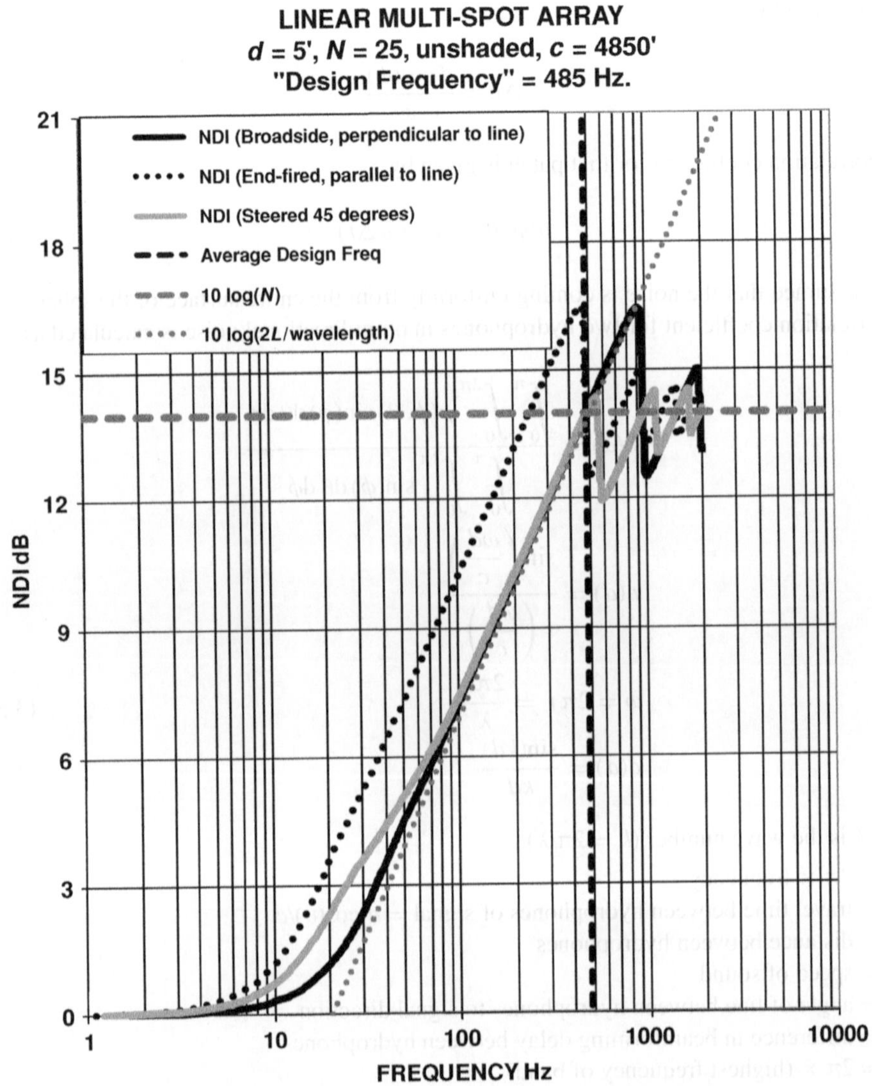

Figure 3.20 Directivity versus frequency and other steered directions for a linear multispot array

array, behaves like a broadside array (with the signal perpendicular to the line of the array) with half the design frequency. Up to the design frequency, the end-fire beam has a higher directivity (about 3 dB over most of the range) than other steered directions. This will become important when trying to understand the directivity of a volumetric array, as the elements inside the volume will be in an end-fired geometry. Also graphed is the equation for a continuous line array (10 log($2L/\lambda$). This is a near-perfect fit over most of the frequency range for the broadside case. Notice also that after the directivity passes 10 log(N), it oscillates about the 10 log(N) level.

3.13 Directivity of Area Arrays

Figure 3.21 shows the directivity of a square array with the same number of hydrophones as the linear array shown in Figure 3.20 versus frequency and steered direction. Notice that the directivity begins climbing at a higher frequency than the line array. Arrays start to have directivity when one of their dimensions approaches a half wavelength. Directivity climbs steadily as the frequency squared, or frequency for the linear array, and reaches 10 log(N) around the same frequency as the linear array. For rectangular arrays, the term design frequency, though still used, is a misnomer since the phone-to-phone separation is not a simple multiple of a constant. For this array, adjacent hydrophones are spaced 5 feet apart horizontally and vertically, but diagonal spacing to adjacent hydrophones is 7.07 ($5\sqrt{2}$) feet and 11.2 feet to a

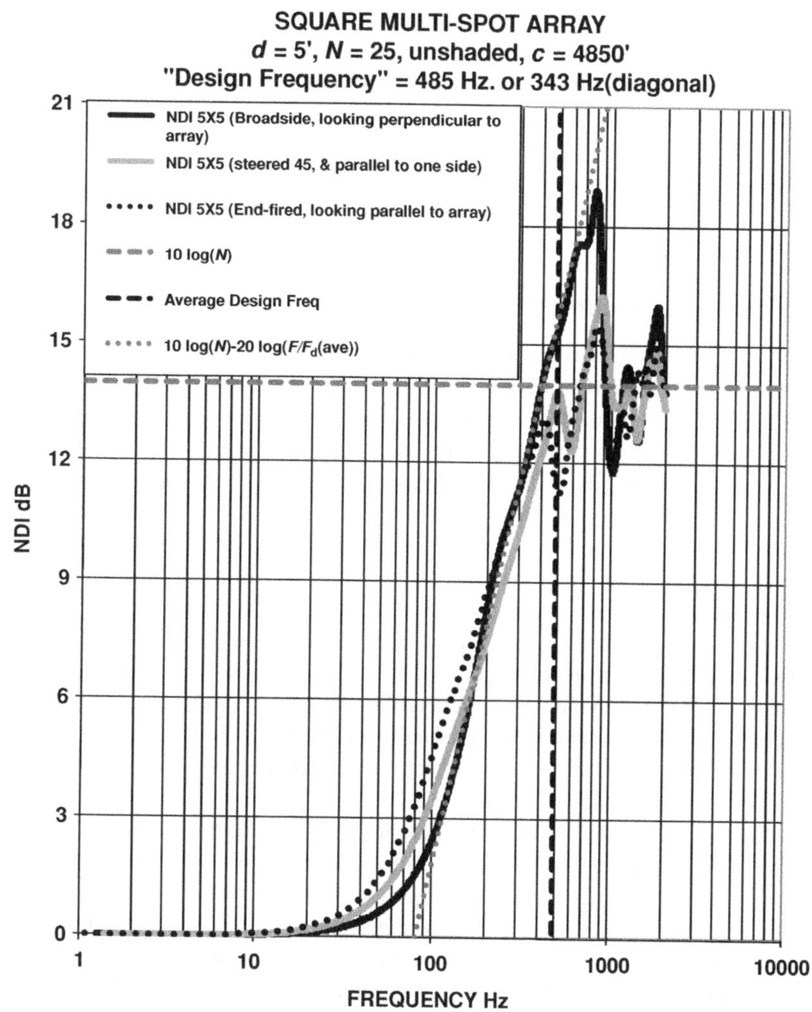

Figure 3.21 Directivity of square planar array versus frequency and steered direction

Table 3.4 Directivity and beam widths for various simple arrays

Type	NDI = DI = 10 log( )	−3 dB beam width (half) = A sin( )	Gain for baffle
Line	$2L/\lambda$	$1.39\lambda/(\pi L)$	3 dB
Rectangle	$5.5LW/\lambda^2$	$1.39\lambda/(\pi L)$ $1.39\lambda/(\pi W)$	3 dB
Circular	$(\pi D/\lambda)^2$	$1.45\lambda/(\pi D)$	
Cylinder (1/3 surface beamformed)		$1.26\lambda/(\pi D)$ $1.39\lambda/(\pi H)$	<3 dB

hydrophone over one and up two, etc. Here, the simplified curve that fits the broadside response over most of its linear region uses a design frequency that is the average of the horizontal and first diagonal frequencies. In general, because of the multiple frequencies that match the various spacings, the directivity is less smooth and the behavior more complex as the array is steered off the normal.

Table 3.4 gives the directivity and beamwidth for various simple array shapes.

3.14 Directivity of Volumetric Arrays

As we saw above, the directivity of a linear array increased as 10 log(f) until the design frequency was reached. Then, the directivity oscillated about 10 log(n) and, for a square array, increased as 20 log(f) and then oscillated about 10 log($n \times n$). Based on this, the expectation may be for a cubical array to increase as 30 log(f) up to 10 log($n \times n \times n$). However, this is not the case, as can be seen in Figure 3.22. The dark dotted curve is a cubical array with constant spacing in the x, y, and z directions (a simple cubic lattice). Notice that this array behaves poorly around 500 Hz. This problem can be corrected if the spacing in the depth direction is one half the spacing in the other two directions. However, this implies that if steered in other directions, this asymmetry cannot be maintained and poor behavior can be expected. The directivity of the cubical array increases by 20 log(f) and does not reach 10 log($n \times n \times n$) until about twice the design frequency. Figure 3.23 shows the comparison in directivity between a baffled planar array and the volumetric rectangular array with half spacing in depth. Notice that adding a baffle to a planar array gives a virtually identical directivity up to twice the design frequency. If given the choice of 5 times as many hydrophones to process or install a baffle plate, the later is usually the better option.

For volumes other than cubical, different, perhaps unexpected, problems arise. Consider a spherical array [1]. If a spherical shell is beamformed and compared to a fully volumetric array, the main lobe becomes wider. The reason is that the aperture is effectively center weighted or shaded because there are more hydrophones being used at the center of the array than anywhere else. Since most of the response is in the main lobe, directivity is worse until the array is greater than about seven wavelengths in diameter and is only 0.4 dB better at a 100 wavelength diameter, at a cost of many times the processing. Finally, if the shell was mounted on a baffle, the breakeven point would extend to about 80 wavelengths.

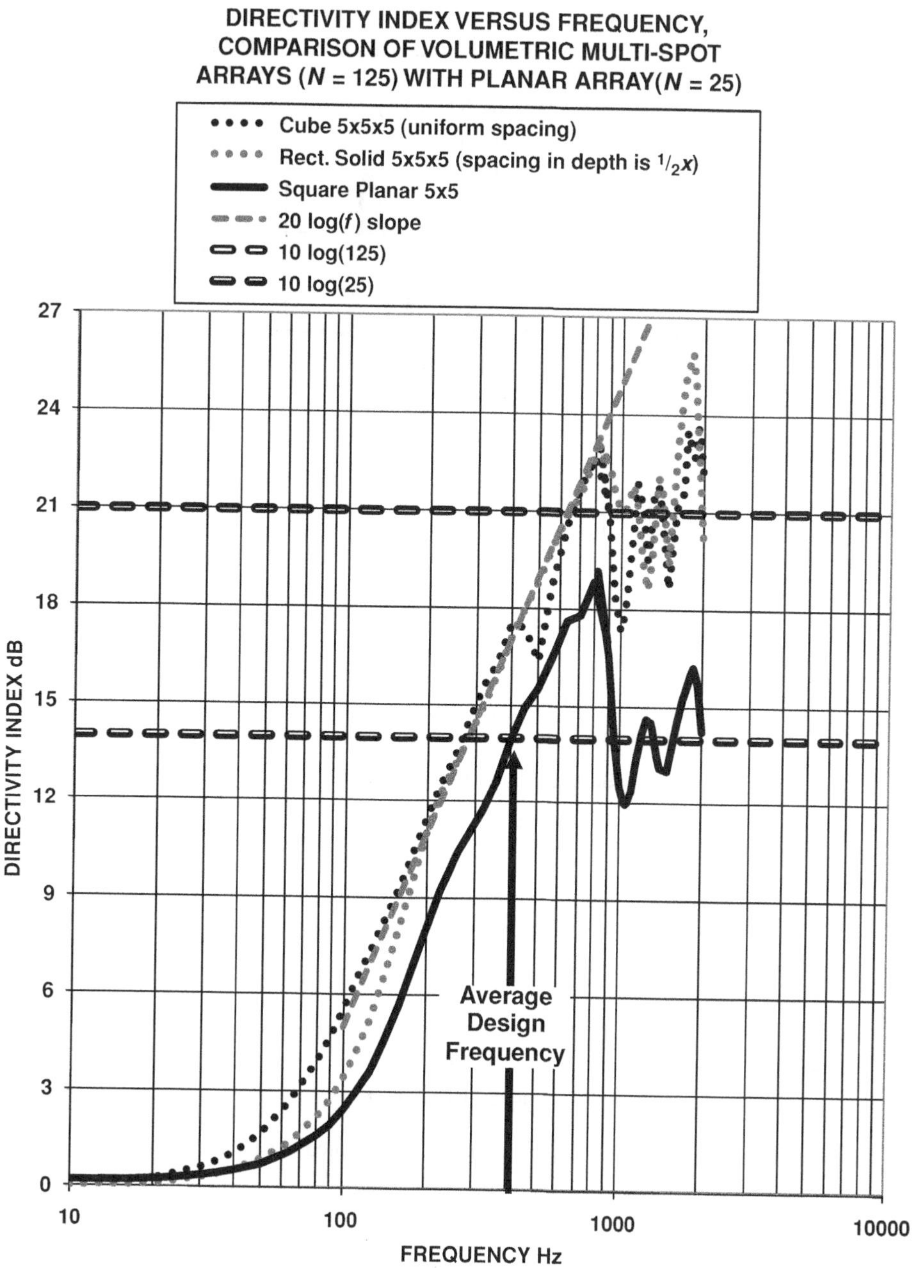

Figure 3.22 Directivity of a volumetric array as compared to a planar array

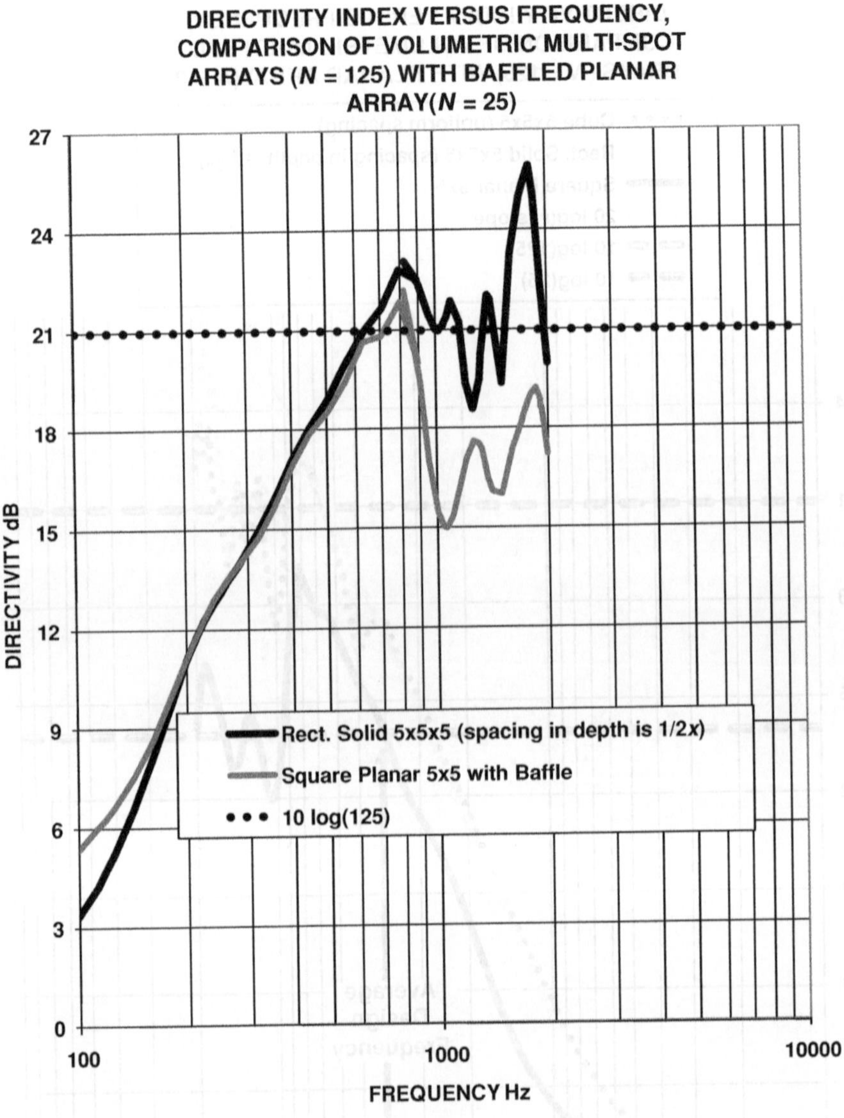

Figure 3.23 Directivity of a volumetric array as compared to a baffled planar array

3.15 Difference Arrays

So far, only arrays in which the outputs of the hydrophones are added with weightings and time delays have been considered. This is because, for low signal-to-noise ratios, it can be shown from information theory that there is no better method of increasing the information content of a signal than to add the outputs of the elements of an array to improve the signal to noise [2]. Consider a pair of hydrophones separated by a distance, d, the outputs of which are subtracted. The output is given by

$$S_{\text{diff}} = S\sin(\omega t) - S\sin\left[\omega\left(t + \frac{d\sin(\theta)}{c}\right)\right]$$
$$S_{\text{dif}}f = S\left\{\sin(\omega t)\left[1 - \cos\left(\omega\frac{d}{c}\sin(\theta)\right)\right] - \cos(\omega t)\sin\left(\omega\frac{d}{c}\sin(\theta)\right)\right\} \tag{3.56}$$

If $\omega\, d/c$ is small, $d \ll \lambda$, the first term vanishes and the output is simply

$$S_{\text{dif}}f \cong -S\,2\pi\frac{d}{\lambda}\sin(\theta)\cos(\omega t) \tag{3.57}$$

The beam pattern is given by

$$R(\theta) = 2\pi\frac{d}{\lambda}\sin(\theta) \tag{3.58}$$

The normalized power response is

$$B(\theta) = [\sin(\theta)]^2 \tag{3.59}$$

This is a classic dipole pattern and is constant for all wavelengths such that $d \ll \lambda$ (Figure 3.23). On the maximum response axis ($\pm$ 90), the signal power out will be

$$S_{\text{out}} = 4\pi^2\left(\frac{d}{\lambda}\right)^2 S^2 \tag{3.60}$$

In the absence of a signal, the output would be given by

$$N_{\text{out}} = 4\pi^2\left(\frac{d}{c}\right)^2 N^2\frac{\displaystyle\int_{-\pi/2}^{\pi/2} 2\pi R^2\,[\sin(\theta)]^2\cos(\theta)\,d\theta}{\displaystyle\int_{-\pi/2}^{\pi/2} 2\pi R^2\cos(\theta)\,d\theta} = \frac{4}{3}\pi^2\left(\frac{d}{\lambda}\right)^2 N^2 \tag{3.61}$$

Given this, the signal to noise out would be

$$S/N_{\text{out}} = 3S/N_{\text{in}} \quad \text{or an increase of } +5\,\text{dB} \tag{3.62}$$

This gain is higher than the sum of two hydrophones at any frequency.

Now consider a pair of hydrophones separated by a distance, *d*. The outputs of the hydrophones are subtracted after a time delay equal to the travel time of the hydrophone separation inserted. The output is then given by

$$S_{\text{diff delay}} = S\sin(\omega t) - S\sin\left[\omega\left(t + \frac{d\sin(\theta)}{c} + \frac{d}{c}\right)\right]$$
$$S_{\text{diff delay}} = S\left\{\sin(\omega t)\left[1 - \cos\left(\omega\frac{d}{c}(\sin(\theta)+1)\right)\right] - \cos(\omega t)\sin\left(\omega\frac{d}{c}(\sin(\theta)+1)\right)\right\} \tag{3.63}$$

If $\omega\, d/c$ is small, $d \ll \lambda$, then the first term vanishes and the output is simply

$$S_{\text{diff delay}} = -S\,2\pi\frac{d}{c}\,[\sin(\theta) + 1]\cos(\omega t) \tag{3.64}$$

The beam pattern is given by

$$R(\theta) = 2\pi\frac{d}{c}\,[\sin(\theta) + 1] \tag{3.65}$$

The normalized power response is

$$B(\theta) = 0.25\,[\sin(\theta) + 1]^2 \tag{3.66}$$

This is a classic cardioid pattern and is constant for all wavelengths such that $d \ll \lambda$ (Figure 3.24). On the maximum response axis (+ 90), the signal power out will be

$$S_{\text{out}} = 16\pi^2\left(\frac{d}{\lambda}\right)^2 S^2 \tag{3.67}$$

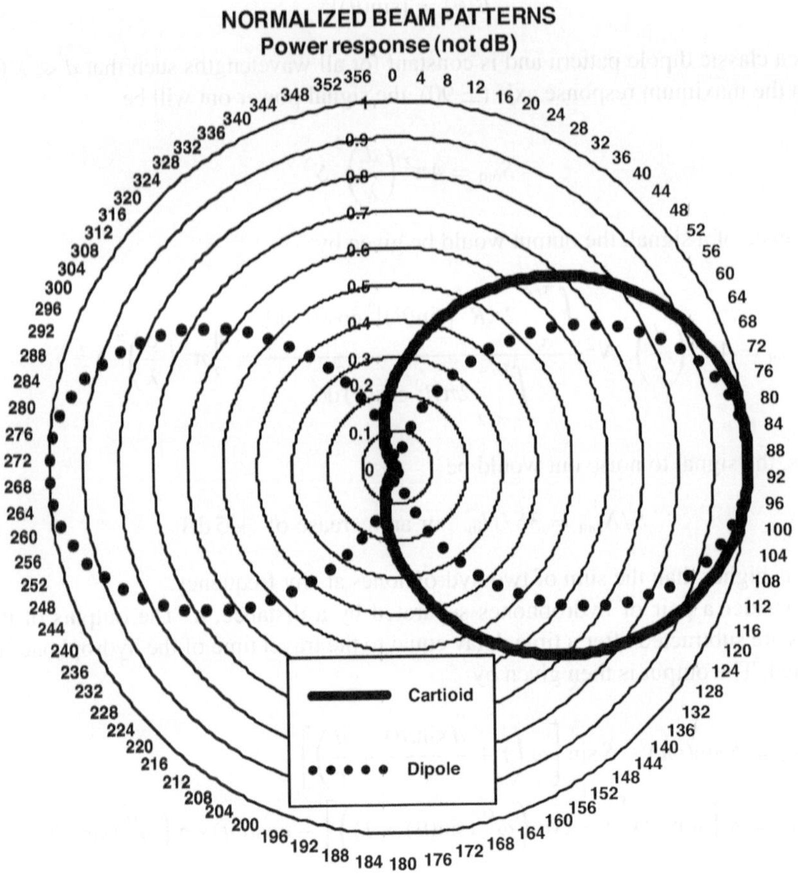

Figure 3.24 Beam patterns for dipole and Cartioid arrangements

In the absence of a signal, the output would be given by

$$N_{\text{out}} = 4\pi^2\left(\frac{d}{\lambda}\right)^2 \frac{\displaystyle\int_{-\pi/2}^{\pi/2} 2\pi R^2\,[\sin(\phi)+1]^2\cos(\phi)\,d\phi}{\displaystyle\int_{-\pi/2}^{\pi/2} 2\pi R^2\cos(\phi)\,d\phi} = \frac{16}{3}\left(\frac{d}{\lambda}N\right)^2 \tag{3.68}$$

Given this, the signal to noise out would be

$$S/N_{\text{out}} = 3S/N_{\text{in}} \quad \text{or an increase of } +5\,\text{dB} \tag{3.69}$$

This gain is higher than the sum of two hydrophones at any frequency. An important characteristic of this array is a zero response at 180° from the main response.

This type of response pattern can also be derived from velocity sensitive hydrophones. The process followed above is equivalent to taking the derivative of the pressure, which is proportional to the particle velocity. The first practical problem with this type of processing is sensitivity. Since the output of the subtracted hydrophones is $2\pi d/\lambda\ S_{\text{in}}$ and it was assumed that $2\pi d/\lambda \ll 1$, the output could run into intrinsic noise floors. Second, while the sensitivity is low for a far-field signal and noise it is not for near-field signals and noise. In fact, this is the basis for many microphones that allow a singer to use the microphone without massive feedback or cross-talk from other singers or band instruments. In sonar applications, these hydrophones are usually only used for stationary arrays because they respond poorly to near-field flow or platform noise.

3.16 Multiplicative Arrays

Instead of linear beamforming, adding or subtracting elements with weightings and time delays, it is possible to develop arrays by multiplying the outputs of elements together to form multiplicative or correlation arrays. These are sometimes referred to as super directive arrays.

A two-element array can be made to yield directivity equal to that of an n-element linear array. However, this requires a high signal-to-noise ratio. As we have shown above, the output of a linear array (with an even number of hydrophones) is given by

$$R(\theta) = \sum A_j \cos(2ju) \tag{3.70}$$

where

$u = \pi d\ \sin(\theta)/\lambda$
A_j = shading coefficients (assumed to be symmetric about the center)

Figure 3.25 shows the processing necessary to carry out this procedure, where $T_n(x)$ is the Chebyshev polynomial of the first order. In the case where the signal to noise is infinite, averaging over a half period is sufficient. If this is not the case, longer averaging times would be required to reduce the noise contribution.

Other schemes for nonlinear processing include: intraclass [3] and time-averaged product (TAP) array processing [4]. The idea here is to replace the usual linear additive operations

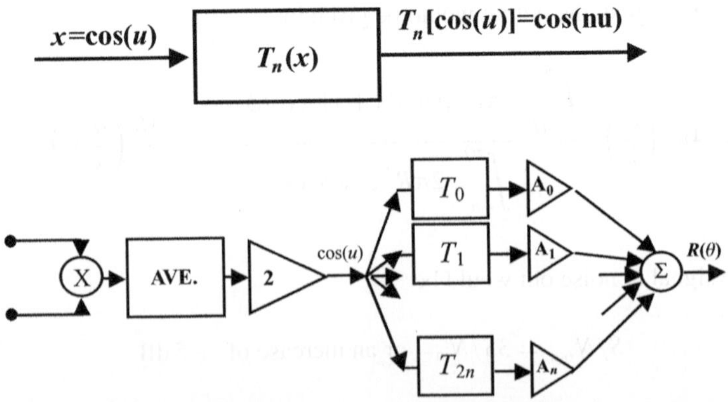

Figure 3.25 Array processing to synthesize directivity of linear multispot arrays using two hydrophones and a nonlinear scheme

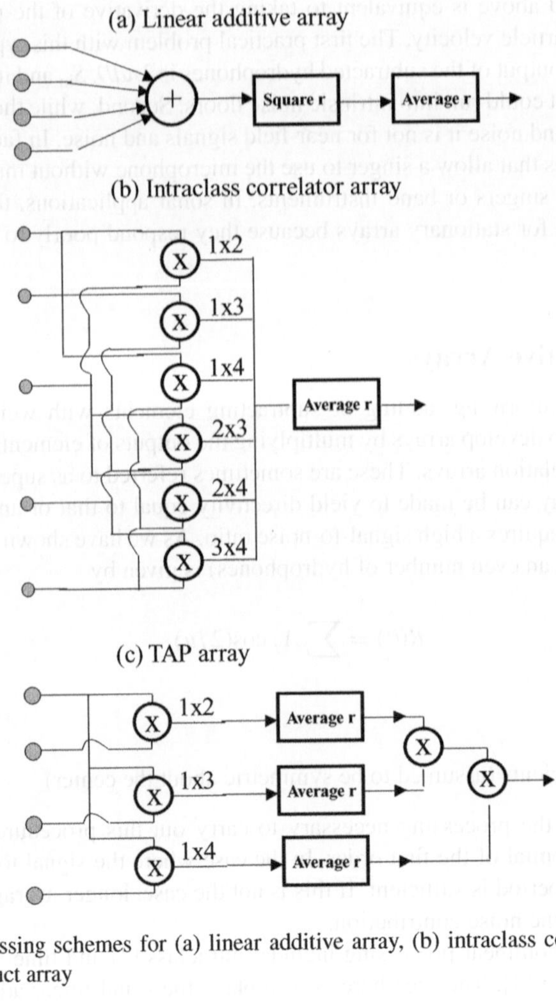

Figure 3.26 Processing schemes for (a) linear additive array, (b) intraclass correlator array, and (c) time-averaged product array

Table 3.5 System performance for linear additive, intraclass, and time-averaged product (TAP) processes

	$(S/N_{out}) = d$	Detection threshold $=10 \log(S/N_{in})$
Add and square law detected	$(n-1)^2\ \omega T(S/N_{in})^2$	$5 \log\{d/[(n-1)^2\ \omega T]\}$
Intraclass correlator	$n(n-1)\omega T(S/N_{in})^2$	$5 \log\{d/[n(n-1)\ \omega T]\}$
TAP	$n!n/2\ \omega T(S/N_{in})^m$	$10/n \log(2d/[n!n\omega T])$

with some form of nonlinear processing. Figure 3.26 shows two examples of such processing compared to the linear case for four hydrophones. Unfortunately, because of the averaging taking place in the beamforming, it is not possible to cleanly separate the array gain and the recognition differential as is done for linear schemes. Table 3.5 shows the system gains for these two types of processors as compared to the normal linear process.

As an example, let $n = 50$, $\omega T = 100$, and $d = 5$. The detection thresholds are -23.41, -23.45, and -13.44 respectively. The normal and intraclass processors are almost identical; the TAP processor is 10 dB worse.

3.17 Sparsely Populated Arrays

When an array with evenly spaced hydrophones is conventionally beamformed at frequencies above the design frequency, side lobes appear with responses approaching that of the desired main lobe (see Figure 3.27). The figure on the left is at 400 Hz and the figure on the right is at 800 Hz. Note the side lobes at 110° and 250° at 800 Hz. As the frequency changes, the position of the strong lobes shifts, making the directivity index oscillate above the design frequency, as the dark curve does in Figure 3.28.

If the spacing is nonuniform or random, the oscillation is frequently damped and the high-level side lobes are reduced. This same principle can be applied in three dimensions. Given this, it is possible to achieve nearly constant directivity versus frequency over a considerable frequency range without major side lobe problems.

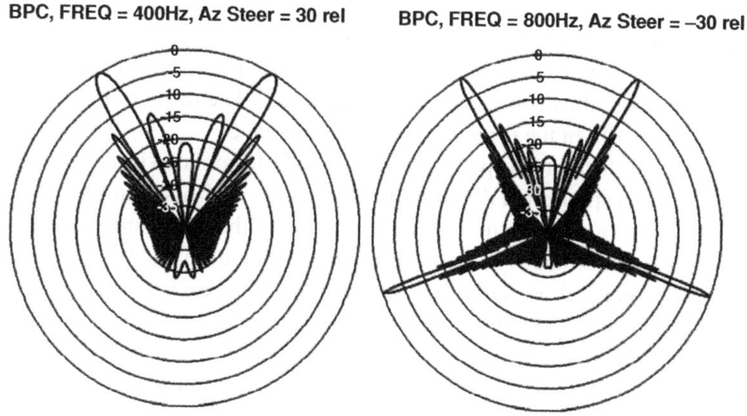

Figure 3.27 Horizontal beam pattern for a line array with a design frequency of 485 Hz beamformed for 30° off forward end-fire

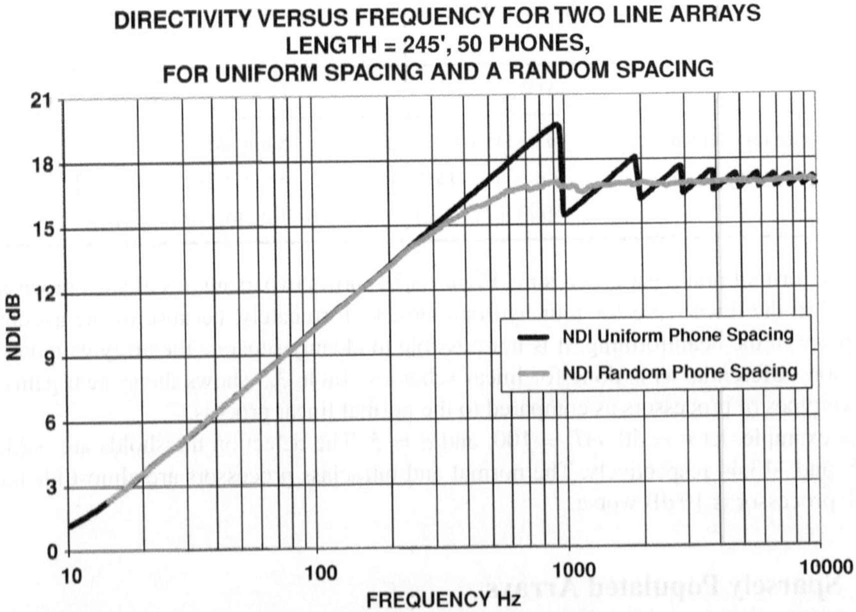

Figure 3.28 Directivity index versus frequency for a line array with uniform spacing and a line array with random spacing

3.18 Adaptive Beamforming

The difference between conventional beamforming and adaptive beamforming is that the latter can adjust to the noise environment, reducing sensitivity in directions of high noise, reverberation, jamming, or simply reducing the gain of noisy hydrophones. Various methodologies of this type of beamforming can be applied to arrays.

An early technique called DICANNE, or Digital Interference Cancelling Adaptive Null Network Equipment [5], formed an estimator beam in the direction of the interference. This interference signal, with appropriate time delays, is subtracted from each element of the array. The resulting outputs were then conventionally beamformed. The result of this processing was a beam pattern having a dip in the direction of the interference.

Modern adaptive beamforming of spatial arrays is usually formulated in terms of a minimum-variance distortionless response (MVDR) optimality criterion. This results in a constrained Wiener filter, which requires the estimation and inversion of the array–space–time covariance matrix, or the cross-spectral density matrix in the frequency domain. Weighting or shading coefficients, which in general are complex, are computed for each hydrophone used in an aperture for each look direction. This is quite computationally intense and the accuracy is limited by the available sample sizes. Adaptive methodologies used include the least mean square, recursive least square, and sample matrix inversion algorithms, or use of the conjugate gradient method.

It is instructive to compute the effect of interference on array gain. Consider an idealized case where the beam power response is one (0 dB) within 5° of the target direction and

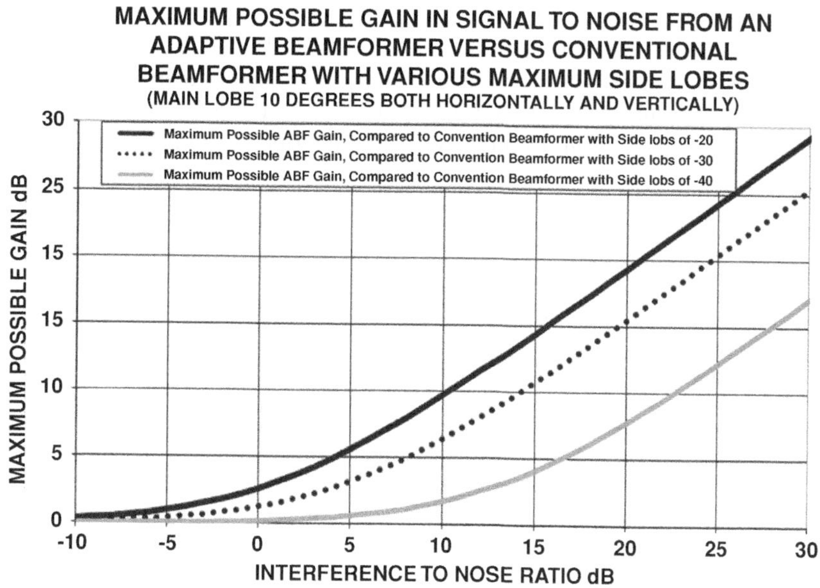

Figure 3.29 Maximum possible gain in the signal to noise from an adaptive beamformer versus a conventional beamformer with various side lobe levels

0.0001 (−40 dB) everywhere else. In an isotropic noise field, the array gain (AG_{iso}), which by definition is also the directivity (DI), is

$$AG_{iso} = 10\log\left(\frac{S}{4\pi N_0}\right) - 10\log\left\{\frac{S}{\pi\left(5\frac{\pi}{180}\right)^2 N_0 + 0.0001 N_0\left[4\pi - \pi\left(5\frac{\pi}{180}\right)^2\right]}\right\} = 27\ \text{dB} \tag{3.71}$$

where N_0 is the isotropic noise intensity per steradian. Next, introduce an interfering source in the side lobe region with an intensity ($I = 4\pi N_0$) that will double the noise received by an omnidirectional hydrophone. In this case, the array gain AG(I) is given by

$$\text{AG(I)} = 10\log\left(\frac{S}{8\pi N_0}\right) - 10\log\left\{\frac{S}{\pi\left(5\frac{\pi}{180}\right)^2 N_0 + 0.0001 N_0\left[4\pi - \pi\left(5\frac{\pi}{180}\right)^2\right] + 0.0001 \times 4\pi N_0}\right\} = 29.8\ \text{dB} \tag{3.72}$$

The array gain increased by 2.8 dB. However, the signal-to-noise ratio for the omnidirectional hydrophone with interference present has been reduced by 3 dB. With the directional array, the signal-to-noise ratio has been reduced by 0.2 dB. Therefore, in this case, a perfect adaptive beamformer could gain a maximum of 0.2 dB. Figure 3.29 shows the maximum possible gain from adaptive beam forming versus the interference-to-noise ratio for several side lobe levels.

References

[1] Anderson, V. C., and Munson, J. C., "Directivity of Spherical Receiving Arrays," *Journal of the Acoustical Society of America*, **35**, 1963, 1162.

[2] Brown, J. L., and Rowlands, R. O., "Design of Directional Arrays," *Journal of the Acoustical Society of America*, **31**, 1959, 1638.

[3] Faran, J. J., and Hills, R., "Application of Correlation Techniques to Acoustic Receiving Systems," Harvard University Acoustical Research Laboratory, Technical Memo, 27 September 1952.

[4] Berman, A., and Clay, C. S., "Theory of Time-Averaged-Product Arrays," *Journal of the Acoustical Society of America*, **29**, 1957, 805.

[5] Anderson, V. C., "DICANNE, a Realizable Adaptive Process," *Journal of the Acoustical Society of America*, **45**, 1969, 398.

4

Active Sonar Sources

Active sonar sources transmit the signal intended to be detected. These sources are called projectors and are frequently arrays of transducers that convert electrical power into directional sound. These same arrays of hydrophones may also be used to receive signals.

4.1 Source Level

If 1 watt of acoustic power is transmitted omnidirectionally, then the intensity at a distance of 1 meter is 1 watt per 4π square meters or 0.08 watts per square meters. For the standard pressure reference of an rms level of 1 microPascal (1 μPa), the intensity in sea water is

$$I_0 = \frac{p^2}{\rho c} = \frac{\left(10^{-6}\right)^2}{1.5 \times 10^6} = 6.67 \times 10^{-19}\ \text{W/m}^2 \tag{4.1}$$

The sound pressure level (SL) for a 1 watt omnisource is given by

$$L_\text{p} = \text{SL} = 10 \log\left(\frac{I}{I_0}\right) = 10 \log\left(\frac{0.08}{6.67 \times 10^{-19}}\right) = 170.8\ \text{dB} \tag{4.2}$$

If additional power is used, this becomes

$$L_\text{p} = \text{SL} = 170.8 + 10 \log(P) \tag{4.3}$$

where

P = acoustic power (W)

SL in dB re: (relative to) 1 μPa, at 1 meter

If this calculation had a reference distance of 1 yard, the constant would have been 171.5 instead of 170.8. If the source is directional, i.e., the power is concentrated, then the intensity in the look direction would be higher by the ratio of the ensonified solid angle, Ω_T, to 4π

Underwater Acoustics: Analysis, Design and Performance of Sonar Richard P. Hodges

steradians. This gain is the transmitter directivity, DI or NDI:

$$\mathrm{DI} = \mathrm{NDI} = 10\log\left(\frac{\Omega_T}{4\pi}\right) \tag{4.4}$$

The resulting SPL would be

$$\begin{aligned} L_\mathrm{p} &= 170.8 + 10\log(P) + \mathrm{NDI} \quad \mathrm{re}: 1\ \mu\mathrm{Pa},\ 1\ \mathrm{m} \\ \mathrm{SL} &= 170.8 + 10\log(P) + \mathrm{DI} \quad \mathrm{re}:\ 1\ \mu\mathrm{Pa},\ 1\ \mathrm{m} \end{aligned} \tag{4.5}$$

or

$$\begin{aligned} L_\mathrm{p} &= 171.5 + 10\log(P) + \mathrm{NDI} \quad \mathrm{re}: 1\ \mu\mathrm{Pa},\ 1\ \mathrm{yd} \\ \mathrm{SL} &= 171.5 + 10\log(P) + \mathrm{DI} \quad \mathrm{re}: 1\ \mu\mathrm{Pa},\ 1\ \mathrm{yd} \end{aligned} \tag{4.6}$$

It is important to remember that P is the acoustic power, not the electrical power, supplied to the transducers. The difference is the conversion efficiency that typically varies from 20 to 80 % depending on the construction of the hydrophone and on how narrowly the hydrophone is tuned in frequency.

Frequently, active transmitters are almost omnidirectional in the horizontal, so as to cover all possible directions of the target. However, because usually only a small region of vertical angles (depression/elevation angles (D/E)) will connect to possible targets with reasonable propagation loss, it is not valuable for it to be vertically omnidirectional. A single transducer can seldom achieve the source level required to achieve the desired performance. Therefore, it is desirable to have multiple transducers for reliability and to prevent cavitation.

4.2 Cavitation

As we will see in Chapter 10, Radiated Noise, cavitation can occur if the local pressure becomes negative, in this case because the signal overcomes the ambient noise. As shown in Table 4.1, a source level of 220 dB results in an rms pressure of 1 atmosphere at 1 yard. Therefore, at shallow depths this might cause cavitation. If bubbles form on the head of the transducer, they will scatter and absorb the signal, as well as change the impedance into which

Table 4.1 Select sound pressure levels and associated parameters for water (impedance of sea water = $\rho c = 1.5 \times 10^6$ kg/m² s)

Sound level	Intensity I(W/m²)	Pressure rms p(N/m²)	Particle velocity u(m/s)	Particle displacement u/ω (m) at 3 kHz
Sea state 3 (50 dB re: 1 μPa)	6.7×10^{-14}	3.2×10^{-4}	2.1×10^{-10}	1.1×10^{-14}
Active source (220 dB re: 1 μPa)	6.7×10^{3}	1.0×10^{5}	6.7×10^{-2}	3.5×10^{-6}

Note: Pressure for active source is 1 atmosphere.

the transducer must work. All three of these effects will degrade performance. The cavitation threshold at 1 atmosphere is nominally 4000 watts per square meters (the 6.7×10^3 value in Table 4.1 is for an rms pressure of 1 atmosphere at 1 yard, not a peak pressure of 1 atmosphere at 1 meter). As the depth, D, of water increases, the threshold intensity increases:

$$I_{\text{thres}}(D) = 4000\left(1 + \frac{D}{33}\right)^2 \ \text{W/m}^2 \tag{4.7}$$

where D is the depth in feet. In reality, this only applies for long pulses and low frequencies. Figure 4.1 shows the frequency dependence being almost constant to 10 kHz, then climbing very rapidly [1]. An expression for this is

$$P_{\text{thres}}(F) = 1 + 0.15F + 0.000\,107F^{2.4} \quad \text{atmospheres} \tag{4.8}$$

$$I_{\text{thres}}(F) = 4000P_{\text{thres}}^2 \ \text{W/m}^2 \tag{4.9}$$

where F is the frequency in kHz. The threshold increases for very short pulses, up to 3 times greater for a 0.5 ms pulse than for a 10 ms pulse, but remains constant for longer pulses

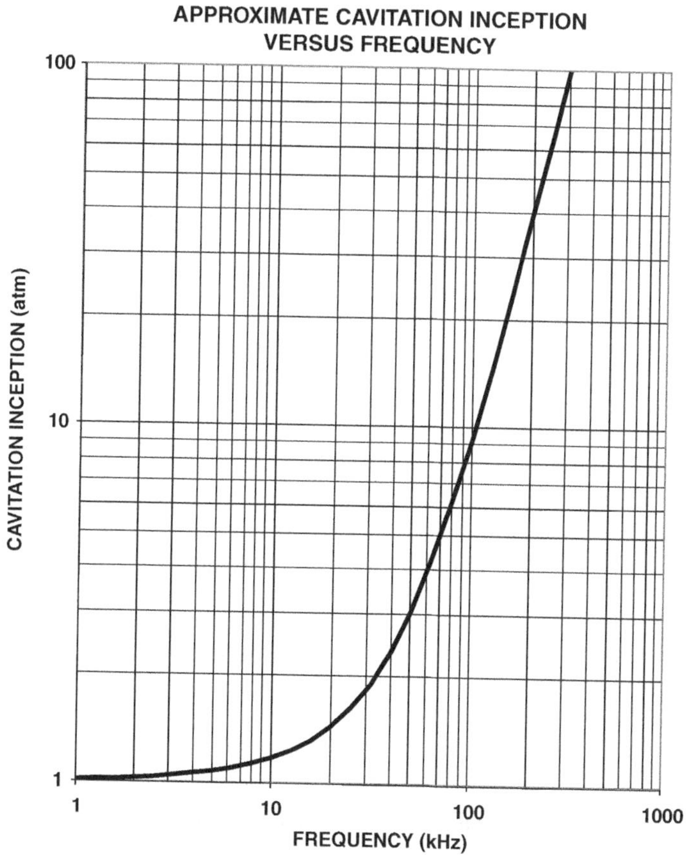

Figure 4.1 Cavitation inception versus frequency

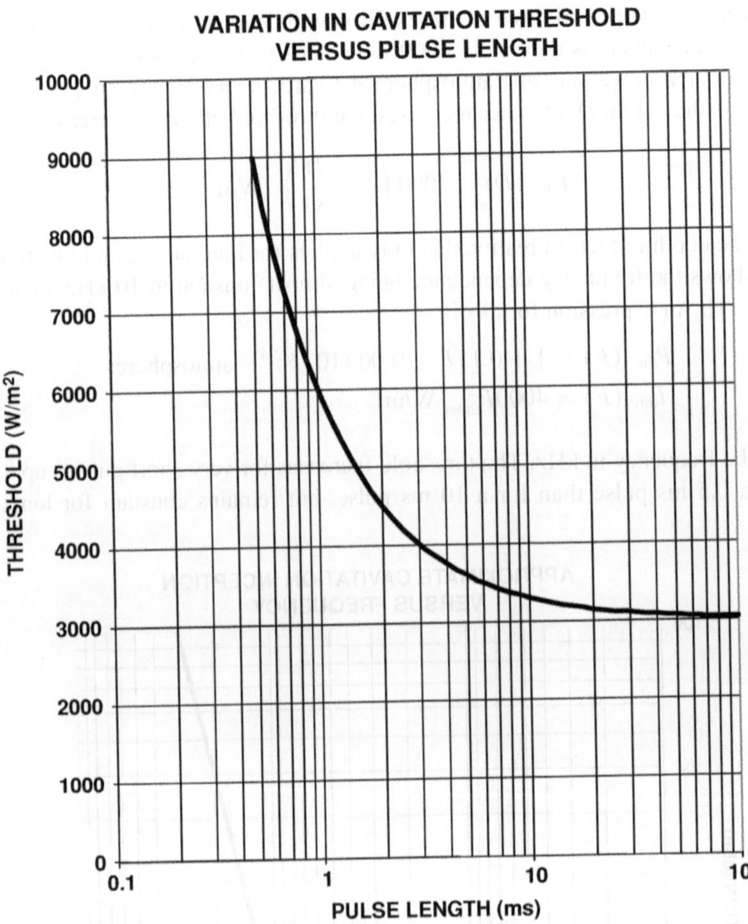

Figure 4.2 Variation in cavitation threshold as a function of pulse length

(Figure 4.2) [2, 3]. An expression for this is

$$I_{\text{thres}}(\tau) = 4000\left(1 + \frac{1}{\tau}\right)\ \text{W/m}^2 \tag{4.10}$$

where τ is the pulse length in ms. Combining all three expressions yields

$$I_{\text{thres}}(D, F, \tau) = 4000\left(1 + \frac{D}{33}\right)^2 (1 + 0.15F + 0.000\,107^{2.4})^2 \left(1 + \frac{1}{\tau}\right) \tag{4.11}$$

In decibels, relative to 1 microPascal at 1 meter (relative to 1 μPa, 1 m), this becomes

$$L_{\text{p}} = \text{PL} = 181.79 + 10\log[I_{\text{thres}}\,(D, F, \tau)] \tag{4.12}$$

Table 4.2 Select sound pressure levels and associated parameters for air (impedance of air = ρc = 415 kg/m^2 s)

Sound level	Intensity I(W/m^2)	Pressure rms p(N/m^2)	Particle velocity u(m/s)	Particle displacement u/ω (m) at 440 Hz
Threshold of human hearing (26 dB re: 1 μPa)	9.6×10^{-13}	2.0×10^{-5}	4.8×10^{-5}	1.7×10^{-11}
Normal conversation (106 dB re: 1 μPa)	9.6×10^{-7}	2.0×10^{-2}	4.8×10^{-5}	1.7×10^{-8}
Rock band/threshold of pain (146 dB re: 1 μPa)	1	2.0×10^{1}	4.8×10^{-2}	1.7×10^{-3}

Note. Human hearing works over a vast power range, 10^{12}. The particle displacement at the threshold of human hearing at middle C is about 6 % of the diameter of a hydrogen molecule, so the ear is able to detect tiny movements.

where the 181.79 constant is equal to 170.8 + 10 log(4π). While Equation 4.12 seems very precise, the experimental data scatters over a fairly large area (a factor of three either way), dependent on such parameters as air content of the fluid and smoothness of the transducer head. While we have computed the mean projector level for a transmitter, there are likely to be hot spots in the near-field.

Revisiting Tables 1.3 and 1.4, Table 4.1 shows select sound pressure levels of interest for water and Table 4.2 shows similar levels with corresponding parameters for air.

4.3 Near-Field Interactions

When large arrays of closely spaced resonance transducers are driven electrically, the signals from one hydrophone will interact with other nearby hydrophones. The signal from one hydrophone, arriving at an adjacent hydrophone, shifts the effective impedance of the medium. If the positive phase of a signal arrives at a hydrophone that is trying to transmit the negative phase, that hydrophone can act as a sink and absorb the energy. Unless this interaction is compensated for, it will reduce the total power output, distort the beam pattern, and produce mismatches with the transmitter, which could damage it, the projectors, or both. There are four methods that have been used to reduce this interaction:

1. Separate the elements. This may distort the beam pattern.
2. Insert a reactance in series with each element. This only works at a single frequency.
3. Make the individual elements large enough that their self-radiation impedance is greater than the mutual radiation impedance between the two elements.
4. The most modern approach uses individual amplifiers on each element to maintain the correct phase and amplitude (e.g., using feedback accelerometers in the transducers).

4.4 Explosive Sources

The most commonly used underwater explosive sources have a fraction of a pound to a few pounds of high explosives. These sources are used for measuring propagation loss or for echo ranging to locate submarines. Somewhat larger charges are used for seismic soundings exploring for oil and gas. Three excellent sources of information on this topic are Cole [4], a three-volume compendium by the Office of Naval Research [5], and Glasstone and Dolan [6].

4.5 Physics of Shock Waves in Water

The characteristics of shock waves in water were well understood by the end of World War II. Here, we will discuss only high explosives that have detonation speeds which are much larger than the speed of sound in water, as opposed to burning or deflagration of materials like gunpowder that have much lower speeds. The detonation of a high explosive produces a shock front where the pressure over a distance measured in a few molecular widths rises to a high peak pressure. This peak is followed by an exponential decay of the pressure. An explosion underwater generates a gas bubble. As the bubble rapidly expands, the internal pressure drops to eventually become less than the ambient pressure as momentum carries the bubble past the equilibrium point. The bubble then collapses inward, only to rebound multiple times, creating what are referred to as bubble pulses. As this happens, the buoyancy of the bubble carries it upward, as shown in Figure 4.3.

Explosives are relatively efficient at producing acoustic energy with about 50 % of the chemical energy being displayed as far-field acoustic energy. Table 4.3 shows the portioning of energy for TNT [7].

The equation for a shock wave pressure is given by [8]

$$P = P_0 \exp\left(\frac{-t}{\tau}\right) \tag{4.13}$$

$$P_0 = K\left[\frac{w^{1/3}}{(R - R_0)}\right]^{\alpha} \tag{4.14}$$

where

K and α = are given in Table 4.4 (adapted from Cooper [9])
P_0 = pressure (psi)
w = weight of the explosive (lb)
R = range (ft)
R_0 = radius of the explosive (ft)

A problem occurs at the interface between the explosive and water as R approaches R_0. The pressure must be obtained by solving for the shock wave interaction between the detonation products (P–u isentrope) and the water (P–u Hugoniot). This is about 100 kbar for TNT. Inside about 1.6, R_0 interpolate between this value and Equations 4.13 and 4.14 given above.

For TNT, at distances great compared to the radius of the explosive the equation becomes

$$P_0 = 2.17 \times 10^4 \left(\frac{w^{1/3}}{R}\right)^{1.13} \tag{4.15}$$

where

P_0 = pressure (psi)
w = weight of the explosive in equivalent pounds of TNT
R = range (ft)

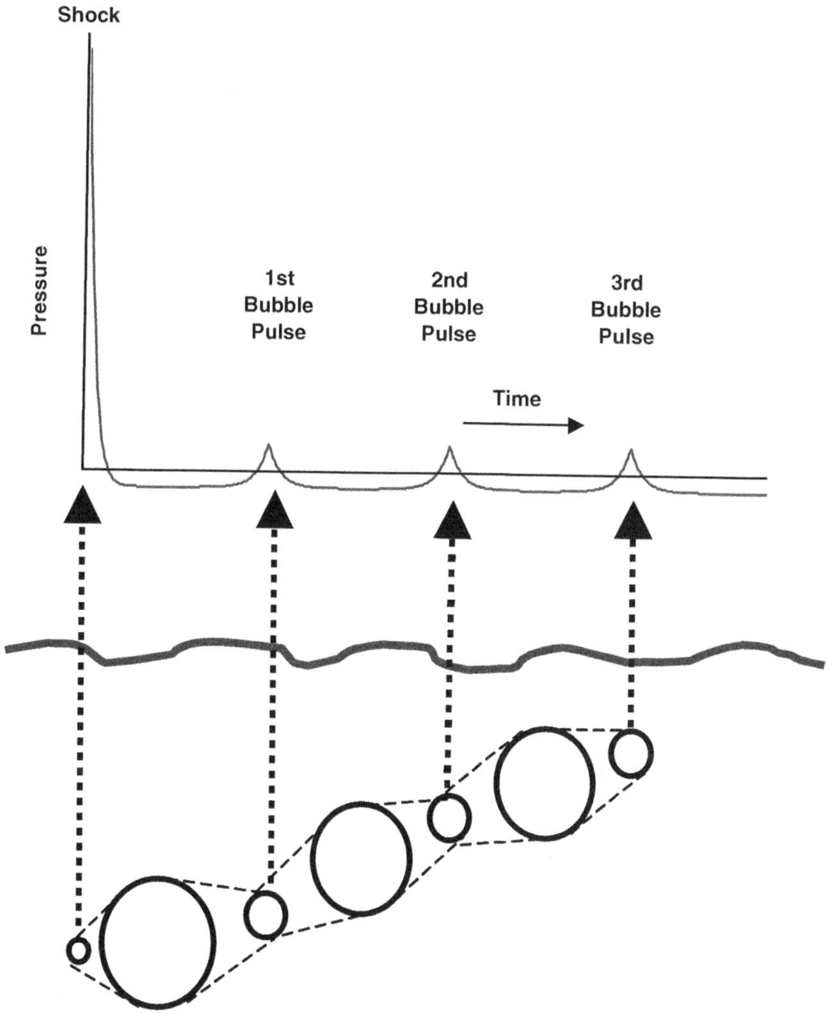

Figure 4.3 Pressure changes, bubble pulsations and depth migration

Table 4.3 Partition of the energy of an explosion of TNT in water

Total radiated acoustic energy[a]	410 cal/g
Shock and first two bubble pulses	
Energy radiated at third bubble pulse	95 cal/g
Shock front dissipation at short range[a]	200 cal/g
Apparent losses (turbulence, viscosity, heat conduction, chemical change in gaseous products)	345 cal/g
Total	1050 cal/g

[a]At $R = w^{1/3}/0.1$ m.

Table 4.4 *K* and α for selected explosives

Explosive	*K*	α	Heat of detonation (kcal/g)
Composition A-3	2.35×10^4	1.13	1.58
HBX-1	2.49×10^4	1.15	1.84
HBX-3(Torpex)	2.68×10^4	1.14	2.11
Pentolite (50 %PETN/50 %TNT)	2.26×10^4	1.13	1.53
PETN	2.32×10^4	1.13	1.65
Tetryl	2.15×10^4	1.15	1.51
TNT	2.17×10^4	1.13	1.41

The time constant associated with the pulse is given by

$$\tau = 58w^{1/3}\left(\frac{w^{1/3}}{R}\right)^{-0.22} \tag{4.16}$$

where

τ = time constant (μs)

Figure 4.4 shows the peak pressure versus range for 1 pound of TNT in water. Figure 4.5 shows the time constant versus range.

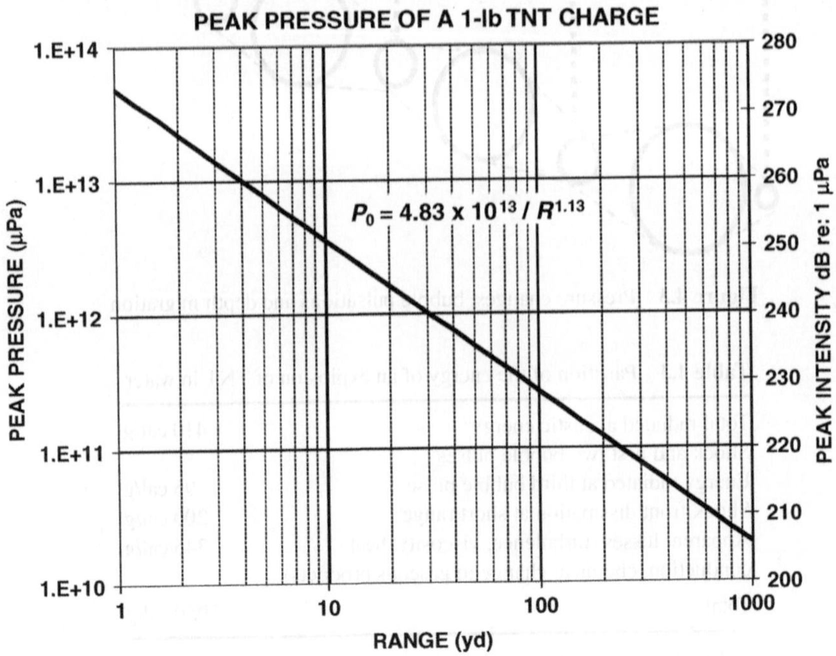

Figure 4.4 Peak pressure versus range for a 1 lb TNT charge

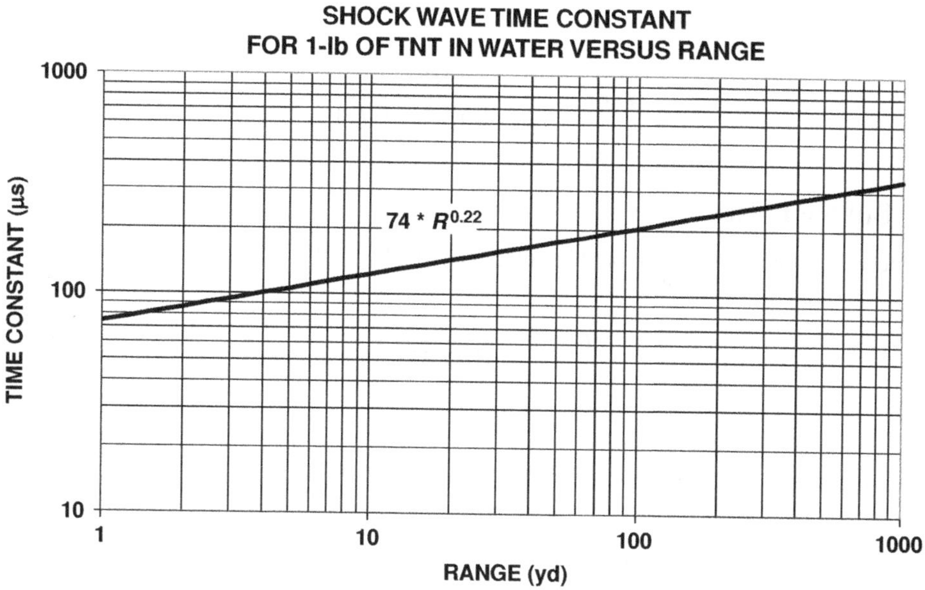

Figure 4.5 Time constant for a shock wave in water generated by a 1 lb TNT charge

Using the pressure versus time relationship, intensity versus frequency can be computed by taking the Fourier transform (see Appendix A):

$$
\begin{aligned}
P(f) &= \int_0^{\infty} P_0 \exp\left(\frac{-t}{\tau}\right) \exp(-i2\pi ft)\,\mathrm{d}t \\
I_0 &= \frac{P\bar{P}}{\rho c}\frac{2}{\tau} \\
I_0(f) &= \frac{2}{\rho c\tau}\left[\frac{\tau}{(1+2i\pi ft)}P_0\right]\left[\frac{\tau}{(1-2i\pi ft)}P_0\right] \\
I_0(f) &= \frac{P_0^2 2\tau}{\rho c\left(1+4\pi^2 f^2\tau^2\right)} \\
I_0(f) &= \frac{2P_0^2}{\rho c\tau\left(\dfrac{1}{\tau^2}+4\pi^2 f^2\right)}
\end{aligned}
\tag{4.17}
$$

The change in pulse length given above is only appropriate for relatively short ranges; after that the frequency-dependent absorption (attenuation) discussed in Chapter 5, Transmission Loss, and appropriate propagation loss models should be used. It is the convention for workers with explosive charge sources to solve the sonar equation in energy flux density instead of intensity, primarily because the energy flux density expression is less sensitive to the change in pulse length with distance:

$$E_0(f) = \frac{2P_0^2}{\rho c\left(1/\tau^2 + 4\pi^2 f^2\right)} \tag{4.18}$$

4.6 Bubble Pulses

The interval in seconds between bubble pulses is given by [10]

$$T_{\text{bubble}} = \frac{kw^{1/3}}{(D+33)^{5/6}} \tag{4.19}$$

where

k = a constant (4.36 for TNT)
w = weight of charge in lbs of TNT
D = depth below sea level of explosion (ft)

The shape of the bubble pulse pressure signature is generally modeled as an exponential rise followed by an exponential decay. If we assume the time constants are the same (T_1), then the spectral shape will be

$$P(f) = \int_0^{\infty} P_1 \exp\left(\frac{-t}{T_1}\right) \exp(-i2\pi ft)\,\mathrm{d}t + \int_0^{\infty} P_1 \exp\left(\frac{t}{T_1}\right) \exp(-i2\pi ft)\,\mathrm{d}t$$

$$P(f) = \frac{-2P_1T_1}{[(1+2i\pi fT_1)](-1+2i\pi fT_1)}$$

$$P(f) = \frac{2P_1}{1+4\pi^2 f^2 T_1^2}$$

$$I_{\text{bubble}} = \frac{2P\bar{P}}{\rho c T_1} \tag{4.20}$$

$$I_{\text{bubble}}(f) = \left(\frac{2P_1}{1+4\pi^2 f^2 T_1^2}\right)^2 \frac{2}{\rho c T_1}$$

$$I_{\text{bubble}}(f) = \frac{8}{\rho c T_1^3}\left[\frac{P_1}{\frac{1}{T_1^2} + 4\pi^2 f^2}\right]^2$$

$$E_{\text{bubble}}(f) = \frac{8}{\rho c T_1^2}\left[\frac{P_1}{\frac{1}{T_1^2} + 4\pi^2 f^2}\right]^2$$

where P_1 and T_1 are values corresponding to a particular bubble pulse, charge weight, and depth. Figure 4.6 shows the peak intensity and energy density for explosive sources without the bubble pulses.

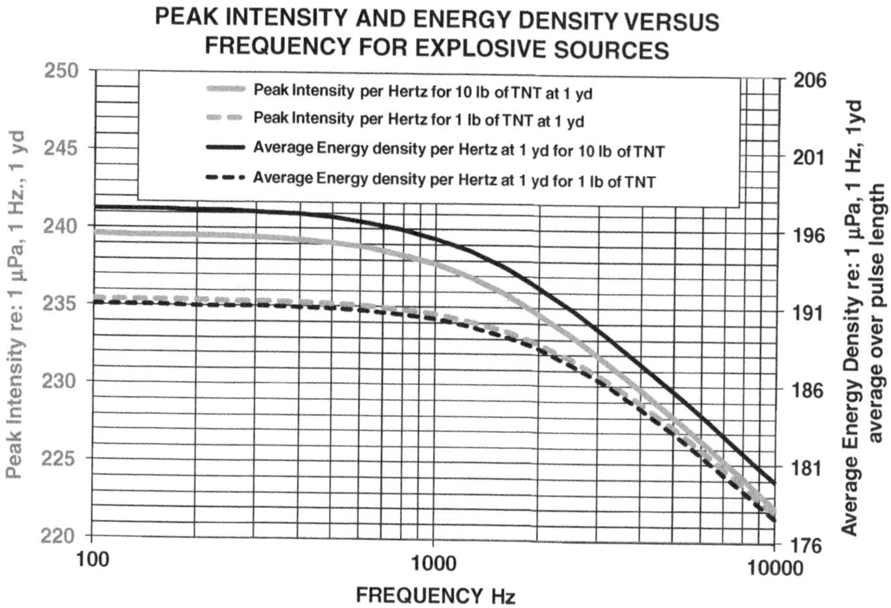

Figure 4.6 Peak intensity and energy density for explosive sources without bubble pulses

4.7 Pros and Cons of Explosive Charges

The advantages of explosive sources are small size, lack of power supply and wiring, broadband of frequencies, omnidirectionality, and short pulse duration (allowing good range discrimination).

The main disadvantage of explosive charges is the requirement to handle high explosives. Many of the same elements from the advantages list are also disadvantages: broadband of frequencies means the charge cannot be tuned to the region of interest and the signal processing is limited; omnidirectionality, which can be somewhat overcome by line charges, results in high reverberation; and short pulse duration limits the recognition differential. Thus, these sources are limited to applications where the merits outweigh the demerits.

4.8 Parametric Acoustic Sources

A parametric acoustic source uses the nonlinearity of the medium to form beams with unique properties. These beams are narrow (for the size of the array) and are nearly side lobe free. Because of the nonlinearity of the medium, if two frequencies are transmitted simultaneously at high amplitude from a directional hydrophone, new frequencies appear at the sum and difference of the two frequencies. The effective length of the "array" is determined by the attenuation of the primary signals. The resulting beam pattern for the new frequencies resembles an end-fired array with exponential shading. Typically, high primary frequencies are chosen so the hydrophone can be directional. The primary signals and the sum frequency decay rapidly because of attenuation, leaving the difference frequency for use. Thus, a small array operating at 20 and 21 kHz can produce a 1 kHz transmission that is very narrow relative to what it would be if the array simply transmitted 1 kHz. The down side is that the power at the difference frequency is much less than the power at the original transmitter. The original concept for this was done by Westervelt [11] with additional important papers by Muir and Willette [12] and by Moffett and Mellen [13].

References

[1] Urick, R. J., *Principles of Underwater Sound for Engineers*, New York: McGraw-Hill, 1967.

[2] Strasberg, M., "Onset of Ultrasonic Cavitation in Tap Water," *Journal of the Acoustical Society of America*, **21**, 1959, 163.

[3] Mason, W. P., "Physics of Acoustic Cavitation in Liquids," in *Physical Acoustics*, Vol. 1, Part B, Chapter 9, New York: Academic Press Inc., 1964.

[4] Cole, R. H., *Underwater Explosions*, Princeton, NJ: Princeton University Press, 1948.

[5] "Underwater Explosions Research," A Compendium of British and American Reports, Washington, DC: U.S. Navy Office of Naval Research, 1950.

[6] Glasstone, S., and Dolan, P. K., "The Effects of Nuclear Weapons," Washington, DC: U.S. Department of Defense and U.S. Department of Energy, 1977.

[7] Arons, A. B., and Yennie, D. R., "Long Range Shock Propagation in Underwater Explosion Phenomena I," U.S. Navy Department Bureau of Ordnance, NAVORD Report 478, 1949.

[8] Arons, A. B., "Underwater Explosion Shock Wave Parameters at Large Distances from the Charge," *Journal of the Acoustical Society of America*, **26**, 1954, 343.

[9] Cooper, P. W., *Explosives Engineering*, Weinheim, Germany: Wiley-VCH, November 1996.

[10] Arons, A. B., "Secondary Pressure Pulses Due to Gas Globe Oscillations in Underwater Explosions II: Selection of Adiabatic Parameters in the Theory of Oscillation," *Journal of the Acoustical Society of America*, **20**, 1948, 277.

[11] Westervelt, P. J., "Parametric Acoustic Array," *Journal of the Acoustical Society of America*, **35**, 1963, 535–537.

[12] Muir, T. G., and Willette, J. G., "Parametric Acoustic Transmitting Arrays," *Journal of the Acoustical Society of America*, **52**, 1972, 1481–1486.

[13] Moffett, M. B., and Mellen, R. H., "Model for Parametric Acoustic Sources," *Journal of the Acoustical Society of America*, **61**, 1977, 325–337.

5

Transmission Loss

As an acoustic signal travels through the ocean, it is weakened and distorted by several mechanisms. In this chapter and the next, we will discuss the physical basis for these effects and take a brief look at a number of the currently (2009) popular models used in practice to accomplish the evaluation of environmental effects on sound propagation. It is important for the practitioner to be familiar with the mechanisms behind these models so as to have a better appreciation for those cases where the models break down.

Transmission loss or propagation loss is defined as the loss in intensity (power per unit area) in decibels (dB) between a point of interest and a reference distance (usually 1 meter or 1 yard). If I_0 is the intensity at the reference distance, then the transmission loss, TL or Nw, at a distance, R, and depth, D, is given by

$$\mathrm{TL} = \mathrm{N_W} = -10 \log \left(\frac{I(R, D)}{I_0} \right) = -20 \log \left(\frac{|p(R, D)|}{|p_0|} \right) \tag{5.1}$$

where

$I(R,D)$ = acoustic intensity at a range, R, and depth, D, of interest
I_0 = acoustic intensity at the reference distance
$p(R,D)$ = sound pressure at a range, R, and depth, D, of interest
p_0 = sound pressure at the reference distance

A time-averaging process is implied since instantaneously the pressure or the intensity could be zero, resulting in $\pm\infty$ as a loss. For continuous signals, this means measuring long enough to smooth out fluctuations. However, for short-duration signals, a definitional problem arises. The phenomenon of multipath, which will be discussed later in this chapter, may result in multiple versions of the signal arriving with little or no overlap in time because each version traveled different paths with almost certainly different intensities. There are several options for managing the effects of multipath: add up all the paths, process each path individually providing multiple TLs for each point of interest; or only sum each path over a specified time window resulting in multiple, window-dependent values. Which option is chosen depends on what is being calculated. If the system of interest can display or detect individual paths, then each path can be processed independently. If the system of interest has a fixed, relatively short

Underwater Acoustics: Analysis, Design and Performance of Sonar Richard P. Hodges

duration time window, then the third choice is most appropriate. Otherwise, for a long time integration window, the first option of summing all paths is most appropriate.

As with all parameters, failure to state the reference distance, particularly if it is not 1 m or 1 yd, can result in huge errors. While many authors will refer to transmission loss as dimensionless, it is not, as the TL value depends on the reference distance chosen.

Conditions that affect propagation loss are spreading, absorption, scattering (either in the volume or at the surface or bottom boundaries), multiple paths combining, and wave effects, e.g., duct leakage.

Throughout this chapter, acoustic propagation will be discussed as though it is both a particle and a wave, much like quantum mechanics. In all cases, the wave model of sound will provide the correct answer, but for some cases the ray model will be adequate. In these cases, the ray model may be much less computationally intense and can provide visual insights not always obvious in the wave domain.

5.1 Sound Speed Profile in the Sea

Sound speed profiles are frequently measured using bathythermographs (BT) or expendable bathythermographs (XBT), which measure temperature versus depth. The speed of sound is computed by assuming a value for salinity and converting the depth to a pressure value. More modern measurement devices also measure the conductivity (CTD or XCDT), allowing for a better estimate of the salinity.

As we saw in Chapter 1, Introduction to Sonar, the speed of sound increases with temperature, pressure or depth, salinity, and decreases with air content. The first 20 meters of water absorbs almost all of the sun's energy that hits the surface of the water. Mixing caused by waves distributes the energy downward. If the wind or wave height is significant, this mixing will form an isothermal layer, called a mixed layer, surface layer, or surface duct, which tends to trap sound. Below this layer, the temperature usually declines steadily, in a region called the theromocline. The thermocline is split into two regions: the seasonal and main thermoclines. Seasonal thermoclines vary considerably from month to month, whereas the main thermocline is almost constant over the year. The thermocline begins near the surface (below the surface layer if any is present) and extends down to around a thousand meters.

Figure 5.1 shows a typical deep water sound speed profile (SSP). The upper portion of the SSP moves higher in the summer and lower in the winter due to seasonal temperature changes. Ninety percent of the oceans' waters are below the thermocline. Deep in the major oceans, the bottom temperature is between 0 and 3 degrees Celsius (32–38 degrees Fahrenheit). Ocean water, with an average salinity of 35 parts per thousand (ppt), freezes at −1.94 °C (28.5 °F). At very high Northern and Southern latitudes, ocean water can reach these low temperatures and freeze. When forming ice, water tends to reject dissolved salts so that sea ice has only a fraction of a part per thousand of salt. The excess salt dissolves in the water below and makes it denser. The sinking of this cold salty water to the bottom helps maintain the return currents in the ocean, replacing the warm water transported north by surface currents like the Gulf Stream and the Kuroshio Current. In the summer, melting sea ice creates a fresh water layer in the Arctic, resulting in a very effective surface duct.

The Naval Oceanographic Office at Stennis Space Center in Mississippi maintains a web site (http://www.navo.navy.mil/) of vast collections of downloadable data from which Figure 5.1

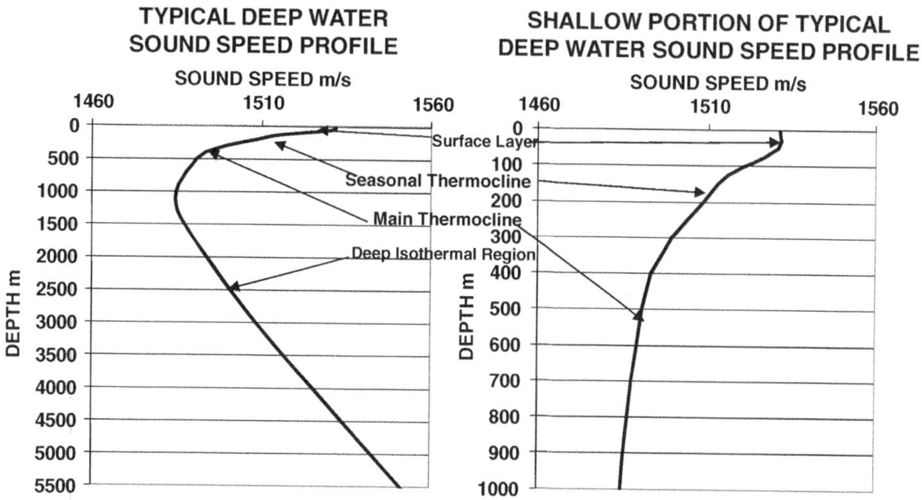

Figure 5.1 Typical deep water sound speed profile (SSP)

was taken. These data include historical SSP and bottom depth data available for most of the world. Generalized Digital Environmental Model Variable Resolution (GDEMV) and Digital Bathymetric Data Base Variable Resolution (DBDBV) are the databases of interest. Table 5.1 gives the SSP data extracted from this site for a specific latitude and longitude in the Philippine Sea during the month of January. The data contain depth values in meters, temperature in degrees Centigrade, salinity in parts per thousand, and sound speed in meters per second.

The U.S. Navy also provides near real-time estimates of SSPs gridded over an area from the Modular Ocean Data Assimilation System (MODAS) program. The MODAS program is a modular toolkit for estimating present and future conditions in the oceans. It presently consists of over 100 individual programs designed to:

1. Acquire and quality-control input data of various types including satellite remote sensed information.
2. Use satellite data to refine climatological temperature and salinity data.
3. Merge in situ measurements with a "first guess" field to produce a "best guess" of the present conditions in the ocean.
4. Provide a short-term forecast of the ocean, including currents and tides.

This system was designed to meet the U.S. Navy's need to produce rapid estimates of present and near-term ocean conditions, often in situations where little or no in situ data was available. MODAS modules are presently (2008) running at various Navy facilities.

5.2 Snell's Law and Transmission across an Interface

Snell's law states that sound and electromagnetic waves like light change direction when they pass between media with different propagation speeds nonnormal to the interface.

Table 5.1 SSP data for the Philippine Sea in January. (Extracted from GDEMV)

Latitude: 18.67 Longitude: 129.67 Valid days: 001–031
Points in profile: 54 min. depth: 0 max. depth: 5804
Version: OAML GDEMV Reader 1.2
Classification: Public Domain

Depth	Temp.	Salinity	Sound speed	Depth	Temp.	Salinity	Sound speed
0	26.22	34.68	1537.3	275	16.79	34.74	1517.11
5	26.22	34.68	1537.38	300	15.93	34.67	1514.81
10	26.22	34.68	1537.46	325	15.09	34.61	1512.52
15	26.22	34.68	1537.55	350	14.25	34.55	1510.16
20	26.22	34.68	1537.63	375	13.41	34.49	1507.74
25	26.21	34.69	1537.7	400	12.57	34.44	1505.26
30	26.21	34.69	1537.78	450	11.14	34.36	1501.03
35	26.17	34.7	1537.78	500	9.71	34.29	1496.61
40	26.15	34.7	1537.82	550	8.56	34.25	1493.1
45	26.13	34.71	1537.87	600	7.42	34.22	1489.51
50	26.11	34.72	1537.92	650	6.62	34.23	1487.21
55	25.99	34.72	1537.73	700	5.82	34.24	1484.86
60	25.88	34.74	1537.58	800	4.78	34.3	1482.35
65	25.78	34.74	1537.43	900	4.15	34.35	1481.45
70	25.66	34.75	1537.25	1000	3.78	34.41	1481.65
75	25.56	34.76	1537.11	1100	3.52	34.48	1482.31
80	25.29	34.77	1536.57	1200	3.29	34.51	1483.04
85	25.02	34.78	1536.03	1300	3.05	34.55	1483.75
90	24.76	34.79	1535.51	1400	2.81	34.58	1484.44
95	24.5	34.8	1534.98	1500	2.56	34.58	1485.04
100	24.24	34.82	1534.46	1750	2.19	34.59	1487.67
125	22.46	34.85	1530.5	2000	1.98	34.6	1491.01
150	20.98	34.88	1527.09	2500	1.79	34.63	1498.77
175	20.15	34.88	1525.27	3000	1.73	34.66	1507.17
200	19.33	34.89	1523.43	4000	1.57	34.67	1523.98
225	18.49	34.85	1521.41	5000	1.6	34.68	1541.93
250	17.64	34.81	1519.31	5804	1.77	34.66	1557.14

Figure 5.2 shows an example of incident and transmitted rays and wave fronts for this case. Here, the speed is lower in the medium on the right side, medium 2. Since the frequency remains constant, the wave fronts are closer together. By simple trigonometry, Snell's law can be derived as follows:

$$C_1 \Delta t = \Delta Y \cos \theta_1 \tag{5.2}$$

$$C_2 \Delta t = \Delta Y \cos \theta_2 \tag{5.3}$$

Dividing yields

$$C_1 / C_2 = \cos \theta_1 / \cos \theta_2 = n \tag{5.4}$$

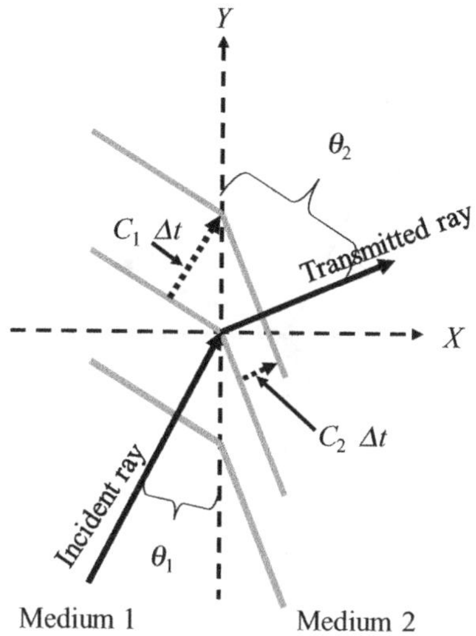

Figure 5.2 Passage of a wave through an interface at oblique incidence

where

n = the index of refraction

This is more commonly written as

$$\frac{\cos\theta_1}{c_1} = \frac{\cos\theta_2}{c_2} \tag{5.5}$$

By a similar argument, it can be shown that the reflected ray (although present in this example is not shown in Figure 5.2) will be reflected at the same angle as the incident ray, but in the upper left quadrant (Figure 5.3). If $c_2 > c_1$ in Equation 5.5, there will be a region of angle θ_1 for which there is no real solution for θ_2 because $c_2/c_1 \cos(\theta_1)$ could be larger than 1. The angle at which this occurs is called the angle of total reflection because at or beyond this angle, for two fluid media, all the energy is reflected and none is transmitted. Later, we will learn that this is really only true if the thickness of medium 2 is very large compared with a wavelength.

5.3 Reflection and Transmission Coefficients

Figure 5.3 is an example of the incident, transmitted, and reflected rays at a fluid interface. In an x,y coordinate system, the three rays can be written as

$$p_\mathrm{i} = p_\mathrm{i}\left[t - \frac{(x\sin\theta_\mathrm{i} + y\cos\theta_\mathrm{i})}{c_1}\right] \tag{5.6}$$

$$p_\mathrm{r} = p_\mathrm{r}\left[t + \frac{(x\sin\theta_\mathrm{r} - y\cos\theta_\mathrm{r})}{c_1}\right] \tag{5.7}$$

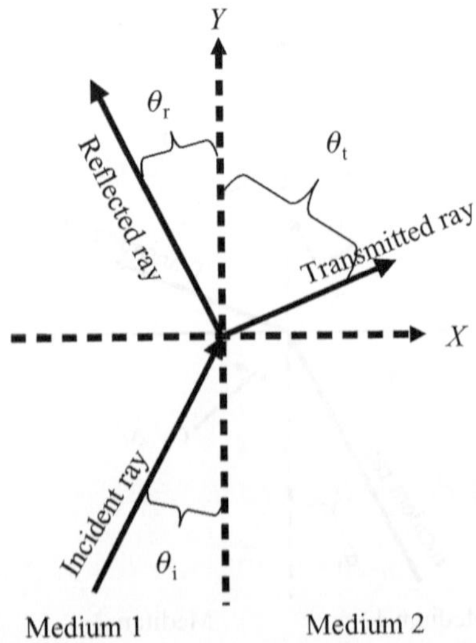

Figure 5.3 Passage of a wave through an interface at oblique incidence with a reflected ray as shown

$$p_t = p_t\left[t - \frac{(x\sin\theta_t + y\cos\theta_t)}{c_2}\right] \tag{5.8}$$

The sign change in Equation 5.7 reflects the direction of the ray, left or right. By applying the pressure boundary condition, which states that the pressure must be continuous across the interface ($x = 0$), we get

$$p_i\left[t - \frac{(y\cos\theta_i)}{c_1}\right] + p_r\left[t + \frac{(-y\cos\theta_r)}{c_1}\right] = p_t\left[t - \frac{(y\cos\theta_t)}{c_2}\right] \tag{5.9}$$

This must be true for all values of y.

If we define a new time $T = t - y\cos(\theta_i)/c_1$, then

$$p_i(T) + p_r\left[T + y\left(\frac{\cos\theta_i - \cos\theta_r}{c_1}\right)\right] = p_t\left[T - y\left(\frac{\cos\theta_t}{c_2} - \frac{\cos\theta_i}{c_1}\right)\right] \tag{5.10}$$

given $\theta_i = \theta_r$ and therefore

$$\frac{\cos\theta_t}{c_2} = \frac{\cos\theta_i}{c_1} \tag{5.11}$$

Using Snell's law again,

$$p_i(T) + p_r(T) = p_t(T) \tag{5.12}$$

or in terms of reflection and transmission coefficients

$$1 + R = T \tag{5.13}$$

where R is the pressure ratio of reflected to incident and T is the pressure ratio of transmitted to incident. Applying the additional boundary condition that the normal component of the particle velocity must be continuous across the interface gives

$$u_i \sin(\theta_i) + u_r \sin(\theta_r) = u_t \sin(u_t) \tag{5.14}$$

and $u_j = p_j/(\rho_j c_j)$, and $\rho_j c_j$ is the impedance Z_j, then

$$\frac{p_i \sin(\theta_i)}{\rho_i c_i} + \frac{p_r \sin(\theta_r)}{\rho_r c_r} = \frac{p_t \sin(\theta_t)}{\rho_t c_t} \tag{5.15}$$

Given $\theta_i = -\theta_r$, $\rho_i{}^* c_i = \rho_r{}^* c_r = Z_1$, and $\rho_t{}^* c_t = Z_2$,

$$\frac{p_i \sin(\theta_i)}{Z_i} - \frac{p_r \sin(\theta_i)}{Z_1} = \frac{p_t \sin(\theta_t)}{Z_2} \tag{5.16}$$

Dividing through by p_i and setting $R = p_r/p_i$ and $T = p_t/p_i$ gives

$$1 - R = \frac{Z_1 \sin(\theta_t)}{Z_2 \sin(\theta_i)} T \tag{5.17}$$

Using $1 + R = T$, T and R are

$$T = \frac{2\,Z_2 \sin(\theta_i)}{Z_2 \sin(\theta_i) + Z_1 \sin(\theta_t)} \tag{5.18}$$

$$R = \frac{Z_2 \sin(\theta_i) - Z_1 \sin(\theta_t)}{Z_2 \sin(\theta_i) + Z_1 \sin(\theta_t)} \tag{5.19}$$

R is known as the Rayleigh reflection coefficient. For normal incidence, where $(\theta_i = \theta_t)$,

$$T(\text{normal}) = \frac{2Z_2}{Z_2 + Z_1} \tag{5.20}$$

$$R(\text{normal}) = \frac{Z_2 - Z_1}{Z_2 + Z_1} \tag{5.21}$$

T and R are pressure ratios, not intensity. In intensity, space equivalent parameters are τ and r, where $\tau + r = 1$ by conservation of power and

$$\tau = \frac{4 Z_1 Z_2 \sin(\theta_i) \sin(\theta_t)}{[Z_1 \sin(\theta_t) + Z_2 \sin(\theta_i)]^2} \tag{5.22}$$

$$r = \frac{[Z_2 \sin(\theta_i) - Z_1 \sin(\theta_t)]^2}{[Z_1 \sin(\theta_t) + Z_2 \sin(\theta_i)]^2} \tag{5.23}$$

Equation 5.22 gives perfect transmission when $Z_1 \sin(\theta_t) = Z_2 \sin(\theta_i)$. Using Snell's law, we can solve for the incident angle at which this occurs θ_{int}, the angle of intromission:

$$\cos(\theta_{int}) = \frac{\frac{Z_2^2}{Z_1^2} - 1}{\frac{Z_2^2}{Z_1^2} - \frac{c_2^2}{c_1^2}} = \frac{\frac{Z_1^2}{Z_2^2} - 1}{\frac{\rho_1^2}{\rho_2^2} - 1} \tag{5.24}$$

Such an angle will not exist for all interfaces, e.g., if $\rho_1 = \rho_2$.

Let us consider a signal hitting a water–air interface at normal incidence, first from the water into air. Using Equations 5.18 and 5.19, Z_2, the acoustic impedance of air (420 Pa s/m) and Z_1, the acoustic impedance of water (1.5×10^6 Pa s/m) gives $T = 0.000\,56$ and $R = -0.999\,44$. Using Equation 5.22, the intensity or power is $\tau = 0.001\,12$, or just over a tenth of a percent of the intensity in the water. Note that R resolves to a negative value, which means that the reflected wave from the surface of the ocean will have its phase shifted 180°. An investigation of R versus angle shows that R slowly approaches -1 at 0°, so the phase reversal is independent of arrival angle.

Reversing the problem, consider a signal hitting a water–air interface at normal incidence from the air to the water, such as the radiated noise of an aircraft. Using Equations 5.18 and 5.19, Z_2, the acoustic impedance of water (1.5×10^6 Pa s/m), and Z_1, the acoustic impedance of air (420 Pa s/m), yields $T = 1.9994$, which implies that the pressure in the water is almost double what it is in the air, and $R = 0.999\,44$. Using Equation 5.22, the intensity or power is $\tau = 0.00112$, or just over a tenth of a percent of the intensity in the air. Despite the pressure being double, the intensity difference, -29.5 dB, is huge because ρc for water is so much higher than for air and intensity is $P^2/(\rho c)$. While the total reflection angle for water-to-air was 0°, the angle for air-to-water will be $\mathrm{acos}(c_{air}/c_{water})$, or about 76.9°. Therefore, for an airplane flying over water, only the acoustic energy in a cone of $\pm 13.1°$ from straight down will enter the water.

5.4 Transmission through a Plate

A problem that frequently arises is the loss in signal when it passes from one medium to another through a separating membrane. For example, sonars are frequently located inside a hydrodynamic faring. This is an extremely complex problem because solids support two types of elastic waves: longitudinal or compression and shear or transverse. There is no simple method for analyzing the reflection of plane waves obliquely incident on a solid surface. A detailed discussion requires consideration of the partitioning of power into a longitudinal wave traveling in one direction and a shear wave traveling at a lower speed in a different direction. A first approximation, which is often done before really detailed calculations are attempted, is to treat the plate as though it was not rigid but instead just another fluid and in effect ignore the transverse wave. For simplicity, we will only consider normal incidence here. Figure 5.4 illustrates the transmission and reflection of a plane wave through a layer of constant thickness at normal incidence.

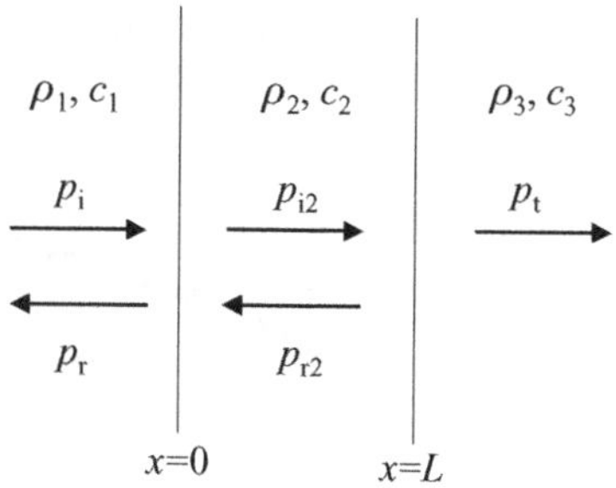

Figure 5.4 Transmission and reflection of a plane wave through a layer of constant thickness at normal incidence

For long, single-frequency plane waves:

$$\begin{aligned} p_{\mathrm{i}} &= P_{\mathrm{i}} \exp\left[i\left(\omega t - k_1 x\right)\right] \\ p_{\mathrm{r}} &= P_{\mathrm{r}} \exp\left[i\left(\omega t + k_1 x\right)\right] \\ p_{\mathrm{t2}} &= P_{\mathrm{t2}} \exp\left[i\left(\omega t - k_2 x\right)\right] \\ p_{\mathrm{r2}} &= P_{\mathrm{r2}} \exp\left[i\left(\omega t + k_2 x\right)\right] \\ p_{\mathrm{t}} &= P_{\mathrm{t}} \exp\left[i\left(\omega t - k_3 x\right)\right] \end{aligned} \tag{5.25}$$

At $x = 0$, the boundary conditions to be satisfied for pressure and particle velocity are

$$\begin{aligned} p_{\mathrm{i}} + p_{\mathrm{r}} &= p_{\mathrm{t2}} + p_{\mathrm{r2}} \\ u_{\mathrm{i}} + u_{\mathrm{r}} &= u_{\mathrm{t2}} + u_{\mathrm{r2}} \end{aligned} \tag{5.26}$$

Similarly, at $x = L$, the boundary conditions are

$$\begin{aligned} p_{\mathrm{t2}} + p_{\mathrm{r2}} &= p_{\mathrm{t}} \\ u_{\mathrm{t2}} + u_{\mathrm{r2}} &= u_{\mathrm{t}} \end{aligned} \tag{5.27}$$

The pressure reflection coefficient and intensity transmission coefficients are

$$r = \frac{\left(1 - \dfrac{Z_1}{Z_2}\right)\cos\left(k_2 L\right) + i\left(\dfrac{Z_2}{Z_3} - \dfrac{Z_1}{Z_2}\right)\sin(k_2 L)}{\left(1 + \dfrac{Z_1}{Z_2}\right)\cos\left(k_2 L\right) + i\left(\dfrac{Z_2}{Z_3} + \dfrac{Z_1}{Z_2}\right)\sin(k_2 L)} \tag{5.28}$$

$$\tau = \frac{4}{2 + \left(\dfrac{Z_3}{Z_1} + \dfrac{Z_1}{Z_3}\right)[\cos(k_2 L)]^2 + \left(\dfrac{Z_2^2}{Z_1 Z_2} + \dfrac{Z_1 Z_3}{Z_2^2}\right)\sin(k_2 L)} \tag{5.29}$$

Figure 5.5 shows the intensity transmission as a function of thickness for steel and plastic in sea water at 20 kHz and 60 kHz at normal incidence. Given this, a sonar

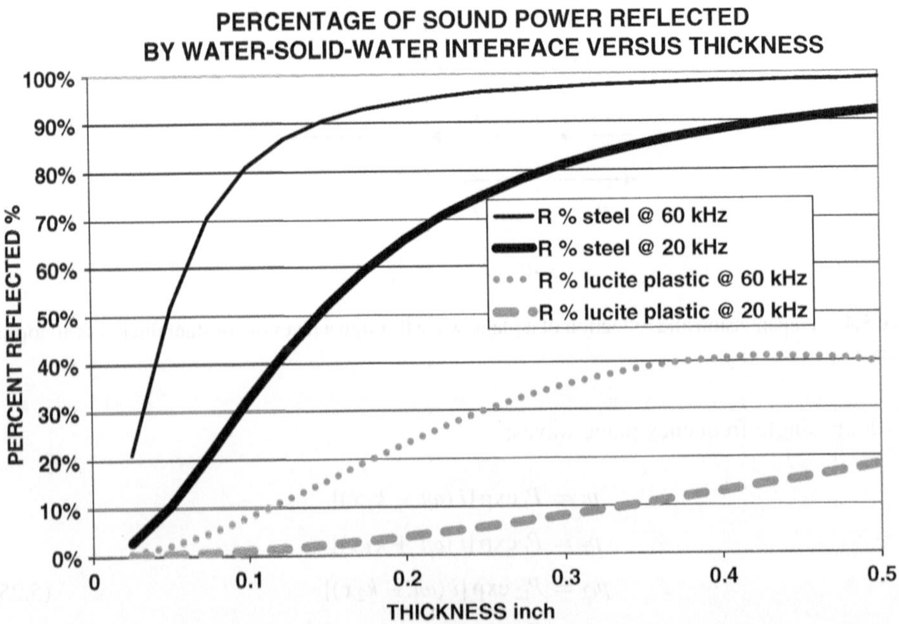

Figure 5.5 Reflection percentage for a water–solid–water interface as a function of thickness

dome for a 60 kHz sonar would have to be very thin steel to avoid very large reflection losses.

5.5 Ray Tracing

Snell's law can be used to visualize how sound propagates. This can be accomplished either by hand or by using one of many RAY TRACE programs. The methodology begins with selecting a point of interest, usually the source of a signal or a receiver. Next, chose a starting angle and, in very small depth steps, follow it as it is bent by the SSP and/or reflected by the boundaries as it moves down range. This process is repeated for many rays until the propagation pattern is established. Rays that connect two points of interest are referred to as eigenrays. According to Snell's law, every point on a ray, $\cos(\theta)/c$, is a constant. Therefore, if a ray begins with a starting sound speed of c_1 at an angle of θ_1, it will never enter a region where the speed of sound c is greater than $c_1/\cos(\theta_1)$, unless it bounces off a nonflat surface or bottom.

The curvature of the ray in a region is defined as

$$R = \frac{\mathrm{d}s}{\mathrm{d}\theta} \tag{5.30}$$

The speed of a sound gradient by definition is $g = \mathrm{d}c/\mathrm{d}z$. If the starting angle is θ,

$$\mathrm{d}z = \mathrm{d}s \sin(\theta) \tag{5.31}$$

Using a derivative of Snell's law gives

$$\frac{d\theta}{dc} = \frac{\cos(\theta)}{c\sin(\theta)} \tag{5.32}$$

Combining Equations 5.31 and 5.32 yields

$$R = \frac{ds}{d\theta}\left[\frac{dz}{ds\sin(\theta)}\right]\left[\frac{d\theta}{dc}\frac{c\sin(\theta)}{\cos(\theta)}\right]\frac{dc}{g\,dz} \tag{5.33}$$

and

$$R = \frac{c}{g\cos(\theta)} \tag{5.34}$$

Since $c/\cos(\theta)$ is a constant for any ray according to Snell's law, if g is constant in a region, then the ray will have constant radius, i.e., it will be an arc of a circle. By inspection of Snell's law, the ray will always bend toward the region of lower sound speed.

Consider the top 50 m of the SSP shown in Figure 5.1 and Table 5.1. It has a positive gradient, i.e., a constantly increasing sound speed with depth. In this region, any sound ray will curve towards the surface. If a ray starts at the surface, it will be trapped in this surface region, called a surface duct, provided its initial angle is less than $\mathrm{acos}[c(0)/c(50)] = \mathrm{acos}(1537.3/1537.92)$, which is 1.63° either up or down. This ray will be horizontal at 50 m in depth and will hit the surface again at a distance x given by

$$x = 2\sin(\theta)R = 7055.8 \text{ m} \tag{5.35}$$

where x is called the cycle distance. All of the rays between 0 and 1.63° will hit the surface at least once in this horizontal distance. Figure 5.6 shows the paths of these rays that are trapped in the surface duct. Rays are plotted from −1.63 (up) to 1.63 (down) every 0.1°.

Rays that start out more than 1.63° either up or down will leave the surface layer. As they continue downward, they will bend down because the gradient is now negative, until they reach 900 m, which is the sound axis or minimum deep sound speed, where the gradient again reverses sign and the rays start bending up again. If the ray started either up or down at greater than 9.156° ($\mathrm{acos}[c(0)/c]$ (bottom) = acos(1537.3/1557.14)), then the ray will eventually hit the bottom and be reflected and attenuated as it returns to the surface, and so forth. These reflections and attenuations are called bottom bounce paths. For rays between 1.63+ and 9.156° (either up or down), they will return to the surface by refraction alone.

As can be seen in Figure 5.6, these rays will clump together forming what is called a convergence zone, which will repeat at constant range spacing if the environment remains constant. The requirement for the formation of a convergence zone is a deep region where the sound speed is greater than any of the sound speeds at shallower depths. In this case, the layer depth or highest speed point near the surface is at 50 m (1537.92 m/s) and the bottom sound speed is higher (1557.14 m/s). In fact, the entire region from about 4500 m to the bottom has a sound speed that exceeds the layer sound speed. This is referred to as a depth excess. Here, the depth excess is 1298 m. Generally, the improvement or reduction in propagation loss increases with the amount of depth excess. This discussion assumes that the SSP, bathymetry, and

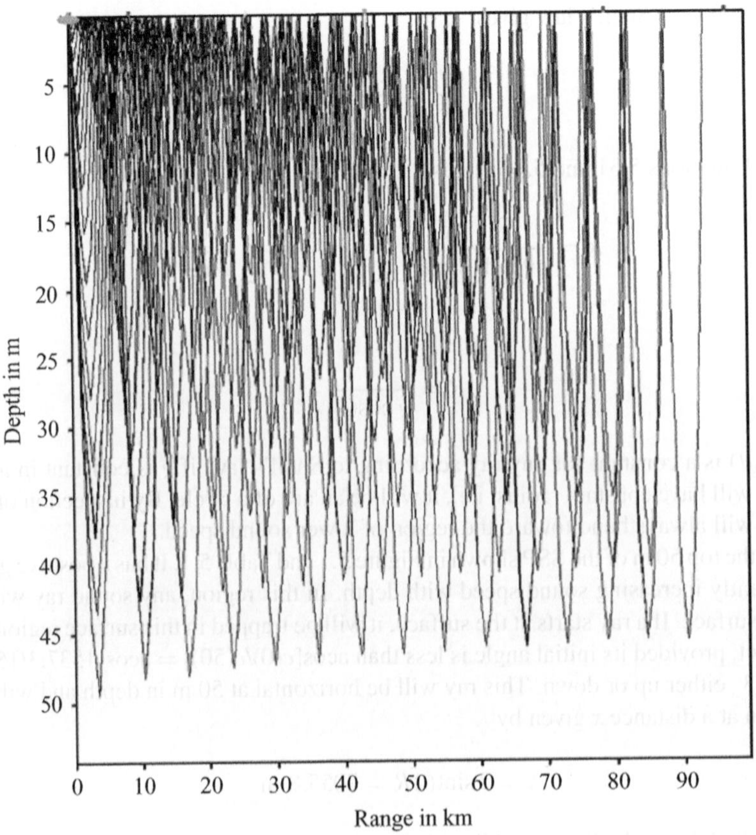

Figure 5.6 Ray trace of the surface layer only

bottom depth are constant with range making it a range-independent calculation, as opposed to a range-dependent calculation, as is generally the case where the SSP and bathymetry vary with location. In fact, if you look carefully at Figure 5.6 you will notice that the lowest rays have vertices at shallower depths as they move to the right. This is because this ray trace was actually run with a range-dependent SSP.

Figure 5.7 shows a ray trace covering the full depth of the sound speed profile shown in Figure 5.1. The plot shows both rays that have bounced off the bottom and rays that have not yet bounced.

Figure 5.8 shows a full field plot of propagation loss versus range and depth. The left side of the graph displays the sound speed profile and the right side of the graph shows the propagation loss using the intensity scale on the far left of the figure.

Figure 5.9 shows propagation loss versus range for two depth combinations for a convergence zone (CZ) environment.

Consider a target just below the layer at 50 m. As the target moves out in range, it suddenly goes from a direct path region where rays from near the surface travel in direct paths to the target to a region where only rays that bounce off the bottom can reach the target. This is the

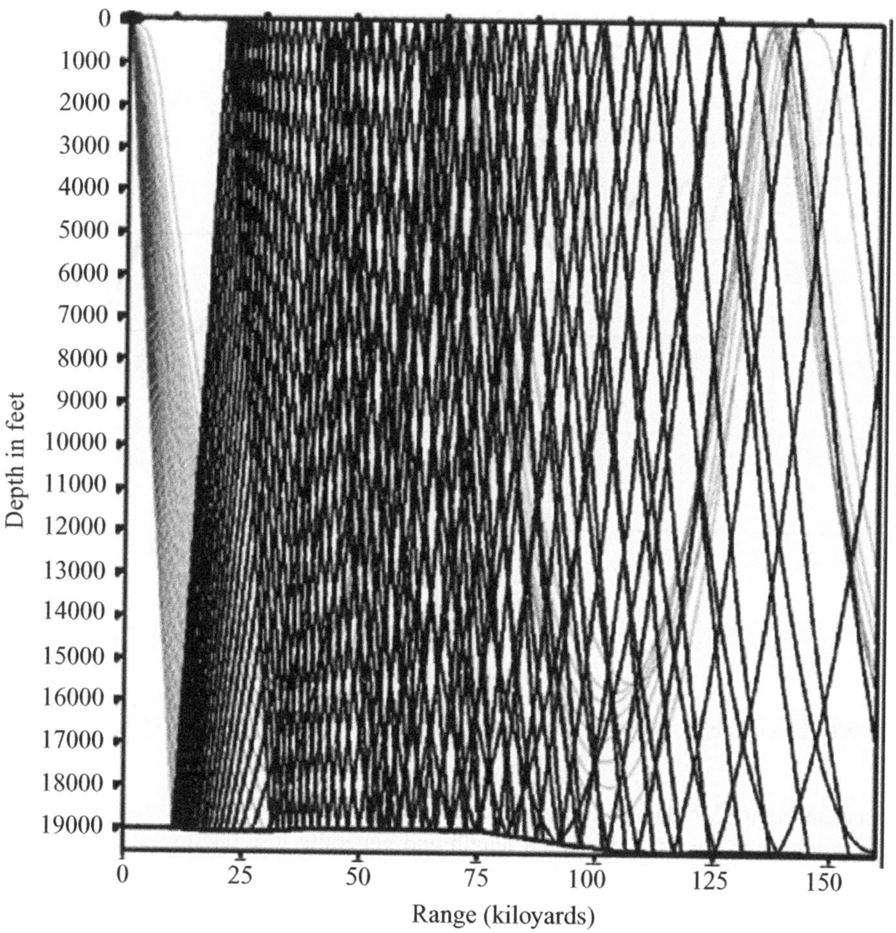

Figure 5.7 Ray trace covering the full depth of the profile shown in Figure 5.1

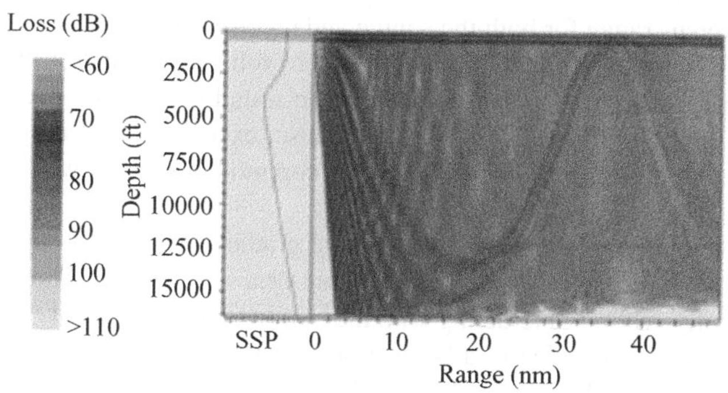

Figure 5.8 A full field plot of propagation loss versus range and depth

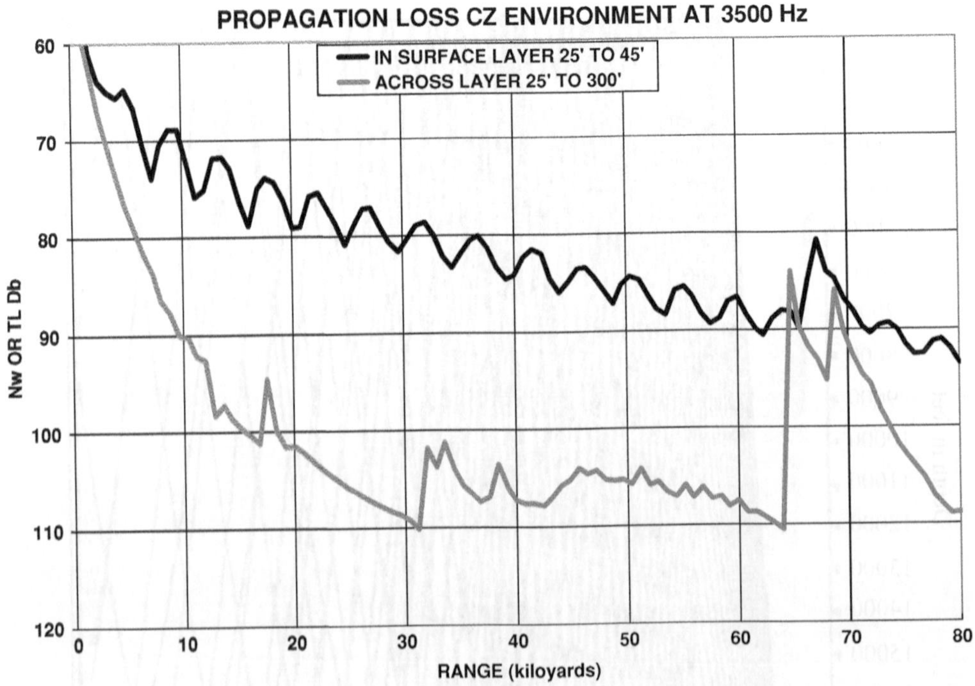

Figure 5.9 Plot of propagation loss versus range for two depth combinations for a CZ environment

gray curve in Figure 5.9. This bottom bounce region is also called the shadow zone because the rays going there have a longer path length and generally lose energy reflecting off the bottom, making the propagation loss higher.

As the target moves further out, it will reach the convergence zone where the propagation loss will suddenly drop because a large bundle of refracted rays will now connect the source and receiver. In this case, the convergence zone has split into two peaks, because the bundle of CZ rays that start upward and reflect off the surface travel further than the bundle of CZ rays that started off downward to reach the surface again. The black curve in Figure 5.9 shows the propagation loss versus range for both the source and receiver in the surface duct.

Figures 5.10 through 5.13 show four shallow water profiles. The first profile, Figure 5.10, is referred to as a negative gradient profile where the sound speed decreases steadily to the bottom. This is a bottom-limited environment because, except at short range, all ray paths interact with the bottom. This type of profile is very common in shallow water in the summer, particularly at lower latitudes.

The second profile, Figure 5.11, shows a positive gradient profile where the sound speed steadily increases with depth, forming a half channel or a surface duct. Many rays cover the entire depth and range without interacting with the bottom, but interact only with the surface. There is frequently lower loss due to a bounce on the surface than a bounce on the bottom unless the sea state is high. Therefore, this profile usually has lower propagation loss versus range than bottom-limited profiles. This type of profile is common in winter conditions in shallow water, particularly at higher latitudes.

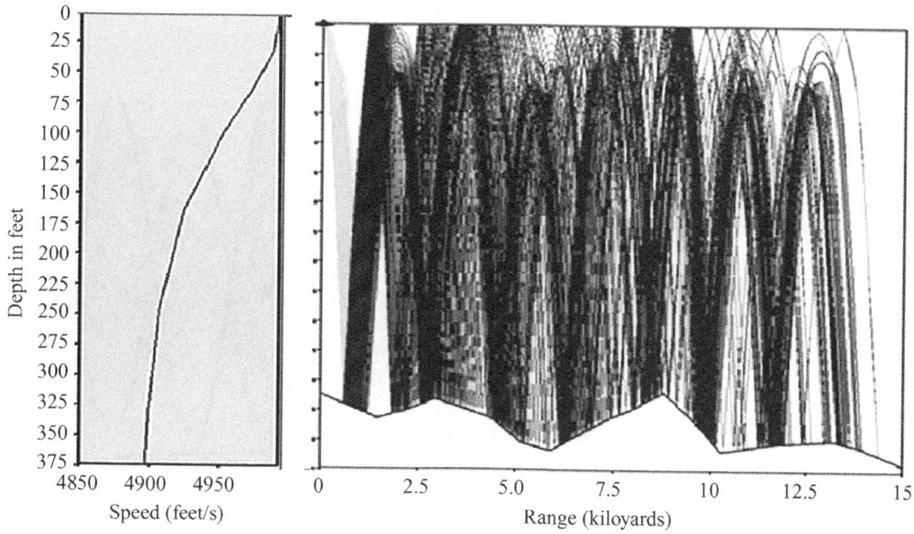

Figure 5.10 Ray trace of a negative gradient profile

The third profile, Figure 5.12, is called a deep or suppressed duct. This profile has a negative gradient followed by a positive gradient, so the rays between the limiting angles will be trapped in the duct due to refraction only, as is shown in the ray trace. This profile usually has very low propagation loss versus range if the source and receiver are in the duct. This type of profile is common in shallow water summer conditions, particularly at higher latitudes. The most spectacular example of such a duct is seen in Figure 5.1 for targets and receivers near the sound axis at 900 m. This type of deep channel is sometimes called the SOund Fixing

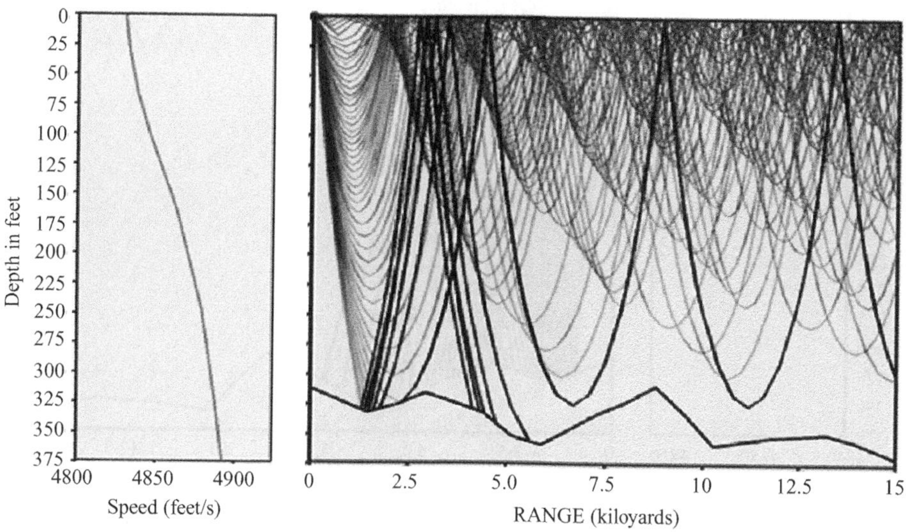

Figure 5.11 Ray trace of a positive gradient profile

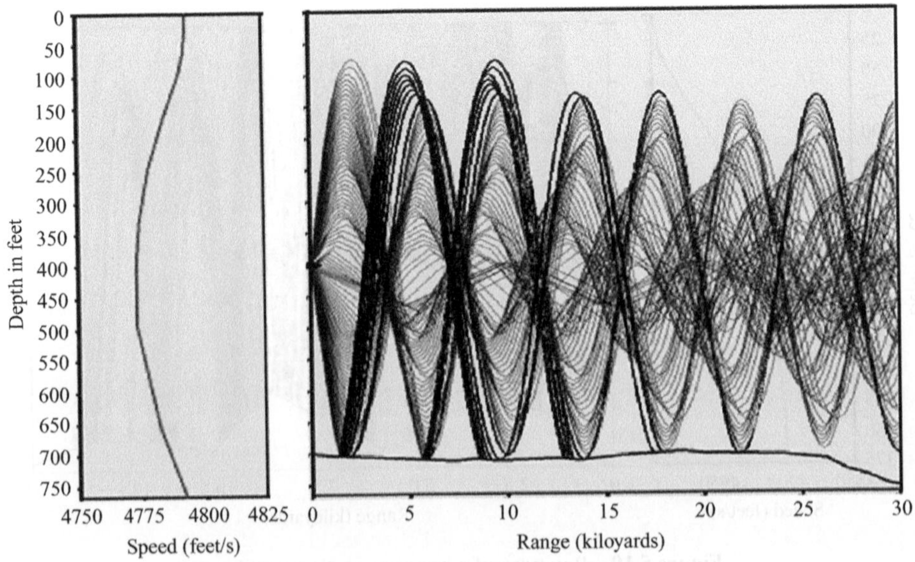

Figure 5.12 Ray trace of a deep or suppressed duct

And Ranging (SOFAR) channel. In World War II, the military used small charges dropped by downed aviators and received at shore stations to localize the aviator. The propagation loss was so good that small charges could easily be detected thousands of miles away. Two or more stations were required to localize the position. Many paths make up the SOFAR channel, each having a different travel time over any range of interest. The short pulse from the explosion

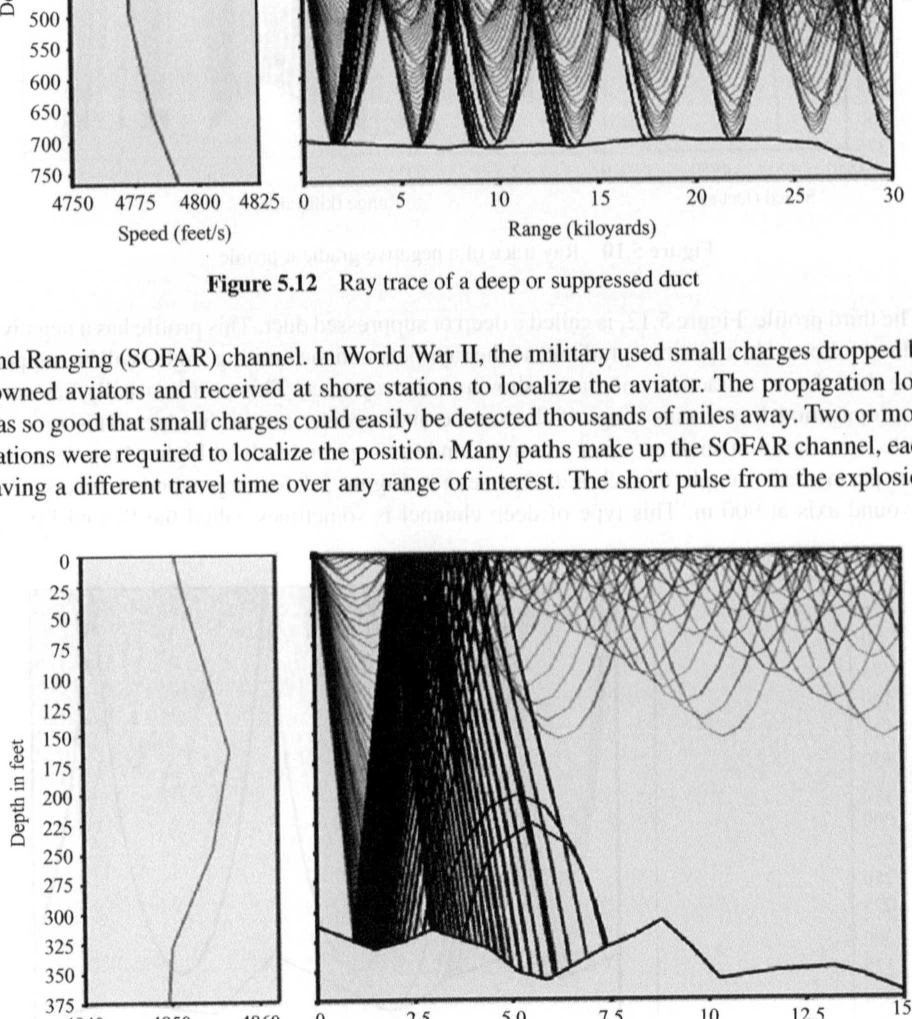

Figure 5.13 Ray trace of a surface duct followed by a negative gradient to the bottom. This type of profile is likely to arise if a storm enters a region that started with the first profile

spreads out in time by an amount around 9–10 seconds per 1000 miles of travel. An interesting observation is that for the trapped refracted rays, the rays that deviate the most in depth are the first to arrive and the rays very near the axis are the last to arrive, even though the former travel a much longer path.

The fourth profile, Figure 5.13, has a surface duct followed by a negative gradient to the bottom. This type of profile is likely to arise if a storm enters a region that started with the first type of profile. The high sea state will mix the surface layer and create a surface duct without affecting the deeper water. Propagation loss may be good for a source and receiver above the layer depth, but for a cross-layer situation (source and receiver on each side of the layer) the loss is likely to be very high because, except for very close paths, the paths will bounce off the bottom many times.

The losses associated with interactions with the bottom and with the surface can be avoided if the receiver, or the receiver and source, for active sonar are placed below the reciprocal depth. The reciprocal depth is that depth in the deep isothermal region where the sound speed is higher than any sound speed above it. This will allow for direct path propagation to all depths and ranges out to half the convergence zone range. These are referred to as reliable acoustic path, or RAP.

5.6 Spreading Loss

In an unbounded, iso-velocity, nonabsorbing ocean, sound would spread out uniformly in all directions, i.e., spherically symmetrical. Because of the conservation of energy, the intensity times the surface area at any range would have to be constant is given by

$$I(R)4\pi R^2 = I_0 4\pi R_0^2 \tag{5.36}$$

where

I_0 = intensity at the reference range (R_0)
$I(R)$ = intensity at range, R

Using Equation 5.1 gives

$$\mathrm{TL} = \mathrm{Nw} = -10\log\left[\frac{I(R)}{I_0}\right] = -10\log\left(\frac{4\pi R_0^2}{4\pi R^2}\right) = -20\log\left(\frac{R_0}{R}\right) \tag{5.37}$$

or $R_0 = 1$ yard or 1 meter:

$$\mathrm{TL} = \mathrm{Nw} = 20\log(R) \tag{5.38}$$

This is called the spherical or inverse square law spreading loss. Here, positive values are used for the transmission loss; many transmission loss programs use the opposite convention, namely negative values.

The oceans are not unbounded. The most obvious boundaries are the surface and the bottom. As sound spreads, if there are no reflection losses or scattering at the boundaries, the power is trapped in a cylinder of height equal to the water depth and radius, R. Here, the sound can no longer spread spherically. In addition, because it is no longer spreading vertically, the transmission loss starts to vary as 1/*R* or 10 log(*R*) in decibels. This is called cylindrical or inverse range law spreading.

Time spreading can also occur for short signals in the SOFAR channel. If the time spreading is proportional to the range, the expected cylindrical spreading can change to become inverse square law spreading if we consider only a time period equal to the original pulse length. If we add up the power over the duration of all the multipaths, the time spreading loss would disappear.

5.7 Absorption of Sound in the Ocean

The terms absorption and attenuation are frequently used interchangeably. However, historically, attenuation has been all of the loss that is proportional to range and thus includes surface and bottom losses. Absorption is frequently taken to be those losses due to mechanisms that convert sound energy to heat in the water. Sound absorption in the oceans is, at most frequencies of interest to sonar, an order of magnitude greater than in fresh water.

5.7.1 Mechanisms of Absorption

In the ocean, there are three mechanisms that cause absorption. The first mechanism was investigated theoretically by Lord Rayleigh [1] who computed the expected absorption due to shear viscosity. When actual measurements became available decades later, this expected value accounted for only one third of the observed absorption in distilled water. The additional absorption observed was due to volume viscosity. The dominant sources of absorption in sea water above 1 MHz are given by

$$\alpha = 16\pi^2 (\mu_s + 0.75\mu_v) f^2/(3\rho c^3) \tag{5.39}$$

where

α = intensity absorption coefficient (cm^{-1})
μ_s = shear viscosity (poise, about 0.01 for water)
μ_v = volume viscosity coefficient (poise, about 0.028 for water)
f = frequency (Hz)
ρ = density (g/cm^2, about 1 for water)
c = speed of sound (1.5×10^5 cm/s)

This yields about $3 \times 10^{-4} \times F^2$ (dB/kyd) for *F* in kHz.

Early measurements showed that the absorption in sea water was two orders of magnitude higher than for fresh water at frequencies below 100 kHz. In 1948, Liebermann proposed [2] that this was due to ionic relaxation, which dominates the absorption at lower frequencies and has the form

$$\alpha_r = af^2 f_r/(f^2 + f_r^2) \tag{5.40}$$

where

f_r = relaxation frequency of the ions

In 1967, Thorpe published an equation for attenuation [3], which has become the most commonly seen equation for attenuation:

$$\alpha = 0.1 f^2/(1 + f^2) + 40 f^2/(4100 + f^2) \tag{5.41}$$

where f *is* in kHz and α is in dB/kyd. (Most authors add the pure water correction $0.003F^2$. This equation is actually based almost entirely on North Atlantic measurements and lacks the temperature, pH, salinity, and pressure dependences of later models.)

The U.S. Navy currently uses the François–Garrison model [4, 5] for attenuation, which includes ionic relaxation for boric acid and magnesium sulfate plus viscosity. It is temperature, salinity, pH, depth, and pressure dependent:

$$\begin{aligned}
\alpha &= \frac{A_1 P_1 F_1 F^2}{F_1^2 + F^2} + \frac{A_2 P_2 F_2 F^2}{F_2^2 + F^2} + A_3 P_3 F^2 \\
A_1 &= \frac{8.86}{c} 10^{0.78\ \mathrm{pH} - 5} \\
P_1 &= 1 \\
F_1 &= 2.8 \left(\frac{S}{35}\right)^{0.5} 10^{4 - 1245/T + 273} \\
c &= 1412 + 3.2T + 1.19S + 0.0167D \\
A_2 &= 21.44 \frac{S}{c} (1 + 0.025T) \\
P_2 &= 1 - 1.37 \times 10^{-4} D + 6.2 \times 10^{-9} D^2 \\
F_2 &= \frac{8.17 \times 10^{8 - 1990/(T + 273)}}{1 + 0.0018 (S - 35)}
\end{aligned} \tag{5.42}$$

For $T < 20\ ^\circ\mathrm{C}$:

$$A_3 = 4.937 \times 10^{-4} - 2.59 \times 10^{-5} T + 9.11 \times 10^{-7} T^2 - 6.5 \times 10^{-8} T^3 \tag{5.43}$$

For $T > 20\ ^\circ\mathrm{C}$:

$$A_3 = 3.964 \times 10^{-4} - 1.146 \times 10^{-5} T + 1.45 \times 10^{-7} T^2 - 6.5 \times 10^{-10} T^3 \tag{5.44}$$

$$P_3 = 1 - 3.83 \times 10^{-5} D + 4.9 \times 10^{-10} D^2 \tag{5.45}$$

where

S = salinity (ppt)
C = speed of sound (m/s)
D = depth (m)

It now appears that three relaxations, together with viscosity, explain the bulk of absorption in the oceans. These three relaxations have been identified as those associated with magnesium sulfate, boric acid, and magnesium carbonate. The absorption coefficient (α) in dB per km has

Table 5.2 Sample temperature, pH, and ionic absorption

Area	Temperature	pH	α
Atlantic	4	8.0	$0.007f^2 + 0.1f^2/(1+f^2) + 0.18f^2/(36+f^2)$
Mediterranean	14	8.3	$0.006f^2 + 0.26f^2/(2+f^2) + 0.78f^2/(144+f^2)$
North Pacific	4	7.7	$0.007f^2 + 0.05f^2/(1+f^2) + 0.09f^2/(36+f^2)$
Red Sea	22	8.2	$0.004f^2 + 0.27f^2/(3.24+f^2) + 1.1f^2/(576+f^2)$

the following form:

$$\alpha = A_1(\mathrm{MgSO_4}) + A_2(\mathrm{B(OH)_3}) + A_3(\mathrm{MgCO_3}) + 0.0003f^2 \tag{5.46}$$

$$A_n = \frac{a_n f^2 f_n}{f^2 + f_n^2} \tag{5.47}$$

where

$$\begin{aligned} a_1 &= 0.5 \times 10^{-D/20} \\ f_1 &= 50 \times 10^{T/60} \\ a_2 &= 0.10 \times 10^{\mathrm{pH}-8} \\ f_2 &= 0.90 \times 10^{T/70} \\ a_3 &= 0.03 \times 10^{\mathrm{pH}-8} \\ f_3 &= 4.5 \times 10^{T/30} \end{aligned} \tag{5.48}$$

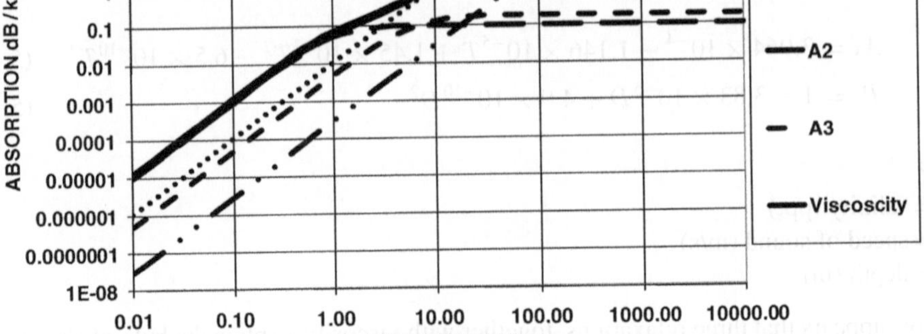

Figure 5.14 Example of absorption in ocean water versus frequency

where

D = depth (km)
F = frequency (kHz)
T = temperature (°C)

The above model is from Mellen [6]. Table 5.2 shows typical temperature, pH, and ionic absorption of various areas of the world. Sophisticated propagation loss models will model absorption as a function of range and depth, taking into account changes in temperature, pH, salinity, and pressure. Figure 5.14 shows the absorption in the North Atlantic versus frequency using the equation in Table 5.2.

References

[1] Rayleigh, Lord, *The Theory of Sound*, Vol II, Chapter XIX., New York: Dover Publications, reprint 1948.

[2] Liebermann, I. N., "The Origin of Sound Absorption in Water and in Sea Water," *Journal of the Acoustical Society of America*, **20**, 1948, 868.

[3] Thorpe, W. H., "Description of the Low-Frequency Attenuation Coefficient," *Journal of the Acoustical Society of America*, **42**, 1967, 270.

[4] François, R. E., and Garrison, G. R., "Sound Absorption Based on Ocean Measurements. Part I: Pure Water and Magnesium Sulfate Contributions," *Journal of the Acoustical Society of America*, **72**, 1982, 896–907.

[5] François, R. E., and Garrison, G. R., "Sound Absorption Based on Ocean Measurements. Part II: Boric Acid Contribution and Equation for Total Absorption," *Journal of the Acoustical Society of America*, **72**, 1982, 1870–1890.

[6] Mellen, R. H., Scheifele, P. M., and Browning, D. G., "Global Model for Sound Absorption in Sea Water," Naval Underwater Systems Center, 1987.

6

Transmission Loss: Interaction with Boundaries

6.1 Sea State, Wind Speed, and Wave Height

The Beaufort wind force scale is used to catagorize weather conditions. The original scale as created by Sir Francis Beaufort related what he saw to the different states of the sea that affected the sailing of a "well-conditioned man-of-war." Wind speed was never actually mentioned in the scale, but rather the force that was exerted on a "man-of-war." The descriptions for Beaufort numbers 0 through 4 describe the wind in terms of the speed it may propel the ship; those for 5 through 9 describe in terms of her mission and her sail carrying ability; and those for 10 through 12 describe in terms of her survival, with 12 being the definition of a hurricane, or that which no canvas could withstand.

The Royal Navy made Beaufort's scale mandatory in 1838, but not until 1912 did the International Commission for Weather Telegraphy seek some agreement on velocity equivalents for the Beaufort scale. A uniform set of equivalents was accepted in 1926 and revised slightly in 1946. By 1955, wind velocities measured in knots replaced Beaufort numbers on weather maps, but there was still a need for eyeball estimates by seamen to fill the gaps in the global observing network. Thus, it became imperative to relate a seaman's guess logged in Beaufort numbers to the wind speed in knots. With this done, the Beaufort's scale became a tool for both the mariner and the meteorologist. See Table 6.1 for a modern day look at the Beaufort scale.

When discussing environmental conditions today, it is common to say something simple like sea state 3 and to assume that this defines both wind speed and wave height. This assumption is not valid. There are frequently used equations for relating wind speed and wave height, but they assume that the wind has been relatively steady for long enough to reach a steady state wave height. Worse than that, there are numerous different definitions of sea state. The most commonly used definitions are from the World Meteorological Organization (WMO) and the U.S. Hydrographic Office (USHO). Table 6.2 shows the wind and significant wave height corresponding to each sea state. The values shown are for medians for the ranges provided in the original sources.

Underwater Acoustics: Analysis, Design and Performance of Sonar Richard P. Hodges

Table 6.1 The modern day Beaufort scale

Beaufort number	WMO wind	Sea state description
0	Calm	Calm, like a mirror
1	Light air	Ripples with appearance of scales, no foam crests
2	Light breeze	Small wavelets, crests of glassy appearance, not breaking
3	Gentle breeze	Large wavelets, crests begin to break, scattered whitecaps
4	Moderate breeze	Small waves, becoming longer numerous whitecaps
5	Fresh breeze	Moderate waves, taking longer form, many whitecaps, some spray
6	Strong breeze	Large waves forming, whitecaps everywhere, more spray
7	Near gale	Sea heaps up, white foam from breaking waves begins to blow in streaks
8	Gale	Moderate high waves of greater length, edges of crests begin to break into spindrift, foam is blown in well-marked streaks
9	Strong gale	High waves, sea begins to roll, dense streaks of foam, spray may reduce visibility
10	Storm	Very high waves with overhanging crests, sea takes white appearance as foam is blown in very dense streaks, rolling is heavy and visibility is reduced
11	Violent storm	Exceptionally high waves, sea covered with white foam patches, visibility still more reduced
12	Hurricane	Air filled with foam, sea completely white with driven spray, visibility greatly reduced

Table 6.2 Relationship of sea state, wind speed, and wave height

	WMO[a]				USHO[b]			
	Wind speed		Significant wave height		Wind speed		Significant wave height	
Sea state	kts	m/s	ft	m	kts	m/s	ft	m
0	1.5	0.75	0	0	0.5	0.25	0	0
1	5.0	2.5	0.5	0.15	2.0	1.0	0.5	0.15
2	8.5	4.4	1.5	0.46	5.0	1.5	2.0	0.6
3	13.5	6.9	3.0	0.91	8.5	4.5	4.0	1.2
4	19.0	9.8	6.0	1.8	19.0	9.8	6.5	3.9
5	24.5	12.6	10.5	3.2	34.0	17.5	10.0	3.0
6	37.5	19.3	16.5	5.0	44.0	22.6	16.0	4.9
7	51.5	26.5	25.0	7.6	51.5	26.5	30.0	9.2
8	59.5	30.6	37.5	11.4	56.5	30.6	>40	>12.2
9	>64	>32.9	>45	>13.7	>64	>32.9	>40	>12.2

[a] World Meterorological Organization [1].

[b] U.S. Hydrographic Office [2].

Throughout this book, the more widely used WMO values will be used because they agree most closely with the Pierson–Moskowitz model for fully developed seas.

The U.S. Navy has a standard conversion of sea state (SS) to wind speed (WS) (knots). If SS < 2.0 then WS = 1.92 + 3.25SS, else

$$\mathrm{WS} = -2.7 + 5.5\,\mathrm{SS}$$

The phase speed of wind waves is given by

$$c = \sqrt{\frac{g\lambda}{2\pi}\tanh\left(\frac{2\pi d}{\lambda}\right)} \tag{6.1}$$

where

g = acceleration of gravity
d = water depth
λ = wave length

If the water is very deep as compared to the wavelength, this reduces to

$$c = 1.25\sqrt{\lambda} \tag{6.2}$$

where c is in meters per second and λ is in meters. In deep water, the group speed is one half of the phase speed.

In the literature, three different wave heights are used. They are

$$\begin{aligned} H &= 0.0186\mathrm{WS}^2 && \text{Significant wave height} \\ h &= 0.012\mathrm{WS}^2 && \text{Average wave height} \\ \sigma &= 0.0046\mathrm{WS}^2 && \text{RMS displacement} \end{aligned} \tag{6.3}$$

where heights are recorded in feet and the wind speed in knots.

6.2 Pierson–Moskowitz Model for Fully Developed Seas

Waves in the ocean are caused by wind (ignoring Tsunamis). As wind blows over a flat calm sea, three processes take place. First, the wind's turbulence produces random fluctuations in pressure causing small waves a few centimeters in size. Second, the wind acting on these small waves causes them to grow larger. The Bernoulli effect amplifies the waves in a very unstable and nonlinear way, resulting in exponential growth [3]. Finally, interactions between the waves produce long-wavelength waves. This interaction transfers energy from short waves generated by the Miles process to waves with frequencies slightly lower than those of the peak of the wave spectrum. Interestingly, as we will see in the equations below, this results in waves moving at speeds greater than the wind speed. The higher the speed of the wind, the greater the area affected (called fetch). The longer the wind blows, the larger the waves become. To understand the interaction of these surface waves with acoustic propagation, ships, oil rigs, breakwaters, and dikes, we need models of the spectra. In 1964, Pierson and Moskowitz [4] developed one of the simplest idealized models for a fully developed sea, meaning the wind has been blowing for about 10 000 wave cycles over an area of 5000 wavelengths on a side and has come close to equilibrium. The model they developed was based on accelerometer measurements on British weather ships in the North Atlantic for time periods when the wind had been blowing steadily for a long period over a large area. The fit to the data they obtained is

$$S(\omega) = \frac{\alpha g^2}{\omega^2} \exp\left[-\beta\left(\frac{\omega_0}{\omega}\right)^2\right] \tag{6.4}$$

where

$S(\omega)$ = spectral density
$\omega = 2\pi f$
f = wave frequency (Hz)
$\alpha = 8.1 \times 10^{-3}$
$\beta = 0.74$
$\omega_0 = g/U_{19.5}$
$U_{19.5}$ = wind speed (m/s) at 19.5 m above the sea surface (the height of the anemometers of the weather ships)

Figure 6.1 shows the spectral densities for various wind speeds. Much of modern wind data is referenced to 10 m, a rough conversion being

$$U_{19.5} = 1.026 U_{10} \tag{6.5}$$

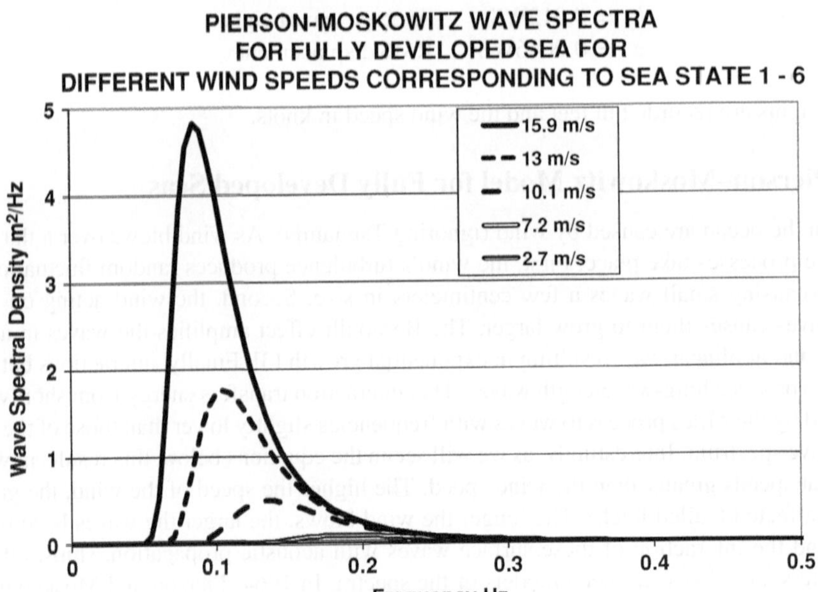

Figure 6.1 Pierson–Moskowitz wave spectral density for various wind speeds

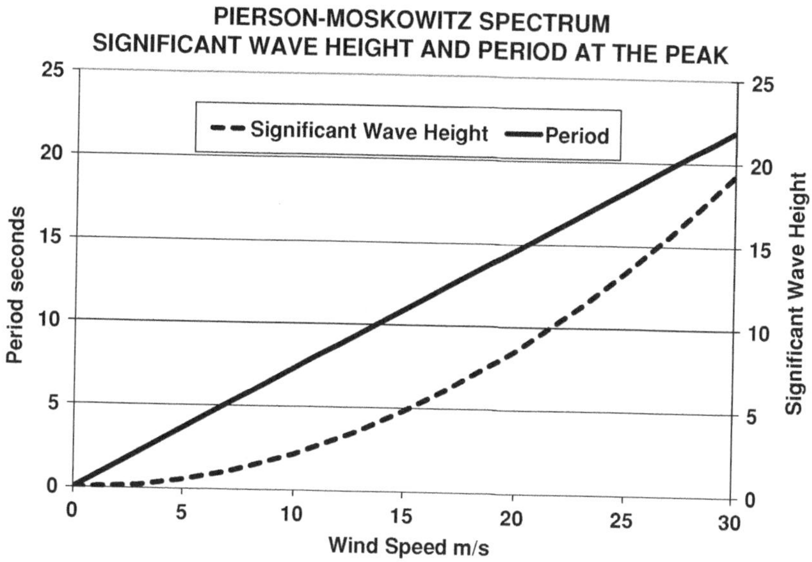

Figure 6.2 Pierson–Moskowitz spectrum of significant wave heights and periods versus wind speed

The peak of the spectrum is

$$\omega_p = 0.887\frac{g}{U_{19.5}} \tag{6.6}$$

The peak wave speed is

$$V_p = \frac{g}{\omega_p} = 1.14U_{19.5} \tag{6.7}$$

Figure 6.2 shows the significant wave height and wavelengths versus wind speed.

6.3 Sea Surface Interaction

Sound reflected from a flat water–air interface is almost perfectly reflected with a 180° phase shift.

6.3.1 Lloyd Mirror Interference

When the surface is relatively smooth, the surface reflected and direct path rays sum at a distance to form constructive and destructive interference patterns. This is called the Lloyd mirror (or image) interference. A good discussion of this topic is given by Casey [5]. Figure 6.3 shows the geometry involved.

Path L_1 is the direct path and L_2 is the surface reflected path, which suffers a 180° phase shift. If we assume straight lines (isovelocity conditions), then the intensity where the two

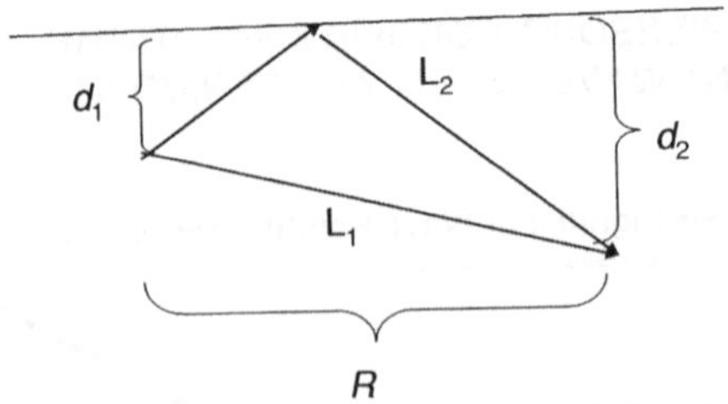

Figure 6.3 Geometry generating Lloyd mirror interference

paths combine have an intensity given by

$$\frac{\int_0^{2\pi/\omega}\left(\frac{\sin(\omega t)}{\sqrt{R^2+(d_2-d_1)^2}}-\frac{\sin(\omega t)}{\sqrt{R^2+(d_1+d_2)^2}}\right)^2 \mathrm{d}t}{\int_0^{2\pi/\omega}[\sin(\omega t)]^2\,\mathrm{d}t} \tag{6.8}$$

There are three regions of interest: (1) in close, (2) where $L_1 \approx L_2$, and (3) the far field where the path lengths are very long.

In region (1), where $L_1 \ll L_2$, the direct path will have a much lower propagation loss and will dominate, producing range squared spreading. Figure 6.4 shows an

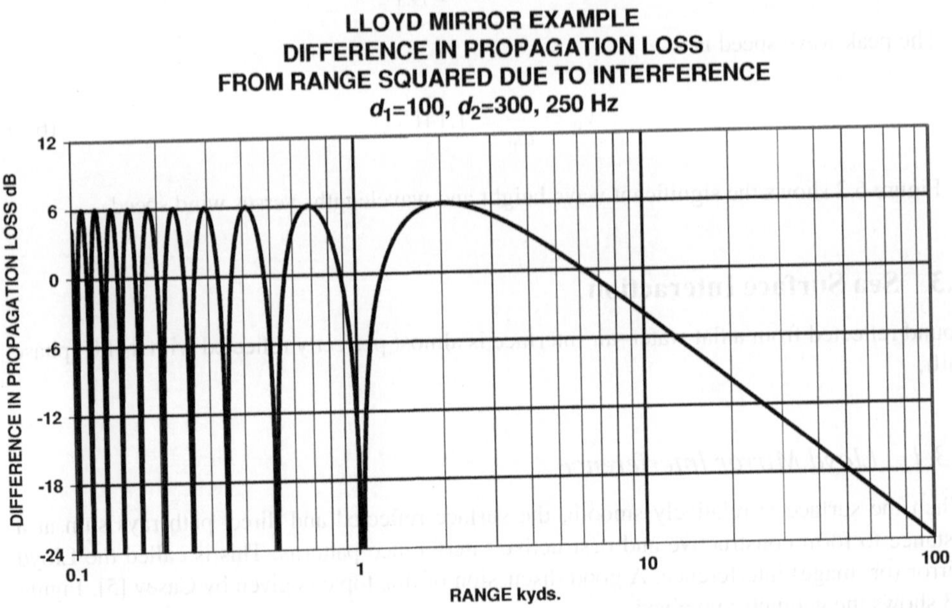

Figure 6.4 Example showing the propagation loss with Lloyd mirror interference relative to range squared spreading

example of propagation loss with Lloyd mirror interference relative to range squared spreading.

In region (2), an interference pattern will occur. The intensity is approximated by

$$I = \frac{I_0}{R^2} 2\left[1 - \cos(\omega\tau)\right] = \frac{I_0}{R^2} 2\left[1 - \cos\left(\frac{2\pi c\tau}{\lambda}\right)\right] \tag{6.9}$$

where I_0 is the reference intensity and

$$c\tau = \sqrt{R^2 + (d_1 + d_2)^2} - \sqrt{R^2 + (-d_1 + d_2)^2} \tag{6.10}$$

To first order, this is

$$\begin{aligned} c\tau &\approx R\left\{\left[1 + \frac{(d_1 + d_2)^2}{2R^2}\right] - \left[1 + \frac{(-d_1 + d_2)^2}{2R^2}\right]\right\} \\ c\tau &\approx \frac{2d_1 d_2}{R} \end{aligned} \tag{6.11}$$

$$I = \frac{I_0}{R^2} 2\left[1 - \cos\left(\frac{4\pi d_1 d_2}{\lambda R}\right)\right] \tag{6.12}$$

If d_1 and d_2 are much greater than the wavelength, the last null will occur at a range of

$$R_{\text{null}} = \frac{2d_1 d_2}{\lambda} \tag{6.13}$$

which, for $d_1 = 100$ and $d_2 = 300$ at 250 Hz, would be just over 1 kyd. The last peak would occur at

$$R_{\text{last peak}} = \frac{4d_1 d_2}{\lambda} \tag{6.14}$$

which is just over 2 kyd for the above case.

Beyond this range, region (3), the level drops off steadily as

$$I = \frac{I_0}{R^4} 8\left(\frac{\pi d_1 d_2}{\lambda}\right)^2 \tag{6.15}$$

For most reasonable depths and frequencies, Lloyd's mirror interference is a short-range phenomenon. This does not prevent similar interference patterns from arising involving other paths such as bottom bounce (see Figures 6.18 and 6.19 later).

6.3.2 *Loss Due to Interaction with the Surface*

As the sea surface becomes rough due primarily to wind, the sound will be scattered in all directions. The term surface loss refers to the decrease in the amount of coherent forward directed (specular reflected) intensity that is reduced. Surface loss is then really a redistribution of the incident power into rays, not corresponding to the expected reflected path. In the discussions of reverberation in Chapter 8, we will consider the "backscattered" intensity, that is the amount sent back in the direction of the incident signal.

6.3.2.1 Rayleigh Surface Loss Model

If the roughness is small as compared to a wavelength, then the surface can be treated as a random scatterer and the reflection coefficient is given by

$$R(\theta) = -\exp(-0.5\Gamma^2) \tag{6.16}$$

where Γ is the Rayleigh roughness parameter defined as

$$\Gamma = 2k\sigma \sin(\theta) \tag{6.17}$$

and
k = acoustic wave number ($2\pi/\lambda$)
θ = incident angle
σ = rms roughness of the surface

The -1 in Equation 6.16 represents the phase reversal on reflection.

In terms of intensity and expressed in decibels, the surface loss (SURFLOSS) would be

$$\text{SURFLOSS} = 300\left\{f^2\sigma^2\,[\sin(\theta)]^2\right\} \tag{6.18}$$

where f is in kHz and σ in meters [6]. Recall that this is only applicable if the roughness is small; for $f\sigma\ \sin(\theta)$ it is less than about 0.25, or SURFLOSS less than about 18 dB. Figure 6.5 shows the Rayleigh surface loss versus frequency with a limit of 18 dB.

6.3.2.2 Eckhart Surface Loss Model

The Eckhart model is based on the Kirchoff approximation, with a small surface slope and a Gaussian distribution of vertical displacements. It gives a surface loss (SL) of

$$\begin{aligned} \text{SL(Eckhart)} &= 20\log(2g) \\ g &= \left[\frac{2f\sigma}{c}\sin(\theta)\right]^2 \end{aligned} \tag{6.19}$$

where

f = frequency
c = speed of sound
σ = rms displacement of the surface
θ = surface grazing angle

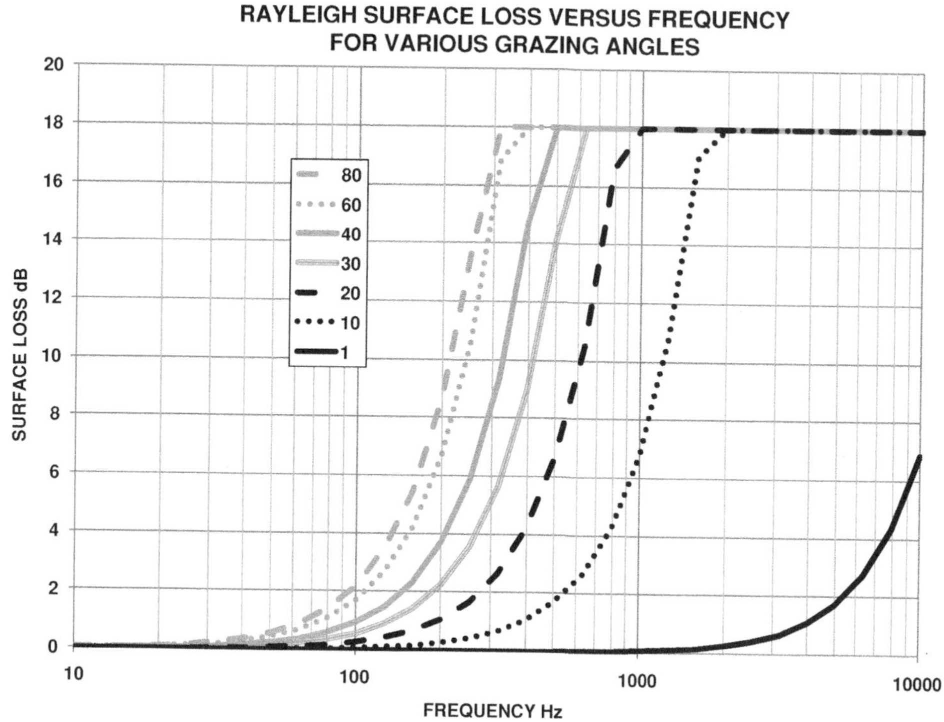

Figure 6.5 Rayleigh surface loss versus frequency and grazing angle for a 15 knot wind speed

Figure 6.6 shows the Eckhart surface loss versus frequency and grazing angle for a 15 knot wind speed.

6.3.2.3 Modified Eckhart Loss Model

The modified Eckhart model represents a theoretical improvement to the original model. The loss is given by

$$\text{SL(ModEckhart)} = -20\log\left[I_0(2g)\exp(-2g)\right] \tag{6.20}$$

where I_0 is the modified Bessel function of zero order. Figure 6.7 shows this model.

The U.S. Navy currently uses the modified Eckhart model with a limit of 11 dB maximum surface loss.

6.3.2.4 Schulkin–Marsh or AMOS Surface Loss Model

The Schulkin–Marsh model [7, 8], also referred to as the AMOS model, is given as

$$\text{SL(Schulkin} - \text{Marsh)} = 10\log\left[1 + \left(\frac{fh}{4.14}\right)^4\right] \quad fh < 4.2691$$

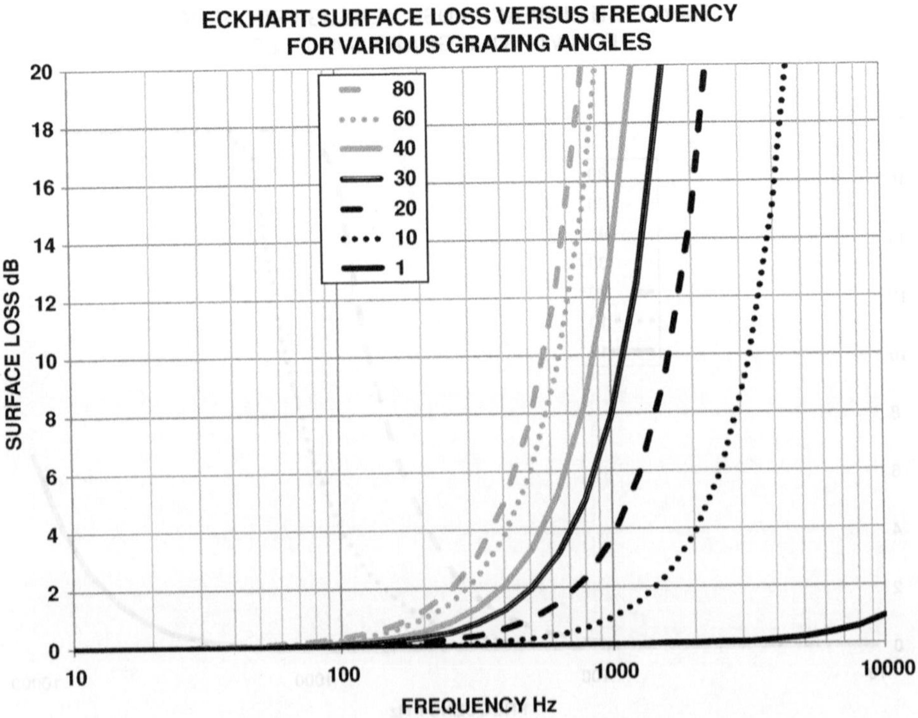

Figure 6.6 Eckhart surface loss versus frequency and grazing angle for a 15 knot wind speed

or

$$\text{SL(Schulkin} - \text{Marsh)} = 1.59\sqrt{fh} \quad fh > 4.2691 \tag{6.21}$$

This model is independent of grazing angle and is based on an empirical fit to surface duct propagation, with the assumption that the distance between bounces is equal to the limiting ray cycle distance.

6.3.2.5 Beckmann–Spizzichino Surface Loss Model

Beckmann and Spizzichino [9] developed a model (BS) for electromagnetic wave scattering that was extended by Leibiger [10] and applied to acoustic surface forward scattering loss. The model is computed as follows:

$$\text{BS} = 20\log(C_v - C_s)$$
$$C_v = \sqrt{1 - \sin(\phi) + \frac{\sin(\phi)\exp(-T)}{\phi\sqrt{\pi\sigma}}} \tag{6.22}$$

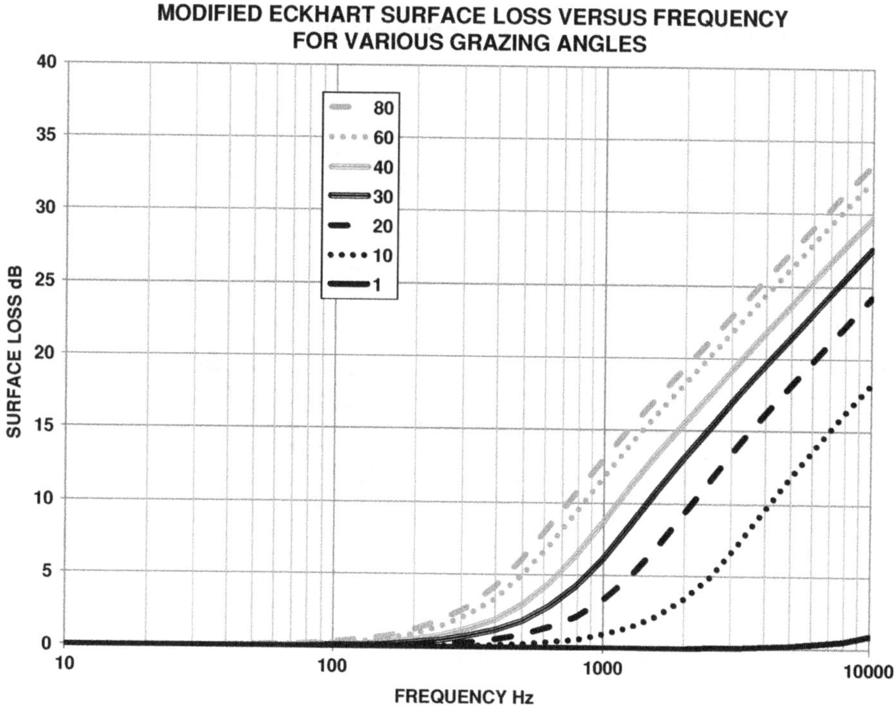

Figure 6.7 Modified Eckhart surface loss versus frequency and grazing angle for a 15 knot wind speed

under the constraint that

$$\frac{\sin(\phi)}{2} \leq \left[\sin(\phi) - \frac{\sin(\phi)\exp(-T)}{\phi\sqrt{\pi\sigma}}\right] \leq 0.99 \tag{6.23}$$

and

$$T = \frac{\sigma\phi^2}{4}$$
$$\sigma = \frac{500}{3 + 2.6S} \tag{6.24}$$

and

$$C_s = 0.3 + \frac{0.7}{1 + 0.01\left(1.16 \times 10^{-5} F S^2\right)^2} \tag{6.25}$$

where

φ = vertical arrival angle (D/E) (rad)
F = frequency (Hz)
S = wind speed (kts)

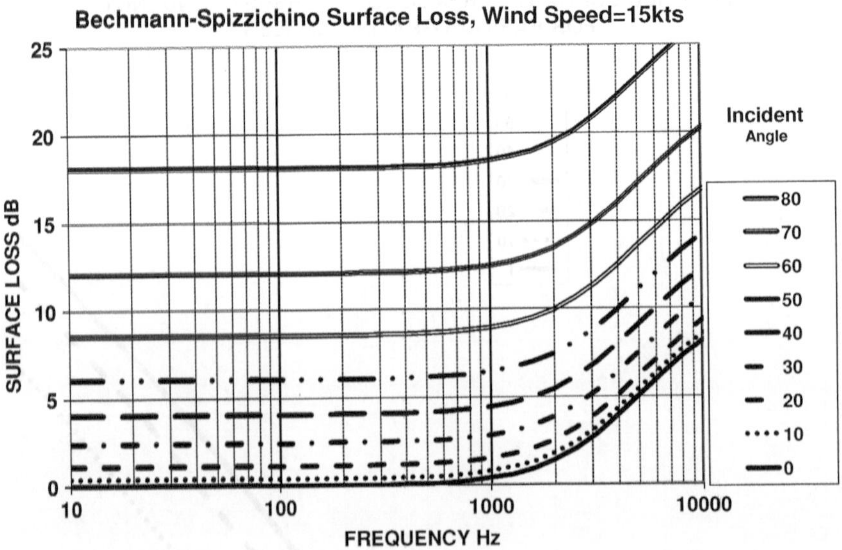

Figure 6.8 Bechmann–Spizzichino sea surface loss versus frequency for a 15 knot wind

Figure 6.8 shows the Bechmann–Spizzichino surface loss (BS above) for a wind speed of 15 knots versus frequency for various incident angles.

6.3.2.6 Surface Model Comparisons

Figure 6.9 shows a wide variation in the predictions of the various models at all grazing angles. Surface loss is most important in a surface duct where paths are interacting with the surface every few thousand yards. A review of the Generalized Digital Environmental Model (GDEM) variable resolution (see Section 5.1) database determined that 56 % of the surface ducts trap rays with grazing angles of 1.5° or less and 94 % of the surface ducts trap rays with 3.0° grazing angles or less. Figure 6.10 shows the same plot as presented in Figure 6.9, but is limited to the surface duct grazing angles. The Schulkin–Marsh (AMOS) curve is a measured result (about 4.8 dB per bounce) and the other curve should be judged in comparison. The only curve that is at all close to the AMOS model is the Bechmann–Spizzichino curve. All of the Eckhart models have losses that are less, by at least 1 dB per bounce, over the entire range of arrival angles.

6.3.2.7 High-Frequency (>10 kHz) Surface Loss

For high frequencies (>10 kHz), there is an additional effect from bubbles near the surface caused by the mixing action of the wind. Dahl [11] at the Applied Physics Laboratory of the University of Washington (APL/UW) has developed the following model:

$$\begin{aligned} \text{SBL(U)} &= \frac{1.26 \times 10^{-3} U^{1.57} f^{0.85}}{\sin(\theta)} \quad U \geq 6 \text{ m/s.} \\ \text{SBL(dB)} &= \text{SBL}(U = 6) \exp\left[1.2(U - 6)\right] \quad U < 6 \text{ m/s.} \end{aligned} \tag{6.26}$$

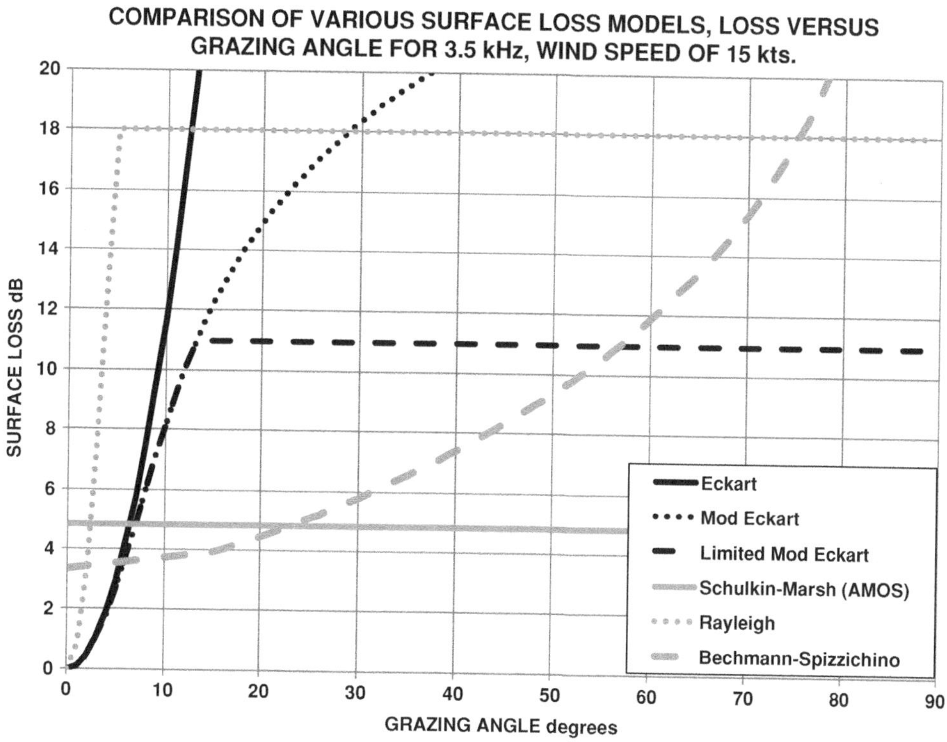

Figure 6.9 A comparison of the surface loss versus grazing angle at 3.5 kHz for a 15 knot wind speed

where

U = wind speed measured 10 m above the sea surface
f = frequency (kHz)
θ = grazing angle of the surface bounce

The 6 m/s division is to take into account the threshold of breaking waves, which are the source of the bubbles. This version of the model is strictly applicable to the isovelocity case and more complex models from the same source treat nonisovelocity cases.

Bubbles in water can have another important effect, namely they change the speed of sound. Recalling from Chapter 1, the speed of sound is given by

$$c = \sqrt{\frac{1}{\rho K}} \tag{6.27}$$

where

ρ = density
K = compressibility (reciprocal of the volume elasticity or bulk modulus)

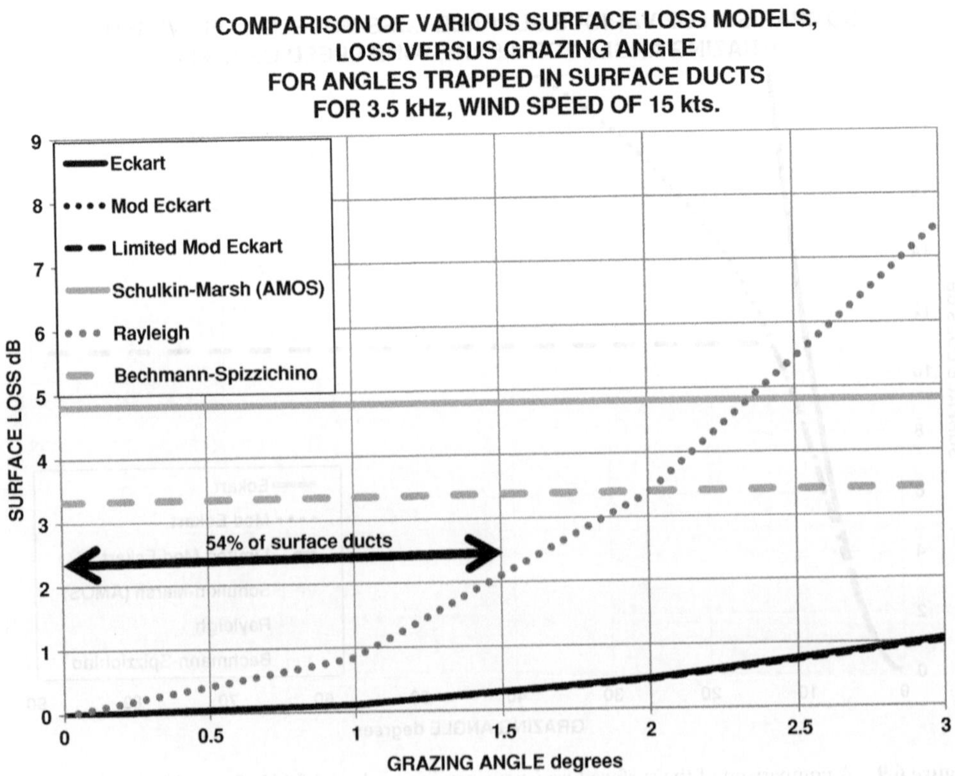

Figure 6.10 A comparison of the surface loss versus grazing angle at 3.5 kHz for a 15 knot wind speed, for grazing angles trapped by surface ducts

If γ is the fraction of air bubbles by volume, then the speed of sound for bubbly water is given by

$$c_{bw} = \left[\frac{1}{(\gamma \rho_a + (1-\gamma)\rho_w)(\gamma K_a + (1-\gamma) K_w)} \right]^{\frac{1}{2}} \tag{6.28}$$

Since $\gamma \ll 1$, $\rho_a \ll \rho_w$ and $K_a \gg K_w$, this can be rewritten as

$$c_{bw} = c_w \left(\frac{1}{1 + \gamma \frac{K_a}{K_w}} \right)^{\frac{1}{2}} \tag{6.29}$$

or

$$c_{bw} = c_w \left(\frac{1}{1 + 25000\gamma} \right)^{\frac{1}{2}} \tag{6.30}$$

If 25 000γ = 3, then the speed of sound is reduced by 50 %, which is for γ = 0.000 12 just over 0.01 %. This is a huge change, as typical SSPs previously shown have total changes of less than 10 %. However, since the bubble density drops very quickly with depth, this effect is typically ignored in computing the speed of sound. In very heavy seas, this can be a significant error. Hall [12] has developed a comprehensive wind-generated bubble model for estimating bubble density and size versus depth, as a function of wind speed:

$$N(a) = N_0 G(a, z) U(W) Y(z, W) \tag{6.31}$$

where N_0 is a constant (1.6×10^{10} m^{-4}),

$$Y(z, W) = \exp\left[\frac{-z}{L(W)}\right] \tag{6.32}$$

$L(W)$ is the e-folding depth constant (1.03 m), and

$$\begin{array}{ll} \left(\dfrac{a}{a_1}\right)^2 & a < a_1 \\ 1 & a_1 < a < a_2 \\ \left(\dfrac{a_2}{a}\right)^p & a > a_2 \end{array} \tag{6.33}$$

where

$a_1 = (34 + 1.24z) \times 10^{-6}$ m
$a_2 = 1.6\, a_1$
$p = p(z) = 4.37 + (z/2.55)^2$
$U(W) = (W/13)^3$
W = wind speed (m/s)

To obtain the total volume of bubbles per unit volume we must integrate over all possible radii as

$$\gamma = \frac{4\pi}{3}\int_0^\infty a^2 N(a)\, da \tag{6.34}$$

For the surface ($z = 0$), the bubble fraction is

$$\gamma(z = 0) = 7.85 \times 10^{-10} W^3 \tag{6.35}$$

For a wind speed of 6 knots (3 m/s), $\gamma(z = 0) = 2 \times 10^{-8}$ and the speed of sound is only changed by –0.03 %. However, at hurricane levels of 64 kts (33 m/s), the speed of sound at the surface is changed by –23 %. It should be recalled that the "standard" equations for computing sound speed are functions of temperature (T), salinity (s), and depth (D), but not bubble entrainment. Therefore, sensors that detect T, s, and D, which are used to compute the speed of sound, could have significant errors if the sea state is high. Figure 6.11 shows the percentage change in the speed of sound due to the entrainment of air in sea water.

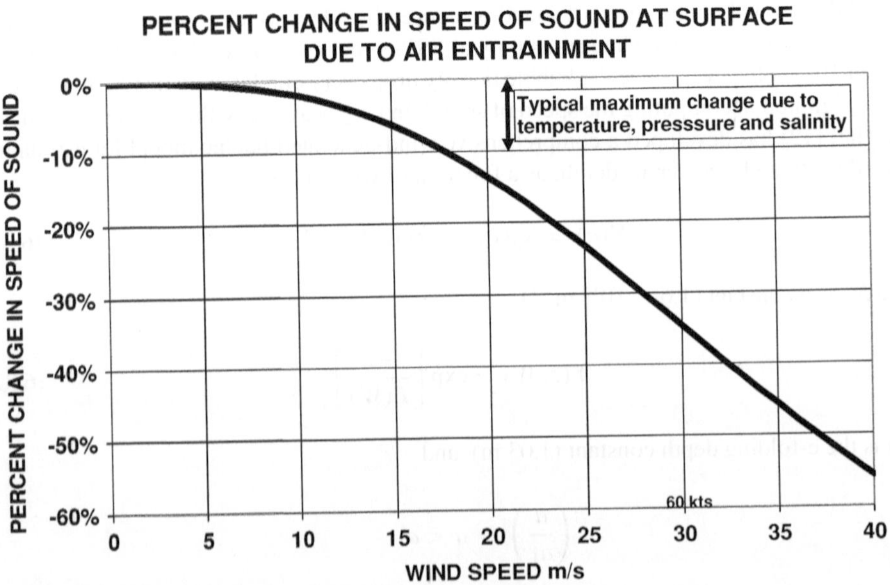

Figure 6.11 Percentage change in the speed of sound due to the entrainment of air in sea water versus wind speed

6.4 Bottom Loss

At first glance, it would seem reasonable to treat bottom loss similar to surface loss as both have a component caused by the roughness. However, as we have seen above, the surface reflects virtually 100 % of the sound incident on it because of the large ρc mismatch between water and air. This is not the case for typical ocean bottoms. Significant energy penetrates into the bottom and, depending on the frequency, it may be necessary to know the details for many meters into the bottom in order to predict what is returned.

Equation 5.19 at normal incidence with Z_2 the acoustic impedance of air (420 Pa s/m) and Z_1 the acoustic impedance of water (1.5×10^6 Pa s/m) gives a reflection power ratio of $r = -0.9989$ or 99.89 % is reflected. For an ocean bottom with a layer of silt, the mismatch is about 13 % or 360 000 %. This will give, for normal incidence, an $r = 0.004$, i.e., only 0.4 % (or –24 dB) is reflected at the interface and most of the energy penetrates into the bottom. This is the basis for a simple bottom loss model [13]. If at a deeper depth, the sound encounters a layer of rock, the ρc mismatch will be large, about 625 %, and the reflection from this layer will be 57 % of the incident power. However, some energy will have been absorbed while traversing the silt. This absorption is frequency dependent, generally increasing with frequency, as does absorption in sea water. At low frequencies (<1 kHz), the reflection characteristics of ocean bottoms can comprise a combination of many layers of material, with reflections occurring at boundaries and absorption occurring in the layers. For monochromatic sound, these layers may give rise to constructive and destructive interference, similar to Fizeau fringes observed with light, while for very high frequencies (>10 kHz), the characteristic may be determined by the top layer of the bottom alone.

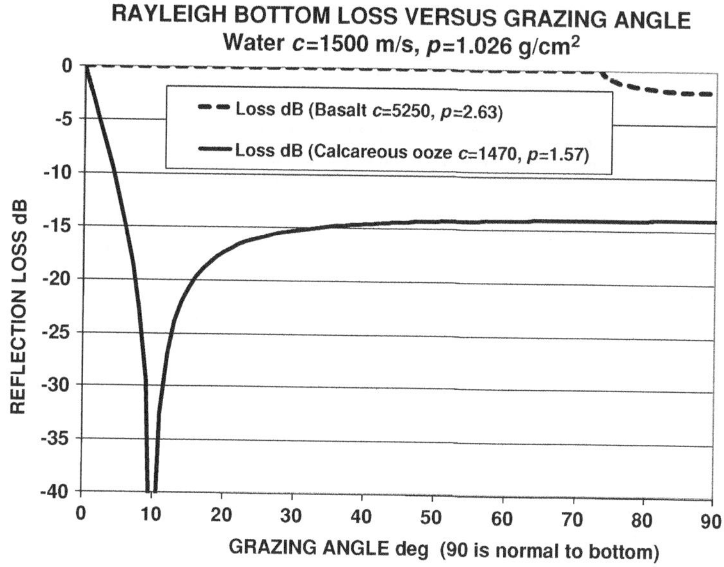

Figure 6.12 Rayleigh bottom loss versus grazing angle examples

6.4.1 Simple Rayleigh Bottom Loss Model

In the reflection loss calculated in Section 6.2, the inputs were density and sound speed at the interface between the water and bottom. As such, it is frequency independent and assumes that the transmitted energy never reappears. Figure 6.12 shows examples of Rayleigh loss. The underlying assumptions here are: (1) scattering is not significant (i.e., smooth bottom) and (2) the only energy reflected is that coming from the water–bottom interface. The remainder of the transmitted energy never returns.

6.4.2 U.S. Navy OAML Approved Models of Bottom Loss

The U.S. Navy has adopted, as of 2007, three separate bottom loss models: LFBL (Low-Frequency Bottom Loss) for < 1 kHz, HFBL (High-Frequency Bottom Loss) for 1.5 to 4 kHz, and HFEVA (High-Frequency Environment Acoustic) for > 10 kHz. Documentation for these models is available from http://stinet.dtic.mil/index.html [14].

6.4.3 Low-Frequency Bottom Loss (LFBL) Model: 50 to 1000 Hz

The database for the LFBL model consists of variable resolution data with multiple bottom layers referred to as an *n*-layer model, each with its associated acoustical properties. Unfortunately, the databases supporting this model are classified, making it unavailable to the general user. At present (2007), LFBL 11.1 is the U.S. Navy's operational global database for geoacoustic information. The data model uses a concept of the geographic coverage entity to define the bounds of individual datasets that are in the database. It has four such entities: world, high-resolution Barents and Kara Sea, high-resolution Yellow Sea–East China Sea, and high-resolution South China Sea. The world coverage is a single layer with a spatial resolution

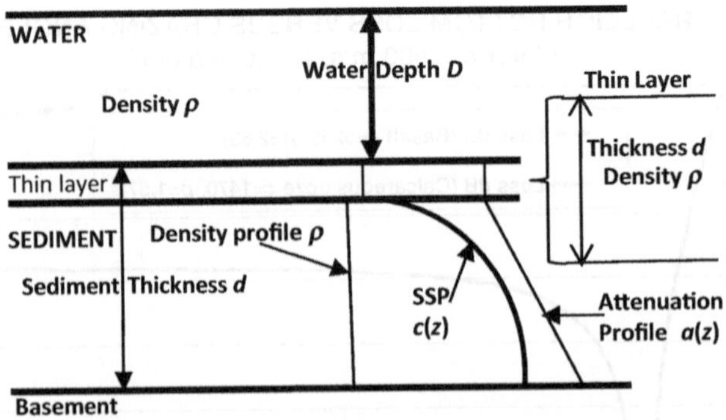

Figure 6.13 LFBL single-layer bottom model

of 5 minutes of arc; all other sets are *n*-layer with a spatial resolution of 12 seconds of arc. For each entity in LFBL, the database stores a corresponding grid cell containing a province identifier. These province identifiers are used to retrieve specific geoacoustic parameters for the cell via a lookup table [14]. The single regions are referred to as the ten parameter regions, as ten geoacoustic inputs are used to define the bottom. The bottom consists of three parts: a thin layer (the so-called stainless steel layer), followed by a sediment layer, and finally a reflective basement as shown in Figure 6.13. Table 6.3 gives a description of the LFBL ten parameters while Table 6.4 gives a description of the *n*-layer parameters.

Multiple layers loss versus angle is extremely variable and, in many cases, very complex, with strong frequency dependences, as can be seen in Figure 6.14.

6.4.4 High-Frequency Bottom Loss (HFBL) Model

This model covers the 1.5 kHz to 4 kHz frequency region and is frequency independent. It is based on bottom provinces (1–9) that were originally determined by a world wide survey, the Maritime Geophysical Survey (MGS). There is one loss versus grazing angle for each

Table 6.3 Description of LFBL single-layer ten parameters

Parameter	Units
Ratio of sediment to water speeds of sound	None
Thin layer thickness	m
Thin layer density	g/cm^3
Sediment surface density	g/cm^3
Sediment sound speed gradient at water interface	s^{-1}
Surface attenuation	dB/m kHz
Attenuation gradient (constant)	dB/m/kHz/m
Attenuation exponent	None
Basement reflection coefficient	None
Two-way travel time	s

Table 6.4 Description of LFBL n-layer parameters

Parameter	Units
Sediment–sediment interface roughness spectral exponent (γ)	N/A
Total number of layers	N/A
Sequential layer numbering scheme	N/A
Layer thickness	s
Compressional wave speed at the top of the layer	m/s
Density at the top of the layer	kg/m^3
Compressional wave attenuation at the top of the layer	dB/m kHz
Shear wave attenuation at the top of the layer	dB/m kHz
Shear wave speed at the top of the layer	m/s
Compressional wave speed at the bottom of the layer	m/s
Density at the bottom of the layer	kg/m^3
Compressional wave attenuation at the bottom of the layer	dB/m kHz
Shear wave attenuation at the bottom of the layer	dB/m kHz
Shear wave speed at the bottom of the layer	m/s
Compressional wave attenuation factor	dB/m (kHz)^n
Compressional wave attenuation gradient	dB/m/kHz/m
Compressional wave attenuation frequency coefficient (n)	N/A
Conductivity	S/m (Siemens/m)
Frequency range	Hz

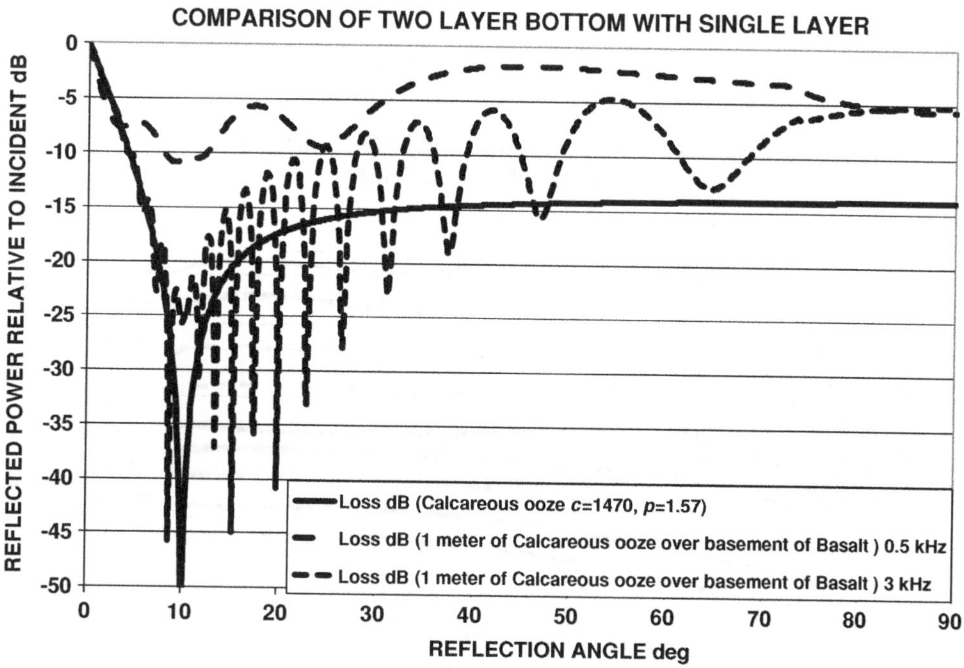

Figure 6.14 Simple physics-based bottom loss models showing constructive and destructive interference for two-layer cases. The solid black curve is the same as in Figure 6.12

Table 6.5 Bottom reflection coefficients for the HFBL model

MGS	C_0	C_1	C_2	C_3	C_4
1	0.94657	0.028424	4.61×10^{-03}	-8.49×10^{-05}	4.25×10^{-07}
2	3.48826	0.049986	3.24×10^{-03}	-6.18×10^{-05}	3.02×10^{-07}
3	5.78675	0.125778	-7.69×10^{-05}	-1.08×10^{-05}	4.40×10^{-08}
4	6.98926	0.389968	-1.02×10^{-02}	1.25×10^{-04}	-5.69×10^{-07}
5	8.04641	0.677356	-2.02×10^{-02}	2.38×10^{-04}	-9.64×10^{-07}
6	9.67454	0.899615	-2.90×10^{-02}	3.75×10^{-04}	-1.70×10^{-06}
7	11.06928	1.281939	-4.41×10^{-02}	5.85×10^{-04}	-2.65×10^{-06}
8	13.0667	1.602316	-5.73×10^{-02}	7.87×10^{-04}	-3.68×10^{-06}
9	15.4861	1.88803	-6.68×10^{-02}	8.90×10^{-04}	-4.05×10^{-06}

province. The survey assigned a province to all deep water (> 600 ft) nonsloping areas in the world. The database was later back-filled with MGS 2 in all shallow water and MGS 9 in all sloping areas. The database that connects areas of the ocean bottom to the MGS province is unfortunately classified. Even though the curves go from 0 to 90° grazing angle, no actual measurements were possible at shallow grazing angles because of the deep water. As a result, the data were extrapolated down to zero. The bottom loss is given by

$$\text{HFBL(mgs)} = C_0 + C_1\theta + C_2\theta^2 + C_3\theta^3 + C_4\theta^4 \quad \text{dB} \tag{6.36}$$

where θ is the grazing angle in degrees. The values for C_i are from Table 6.5 and are a function of the MGS province. Figure 6.15 shows a plot of the HFBL loss versus angle.

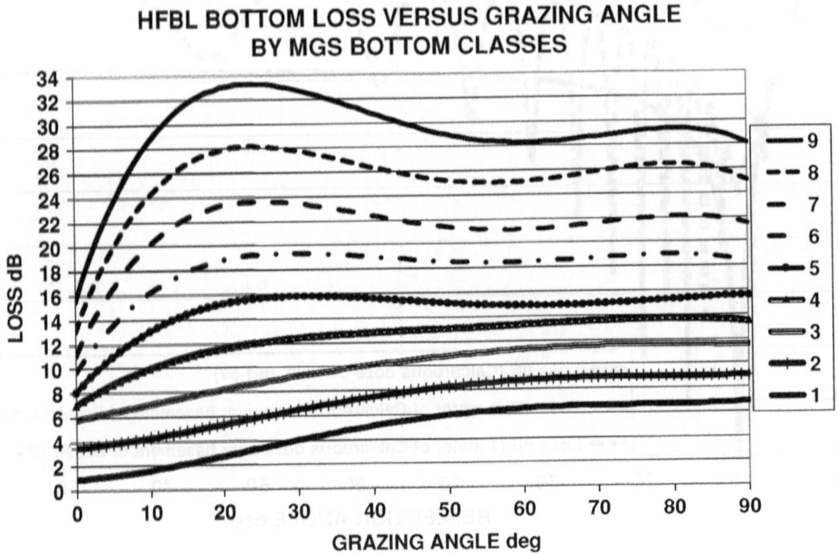

Figure 6.15 High-frequency bottom loss model, 1.5–3.5 kHz, loss versus grazing angle, for Marine Geophysical Survey bottom provinces

6.4.5 High-Frequency Environment Acoustic (HFEVA) Model

This model was developed by the Advanced Physics Laboratory of the University of Washington and is approved for use above 10 kHz. The model is geophysical bottom surface based. Unfortunately, the U.S. Navy's databases for this model are classified. Table 6.6 shows the HFEVA bottom parameters. The bottom loss is computed as follows:

$$\text{HFEVA BL(dB)} = -20 \log \{abs\,[R(\theta)]\}$$

$$R(\theta) = \frac{y-1}{y+1}$$

$$y = \frac{\rho \sin(\theta)}{P(\theta)} \tag{6.37}$$

$$P(\theta) = \sqrt{\left[\kappa^2 - \cos(\theta)^2\right]}$$

$$\kappa = \frac{1}{\nu(1+i\delta)}$$

Figure 6.16 shows bottom loss above 10 kHz versus grazing angle for various bottom types.

6.5 Leakage Out of a Duct, Low-Frequency Cutoff

In a surface duct, such as the one shown in Figure 5.6, ray theory has a large angular spread of rays trapped. Wave theory says that at lower frequencies this energy will leak out. The cutoff frequency is an estimate of the lowest frequency that will be effectively trapped. The functional form comes from radar research [15]:

$$\lambda_{max} = \frac{8}{3}\sqrt{2}\int_0^D \sqrt{n(z) - n(D)}\,\mathrm{d}z$$

$$\lambda_{max} = 0.036\sqrt{g}D^{3/2} \tag{6.38}$$

Since mixed layers are usually of a constant temperature, the gradient (0.017 s^{-1}) is due to pressure only. The cutoff frequency is given by

$$f_{cut-off} = \left(\frac{1000}{D}\right)^{3/2} \tag{6.39}$$

where

D = thickness of the layer (ft)
f_{cutoff} = minimum frequency (Hz)

For example, a mixed layer of 100 ft has a cutoff of about 1 kHz. This is a fairly sharp cutoff, as frequencies below this do not have modes that fit in the duct. At frequencies about this, leakage out of the duct gradually decreases with increasing frequency (Figure 6.17).

Table 6.6 HFEVA model bottom parameters

Sediment name	Bulk grain size ($Mz(\varphi)$	Density ratio (ρ)	Sound speed ratio (ν)	Loss parameter (δ)	Volume parameter (σ_2)	Spectral exponent (γ)	Spectral strength (ω_2)
Rough rock		2.5	2.5	0.01374	0.002	3.25	0.20693
Rock		2.5	2.5	0.01374	0.002	3.25	0.01862
Cobble, gravel, pebble		2.5	1.8	0.01374	0.002	3.25	0.013
Sandy gravel	−1.0	2.492	1.3370	0.01705	0.002	3.25	0.012937
Very course sand	−0.5	2.401	1.3067	0.01667	0.002	3.25	0.010573
Muddy sandy gravel	0.0	2.314	1.2778	0.0163	0.002	3.25	0.08602
Course sand, gravelly sand	0.5	2.231	1.2503	0.01638	0.002	3.25	0.006957
Gravelly muddy sand	1.0	2.151	1.2241	0.01645	0.002	3.25	0.005587
Medium sand	1.5	1.845	1.1782	0.01624	0.002	3.25	0.004446
Muddy gravel	2.0	1.615	1.1396	0.0161	0.002	3.25	0.003498
Fine sand, silty sand	2.5	1.451	1.1073	0.01602	0.002	3.25	0.002715
Muddy sand	3.0	1.339	1.0800	0.01728	0.002	3.25	0.0207
Very fine sand	3.5	1.268	1.0568	0.01875	0.002	3.25	0.001544
Clayey sand	4.0	1.224	1.0364	0.02019	0.002	3.25	0.001119
Course silt	4.5	1.195	1.0179	0.012158	0.002	3.25	0.000781
Sandy silt, gravelly mud	5.0	1.169	0.9999	0.01261	0.002	3.25	0.000518
Medium silt, sand–silt–clay	5.5	1.149	0.9885	0.00676	0.001	3.25	0.000518
Sandy mud	6.0	1.149	0.9873	0.00386	0.001	3.25	0.000518
Fine silt, clayey silt	6.5	1.148	0.9861	0.00306	0.001	3.25	0.000518
Sandy clay	7.0	1.147	0.9849	0.00242	0.001	3.25	0.000518
Very fine silt	7.5	1.147	0.9837	0.00194	0.001	3.25	0.000518
Silt clay	8.0	1.146	0.9824	0.00163	0.001	3.25	0.000518
Clay (all grades)	9.0	1.145	0.9800	0.00148	0.001	3.25	0.000518

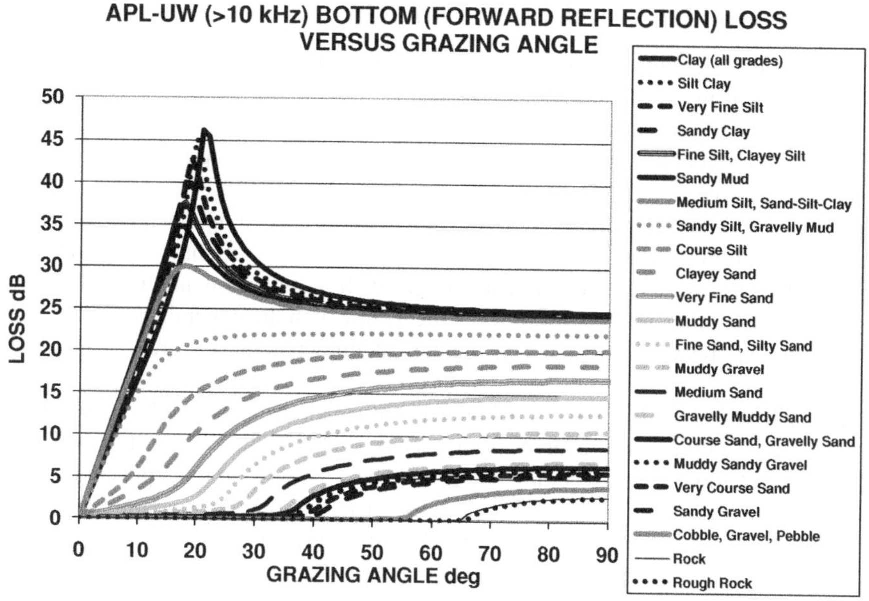

Figure 6.16 Bottom loss above 10 kHz versus grazing angle for various bottom types

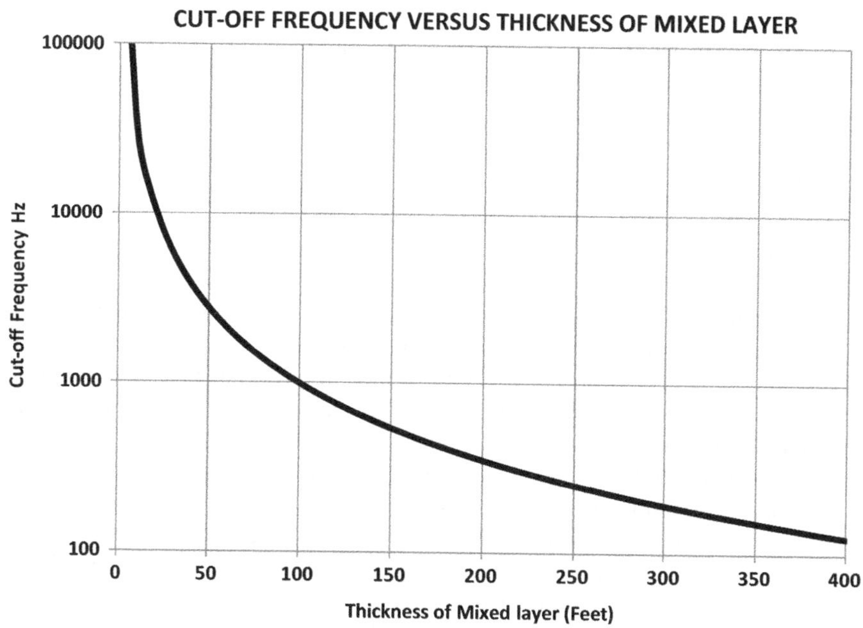

Figure 6.17 Mixed layer cutoff frequency versus layer thickness

6.6 Propagation Loss Model Descriptions

Lord Rayleigh [13] is credited with deriving a reciprocity theorem that applied to a homogeneous, ideal fluid. If the environment is linear and time invariant, then the transmission loss between an omnisource and an omnireceiver are the same – both receiver to source and source to receiver. As a consequence, propagation loss usually needs only to be calculated from sonar to target, not both ways.

If the generalized wave equation can be solved analytically or implemented directly on a computer, it would produce an error-free propagation loss result, of course assuming all the submodels, like surface loss, volume attenuation, as well as environmental inputs, were also error free. However, there are very few environments in which the generalized wave equation can be solved analytically, and, if it can, numerical implementations are computationally intense.

The first approximation for most implemented propagation models is the removal of the effect of the source, or linearization of the wave equation. The result is called the homogeneous wave equation and allows us to treat only the far-field case, somewhat reducing computational requirements. However, the result of this assumption is that all propagation models based on the homogeneous wave equation will have difficulty modeling energy propagation very close to the source. The definition of very close depends on accuracy requirements, but one wavelength is always considered to be near field, while ten wavelengths may be considered a satisfactory far-field range for operational models and normal ASW sound sources.

The next assumption made is a narrowband approximation, considering only a single frequency at a time. This is acceptable for modeling continuous wave (CW) propagation, and is not appropriate for modeling broadband signals. Single-frequency propagation models are run with many frequencies and the results combine to account for broadband effects. The single-frequency homogeneous wave equation is called the Helmholtz equation.

6.6.1 Ray Models

Ray theory makes the assumption that the wavelength of sound is very small. This allows the Helmholtz equation to be rewritten as Snell's law, which can be easily applied in arbitrary environments.

Elementary ray theory states that the energy trapped between two rays will remain between the two rays forever. This leads to an intensity that is inversely proportional to the distance between two bracketing rays. However, when rays cross, this gives an intensity of infinity, and rays cross frequently. The regions where this occurs are called caustics and are areas of high, but not infinite, acoustic intensity. In fact, the lower the frequency, the lower is the intensity at the caustic. Errors in ray theory are greatest at low frequencies where wavelengths are the longest, violating the assumption that they are very short. Ray theory is very attractive because of its ability to produce intuitive visual energy trajectories that are easily understood. A great deal of effort has gone into correcting problems associated with the original assumptions.

To determine the loss between a source and receiver, it is necessary to identify the rays that connect these points; such rays are called eigenrays. For range-independent environments, it is relatively simple to find this family of rays, but for range-dependent environments this is not the case. The first approach to determining this was to find families of rays that are close to eigenrays, by starting a large number of rays at the source and tracing them out to the ranges and depths of interest. Rays that were very close to each other near the source (about a degree apart) could be widely separated at long range. This means that very large numbers of rays must be propagated in order to estimate loss at long range. The other approach is to start with a medium number of rays and then iterate to find eigenrays at long range. In both of these approaches, the resulting eigenrays are actually not exact, but close to being eigenrays, where "close" is an input assumption.

A more modern approach is called the fat beam or Gaussian beam method. Using Gaussian beams allows energy to spread out from rays when they are close together, and the intensity never becomes out of control.

6.6.2 Normal Modes

Normal modes are derived from the Helmholtz equation by initially assuming an incredibly strict assumption of a horizontally stratified medium, frequently referred to as a range-independent medium. Then, sound is propagated as the sum of individual standing waves in the ocean, which do not interact with each other. Each mode has a unique group velocity and corresponding propagation angle and can be decomposed as the product of a range function and a depth function. In the ocean, it is impossible to compute directly the exact depth function of each mode. Two approximations for the depth functions are commonly used.

The first method of approximation is called iteration. This divides the sound speed profile into many thin isovelocity layers. The mode shape in each layer is easy to compute from the upper and lower boundary conditions. For this to work, the boundary conditions must be matched up between layers, i.e., the boundary conditions at the bottom of the nth layer must be the same as the boundary conditions at the top of the $(n + 1)$th layer. Iteration is used to match boundary conditions. This can be computationally intense.

The second method of approximation is to use the WKB (Wentzel, Kramer, and Brillouin) approximation method [16–18]. In this method, the sound speed profile is approximated as piecewise linear instead of piecewise constant. For the linear sound speed profile, analytic functions approximate the depth function of the modes. This allows the process of defining mode depth functions to be orders of magnitude faster than the iteration process. The drawback is that the analytic functions break down at ray turning points and accuracy is compromised.

At this point both approaches are range independent. Range dependence is achieved by dividing the environment into small range bins. Each range bin is assumed range independent at the boundaries; however, how to distribute the energy in each mode to the modes in the next range step needs to be decided. This is referred to as coupling. For example, a mode in one range region might be redistributed into many (all) modes in the next range region. A coupling matrix (mode $\times$ mode) is computed and applied. This too can be computationally intensive. The resulting model can be extremely accurate. These models are one of the very few true benchmark models in existence, and may be the only range-dependent benchmark models.

An alternative to the coupling calculation, in order to achieve computational speed, is to set the diagonal of the coupling matrix to 1, which means that all the energy in mode m in range slice n shows up in mode m in range slice $n + 1$, i.e., there is no redistribution. This is called the adiabatic approximation. The problem here is that mode m may have different characteristics between range steps. For example, if the mode represents a bottom bounce and the bottom slopes significantly, the propagation angles may be significantly different after the bounce. Sloping bottoms add energy to steeper modes in the case of up slopes and the shallower angle modes for down slopes. This will not be the case with the adiabatic assumption. The adiabatic approximation frequently generates significant error if the bottom slope approaches 5° or more.

6.6.3 *Parabolic Equations*

The parabolic equation (PE) solution to the Helmholtz equation makes very different assumptions from normal mode solutions. While the first assumption made in the normal mode approximation is that the environment is horizontally range independent, this assumption is never made in PE models. The two primary assumptions here are that (1) energy is traveling at shallow angles and (2) there is no presence of backscatter. The definition of shallow angles was "about ±20°" at the time of the first use of PE in underwater acoustics, and with propagation of interest traveling along CZ paths, this was not a bad limitation. Since 1974, we have become increasingly interested in steeper angles of propagation, and corrections to the original parabolic approximation have kept up well with requirements. As a result, the parabolic equation is very accurate in most environments to about ± 45°; above that small errors begin to propagate.

Once the shallow angle assumption is made, the Helmholtz equation can be written in a form that is easy to solve by stepping the solution out in range (i.e., if we know the pressure field at all depths at one range, we can compute the pressure field at all depths at a slightly larger range) with one of several available algorithms.

6.6.3.1 Finite Element Branch

The most accurate of these algorithms is the finite element algorithm, which provides excellent results in range-dependent environments and models complex bottom interaction beautifully. The finite element method requires solutions of several large matrices at each range step. This can be very tedious, especially at high frequencies and/or in deep water. One of the primary developers of finite element PE models, The Naval Research Laboratory (NRL), has worked for many years to develop more and more efficient implementations of the finite element PE algorithm, with considerable success.

6.6.3.2 Split-Step Branch

In general, a faster method of solving the parabolic equation is through the use of Fourier transforms to advance the PE solution one range step at a time. The Fourier transform method is very efficient, but requires that the environment be continuous in range and depth, disallowing any rapid changes in the sound speed. This means that if the sound speed at the bottom of the water column is 5000 ft/s and the sound speed at the top of the sediment is 7000 ft/s, we must smooth the discontinuity over a few wavelengths. Also, like the finite element method, combinations of high frequency and deep water cause run times to increase dramatically, although not as dramatically as with the finite element method.

Even with the shallow angle approximation, which is no longer very limiting, and the restrictions on the environment, PE provides extremely dependable and stable results in many complex range-dependent environments.

6.6.4 U.S. Navy Standard Models

There are three standard passive propagation models and two standard mid-frequency active propagation models in use in the U.S. Navy. Complete physics and code documentation of all three can be found in the official documentation for each model.

6.6.4.1 GRAB

The Gaussian RAy Bundle (GRAB) model was originally developed at the Naval Underwater Warfare Center (NUWC). It is applicable to both active and passive sonar. The methodology allows rays to be combined either coherently (taking into account the calculated phase of each ray) or incoherently (power adding rays without regard to the calculated phase). This model is generally bundled in the Comprehensive Acoustic System Simulation (CASS) program, which provides a framework for acoustic analysis supporting the generation of many of the inputs required to feed GRAB and processes the outputs to support analysis. It is a U.S. Navy standard for a wide range of frequencies, from low frequency to torpedo applications. GRAB is not an eigenray model, but instead traces a very dense initial fan of rays. Using Gaussian beams allows energy to spread out from rays so that an exact ray connecting two points is not needed, only a number of close rays to compute intensity. It is currently the favorite model of the U.S. Navy. It has been tested extensively with very good results against both benchmark models and measured datasets. It is robust in deep or shallow water and covers a wide frequency range. The OAML Navy Standard designation has resulted in the production of extensive documentation to be written on GRAB and a lot of verification and validation model testing has been performed by nondevelopers, the people most likely to "break" a model. This model is currently employed in a wide variety of Navy applications and has developed a good reputation. It is not fast, but the speed is independent of frequency, depending only on the ray sampling density. I have personally worked with the model for several years as a user. The developers, Ruth Keenan and Chic Weinburg, are incredibly helpful and knowledgeable and I am proud to count them among my friends.

6.6.4.2 PE

PE, the parabolic equation model, is used for passive propagation modeling only. This is for two reasons. The first reason is speed. The Navy standard PE relies on a Fourier transform method that becomes very slow at the mid-frequency active bands in all but the shallowest of water. The second reason is also speed. Travel time is terribly inefficient to compute using PE. Since the parabolic equation never divides the sound field into modes, it never assigns group velocities to individual modes and cannot efficiently compute these velocities while it computes propagation loss.

Computation of travel time with PE requires an additional loop on frequency, and many PE runs. The number of PE runs required is the product of the bandwidth of the transmitted signal and the time spread of the received signal. If, for example, we wish to model a 3.5 kHz, 50 ms pulse that spreads as much as two seconds due to multipath propagation, then we must run $2(1000/50) = 80$ PE runs. With each individual PE run requiring possibly minutes at 3.5 kHz, 80 PE runs may take hours on 2007 computers.

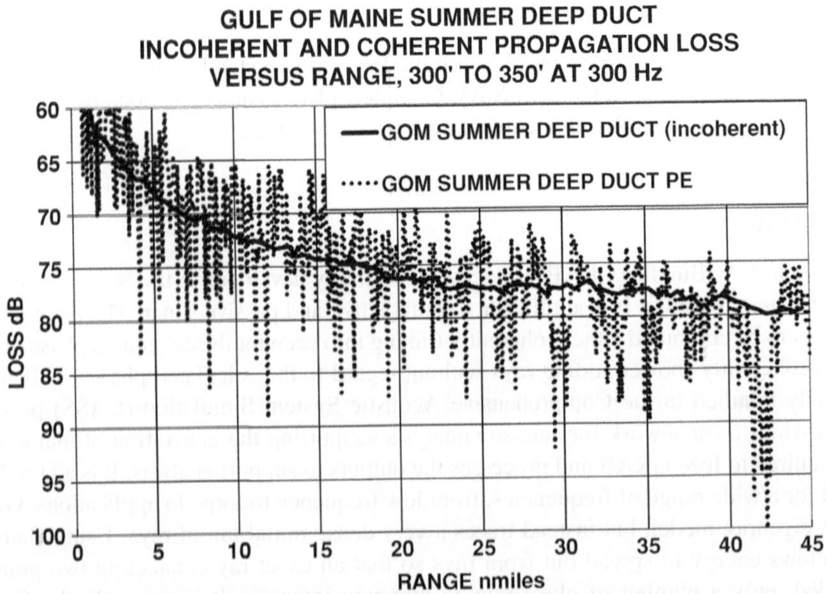

Figure 6.18 Propagation loss predictions using GRAB (incoherent) and PE (coherent) in a deep sound channel

The newest OAML approved version of PE is PE 5.0. It includes the split-step PE, as well as the finite element parabolic equation model (FEPE). FEPE is a highly accurate propagation model, prone occasionally to extreme instability (split-step PE does not have stability problems). Recently, NRL has distributed a new finite element PE model called RAM. This is a complete rewrite of the original FEPE and is purported to be "much better than FEPE." RAM versions of FEPE model have been combined with the Navy standard split-step PE model. The output of PE is a coherent propagation loss; frequently this will be range smooth to remove the large swings in loss. See Figures 6.18 and 6.19 for examples.

6.6.4.3 ASTRAL

The ASTRAL model is an adiabatic range-dependent propagation model. ASTRAL efficiently provides travel times that are used in modeling active system performance, and is therefore used for both passive and active tactical decision aids.

ASTRAL performs some mode averaging, which makes coherent outputs impossible, but is the reason for the models' incredible speed. This model does not compute all modes at very low frequencies; mode averaging has an adverse effect on ASTRAL's accuracy at very low frequency. A rule of thumb that works well for the ASTRAL model is to require 10 wavelengths of water column depth. For example, at 100 Hz, the water column needs to be at least 10 × 50 or 500 feet deep. At 1000 Hz, the water column could be as shallow as 50 feet.

As discussed above, the adiabatic assumption causes the model to generate significant errors if the bottom slope is more than a few degrees, which fortunately is fairly rare, occurring primarily as one approaches the continental shelf.

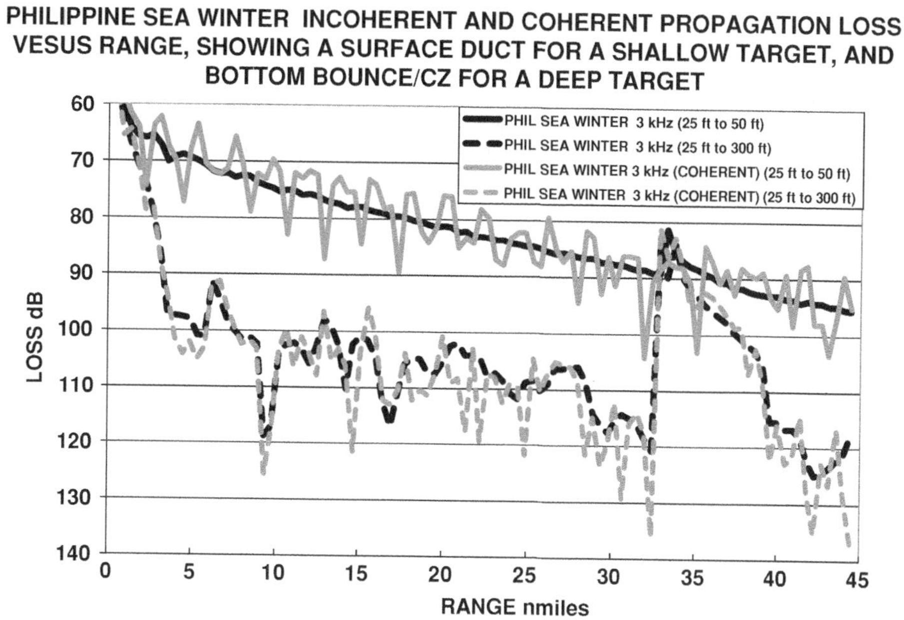

Figure 6.19 Propagation loss predictions using GRAB (incoherent) and PE (coherent) in a deep Philippine Sea in winter

References

[1] Grove, D. G., and Hurt, L. M., *Ocean World Encyclopedia*, New York: McGraw-Hill, 1980.

[2] Fairbridge, R. H., *The Encyclopedia of Oceanography*, New York: Reinhold Publishing Company, 1966.

[3] Miles, J. W., "On the Reflection of Sound at an Interface of Relative Motion," *Journal of the Acoustical Society of America*, **29**, 1957, 226.

[4] Pierson Jr., W. J., and Moskowitz, L. A., "Proposed Spectral Form for Fully Developed Wind Seas Based on the Similarity Theory of S. A. Kitaigorodskii," *Journal of Geophysical Research*, **69**, **1964**, 5181–5190.

[5] Casey, W. M., "Lloyd's Mirror-Image Interference Effects," *Acoustics Today*, **5** (Issue 2), April 2009.

[6] Eckhart C., "The Scattering of Sound from the Sea Surface," *Journal of the Acoustical Society of America*, **25**, 1953, 566–570.

[7] Marsh, H. W., and Schulkin, M., "Underwater Sound Transmission," AVCO Marine Electronics Office, Report, 1962.

[8] Marsh, H. W, and Schulkin, M., "Report on the Status of the Project AMOS (Acoustic Meteorological and Oceanographic Survey)," U.S. Navy Underwater Sound Laboratory, Report 255, 1955.

[9] Beckmann, P., and Spizzichino, A., "The Scattering of Electromagnetic Waves from Rough Surfaces," in *International Series of Monographs on Electromagnetic Waves*, Vol. 4, New York: Macmillan, 1963.

[10] Liebiger, G. A., "A Combined Ray Theory–Normal Mode Approach to Long Range Low Frequency Propagation Loss Prediction (U)," Naval Underwater Systems Center, Technical Memo PA3-0109-71, 1971.

[11] Dahl, P. H., "Revisions and Notes on a Model for Bubble Attenuation in Near Surface Acoustic Propagation," Applied Physics Laboratory of the University of Washington (APL/UW), July 1994.

[12] Hall, M. V., "A Comprehensive Model of Wind Generated Bubbles in the Ocean and Prediction of the Effects on Sound Propagation at Frequencies to 40 kHz," *Journal of the Acoustical Society of America*, **86**, 1989, 1103.

[13] Rayleigh, Lord, *The Theory of Sound*, Vol. II, Chapter XIX, New York: Dover Publishing, reprint 1948.

[14] Steed, C. A., Koehler, K. A., Harvey, D. W., and Northridge, B., "Geophysical Data Base Variable Resolution (GDBV): An Object Oriented Database for Dynamic Geo-Acoustic Data Storage," Naval Research Laboratory, Stennis Space Center, 7 January 2003, available at http://stinet.dtic.mil/index.html, AD Number: ADA417531.

[15] Kerr, D. E., *Propagation of Short Radio Waves*, M.I.T. Radiation Laboratory Series, Vol. 3, New York: McGraw-Hill Book Company, 1951.
[16] Wentzel, G., "Eine Verallgemeinerung der Quantenbedingungen für die Zwecke der Wellenmechanik," *Zeitschrift der Physik*, **38**, 1926, 518–529.
[17] Kramers, H. A., "Wellenmechanik und halbzählige Quantisierung," *Zeitschrift der Physik*, **39**, 1926, 828–840.
[18] Brillouin, L., "La Mécanique Ondulatoire de Schrödinger: Une Méthode Générale de Resolution par Approximations Successives," *Comptes Rendus de l'Academie des Sciences*, **183**, 1926, 24–26.

7

Ambient Noise

In the ocean, ambient noise is the noise associated with a given environment. This noise depends on factors that are generally beyond our control. Ambient noise is defined as the residual noise in the absence of all individual identifiable sources. Potential sources of this noise are turbulence, shipping, wave action, thermal agitation, seismic events, rainfall, marine animals, and ice sheets cracking. For a measurement to be valid, the self noise of the hydrophone must be well below this ambient noise level, which includes the 60 Hz hum and nearby noise sources like individual ships.

Many sources of ambient noise generate continuous frequency bands with Gaussian statistics. This is certainly true for thermal, wave, and rain noise. Shipping noise, however, contains both continuous noise from propeller cavitation and discrete noise from machinery and blade rate components, as discussed in Chapter 10.

7.1 Ambient Noise Models

Prior to World War II, there were essentially no measurements for ambient noise. This was primarily due to two reasons: (1) lack of calibrated hydrophones and (2) an interest in active sonar. For active sonars, the self noise on World War II destroyers was normally well above the ambient noise levels; when self noise was not above ambient levels it was dominated by reverberation. Initially, the advent of acoustic mines and passive sonar was the driving force behind making ambient noise measurements. In the field of underwater acoustics, there are vast amounts of current literature on ambient noise, second in magnitude only to propagation loss literature. Measurements taken during World War II by Knudsen and others [1] resulted in the development of a well-known set of ambient noise versus sea state versus frequency curves, shown in Figure 7.1. See Chapter 6 for a discussion of the relationship between wind speed and sea state.

Knudsen's measurements, which were made primarily in coastal waters with a focus on frequencies above 1 kHz, are of historical interest. Significant, more recent, summary references are those by Wenz [2] (Figure 7.2), Crouch [3], Urick [4], and Sadowski *et al.* [5] (Figure 7.3).

Recent measurements in the Northeast Pacific have shown a substantial increase in ambient noise for frequencies below 300 Hz when compared to data recorded at the same location

Underwater Acoustics: Analysis, Design and Performance of Sonar Richard P. Hodges

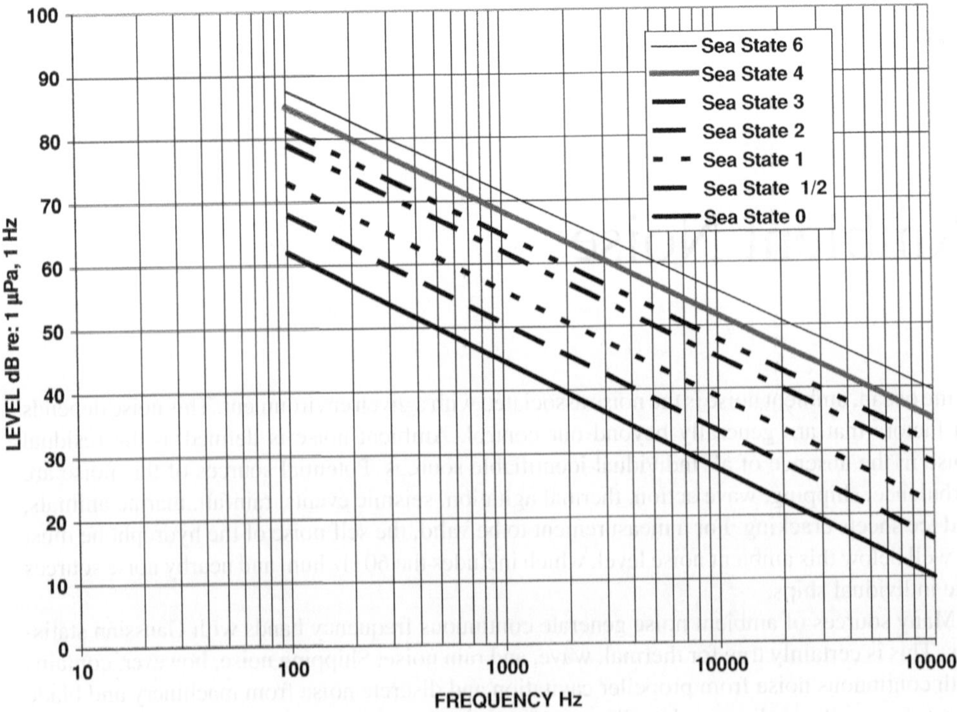

Figure 7.1 Classic "Knudsen" ambient noise curves from World War II

40 years earlier [6]. The recent ambient noise measurements are 10–12 dB higher in the 30–50 Hz region, with the difference gradually decreasing to zero as 300 Hz is approached. A logical explanation for this is the increase in both the number (doubled) and size (quadrupled) of commercial shipping at sea during this time period.

7.2 Seismic Noise

The earth is in a constant state of seismic activity. Individual large quake and volcano eruptions are major transient sources over a large region of frequencies. Microseismic activity has a nearly regular period of about 0.14 Hz and a vertical amplitude on the order of 10^{-6} m. The equivalent acoustic pressure is

$$P = \rho c u = 2\pi \rho c f a \tag{7.1}$$

where

ρ = density of water
c = speed of sound in water
f = frequency
a = amplitude of the displacement

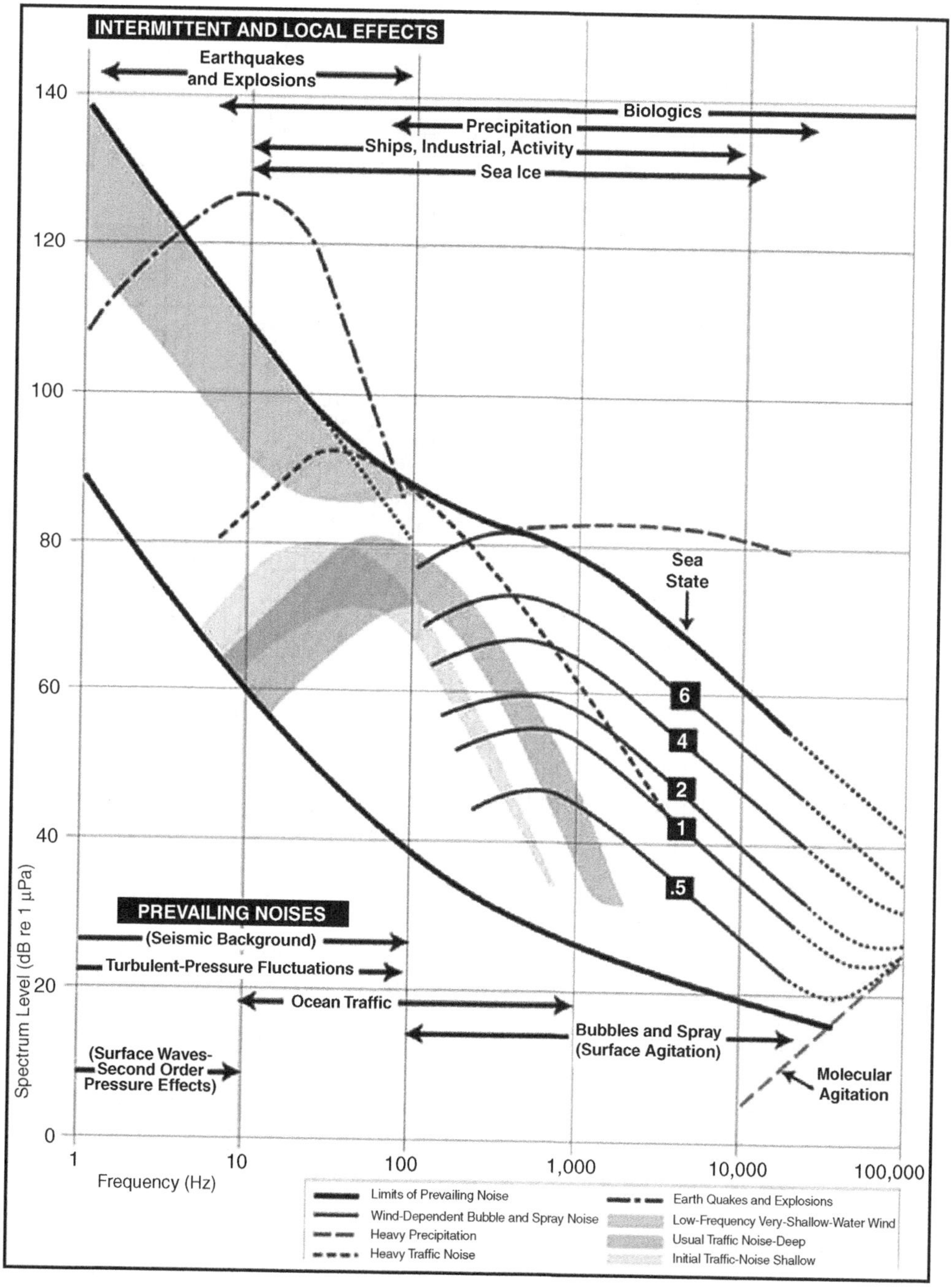

Figure 7.2 Typical sound levels of ocean background. Reprinted from Wenz [2] with permission of The American Institute of Physics

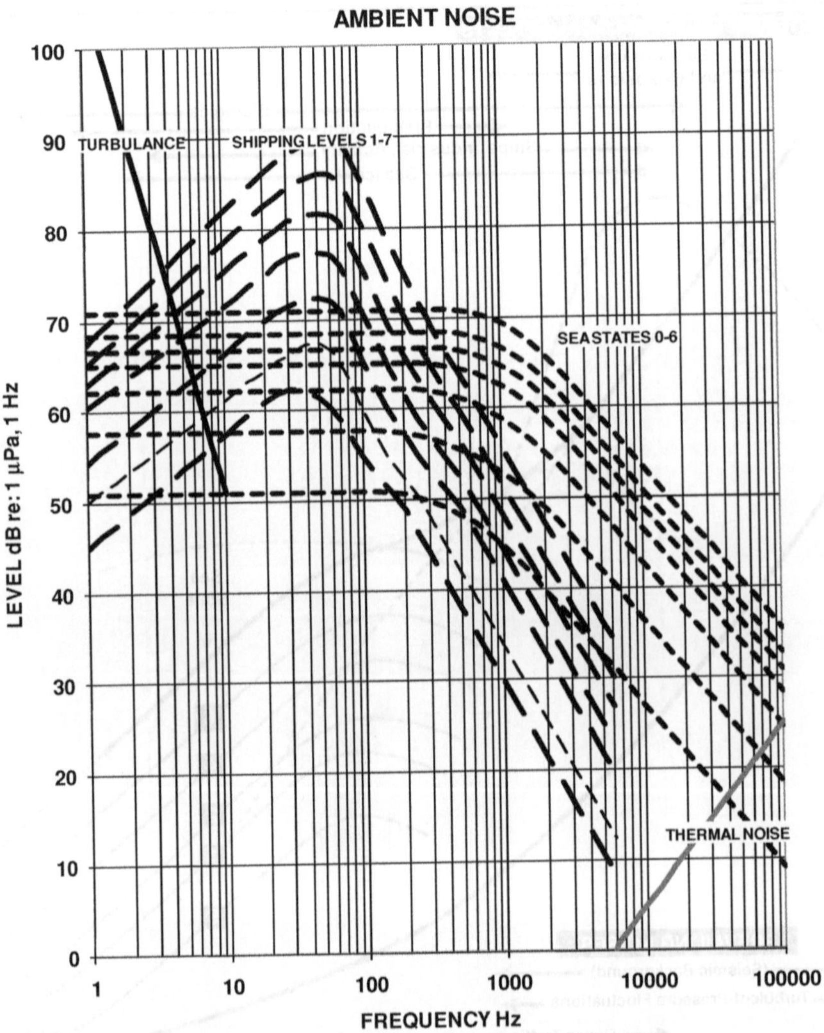

Figure 7.3 Ambient noise standards for acoustic modeling and analysis [5]

The result is 124 dB relative to 1 μPa, the dominant source of energy in the extremely low-frequency (ELF) region below 1 Hz. Also in this frequency region are lunar and solar tidal cycles, as well as the periods of waves and swells in the ocean (see Figure 6.13).

7.3 Ocean Turbulence

In general, turbulence is the dominant source of noise in the very low frequency (VLF) region from 1 to 10 Hz. A theoretical derivation of this was done by Wenz [2]:

$$\mathrm{Ln}_{\mathrm{turb}} = \mathrm{NL}_{\mathrm{trub}} = 107 - 30\log(f) \text{ dB re: 1 } \mu\text{Pa, 1 Hz, } f \text{ in Hz} \tag{7.2}$$

More recently, a result based on measured data was developed by Sadowski *et al.* [5]:

$$Ln_{turb} = NL_{turb} = 108.5 - 57.6\log(f) \text{ dB re: 1 } \mu\text{Pa, 1 Hz, } f \text{ in Hz} \quad (7.3)$$

7.4 Shipping Noise

In general, ambient noise from 10 Hz to the mid-hundreds of Hz is dominated by noise radiated from distant ship traffic. In extremely busy areas, like the English Channel, shipping noise may dominate up to the mid-thousands of Hz frequencies. Table 7.1 shows shipping noise level versus frequency versus shipping level (SL) [5]. The italicized numbers were extrapolated to cover the full frequency region.

7.5 Wave Noise

From the mid-hundreds of Hz to 50 kHz, wave and wind generated noise dominates. This noise is commonly referred to as sea state noise. Table 7.2 shows the wave noise level versus frequency versus sea state (SS) [5]. The italicized numbers were extrapolated to cover the full frequency region.

7.6 Thermal Noise

In the frequency region above wind and wave generated noise, greater than 50 kHz, thermal noise dominates. Thermal noise is simply the molecular bombardment of the receiver. Mellan [7] was the first to theoretically derive the level and frequency dependence of this noise source and found that

$$Ln_{thermal} = NL_{thermal} = 20\log(f) - 75 \text{ dB re: 1 } \mu\text{Pa, 1 Hz, } f \text{ in Hz} \quad (7.4)$$

Ezrow [8] determined experimentally that this equation is accurate to within 4 %.

7.7 Rain Noise

There is a long history of studying the noise produced by rain falling on the ocean. The rain impacting the surface of the ocean generates noise by three mechanisms: impact, oscillation of the droplet after impact, and oscillation of the entrained air carried below the surface. A classic paper by Franz [9] describes each of these mechanisms in detail and derives theoretical models for this type of noise. There is a great deal of literature containing measurements of ambient noise from rain (e.g., by Heindsmann *et al.* [10] and Bom [11]). Figure 7.4 shows curves fit to measured data and are probably adequate for acoustic modeling and analysis purposes. Note that the curves are quite flat over a wide frequency region, 100 Hz to 3 kHz. Table 7.3 shows the noise level generated by rain versus frequency and sea state (Sadowski *et al.* [5]). The italicized numbers were extrapolated to cover the full frequency region.

Table 7.1 Shipping noise level

Frequency	SL 1	SL 2	SL 3	SL 4	SL 5	SL 6	SL 7
1	45.0	50.1	54.4	60.4	62.8	65.0	67.4
1.25	46.2	51.2	55.6	61.5	64.1	66.4	68.8
1.6	47.5	52.3	56.8	62.6	65.3	67.7	70.3
2	48.7	53.4	58.0	63.7	66.6	69.1	71.7
2.5	49.9	54.5	59.2	64.8	67.8	70.4	73.2
3.2	51.2	55.6	60.4	65.9	69.0	71.8	74.6
4	52.4	56.7	61.6	67.0	70.3	73.1	76.1
5	53.7	57.8	62.8	68.1	71.5	74.5	77.5
6.4	54.9	58.9	64.0	69.2	72.8	75.8	79.0
8	56.2	60.1	65.2	70.4	74.0	77.2	80.5
10	57.3	61.1	66.3	71.4	75.2	78.5	81.9
12.5	58.7	62.3	67.6	72.5	76.5	79.8	83.3
15	59.7	63.3	68.5	73.6	77.4	80.8	84.6
16	60.0	63.5	68.8	74.0	77.8	81.3	84.9
20	61.0	64.4	69.9	74.6	78.9	82.5	86.2
25	61.9	65.5	70.9	76.1	79.9	83.5	87.4
30	62.3	66.1	71.4	76.9	80.9	84.5	88.1
31.5	62.3	66.4	71.6	77.0	81.0	84.8	88.4
40	62.2	67.3	72.2	77.3	81.5	85.5	89.3
50	61.3	67.4	72.4	77.3	81.7	86.0	89.7
60	60.2	66.7	72.1	77.1	81.5	86.0	89.9
63	59.7	66.2	71.8	76.9	81.4	85.8	89.7
70	58.8	65.1	71.0	76.0	80.8	85.3	89.5
80	57.4	63.4	69.3	74.4	79.2	83.8	88.6
90	56.1	61.8	67.6	72.6	77.5	82.2	87.0
100	54.9	60.1	65.7	71.0	76.0	80.8	85.6
125	52.7	57.0	62.4	67.8	72.6	77.4	82.3
150	51.4	54.9	60.2	65.3	69.9	74.6	79.4
160	51.0	54.3	59.5	64.4	69.0	73.7	78.6
200	48.9	52.2	57.2	61.6	66.0	70.5	75.3
250	46.2	50.4	54.9	59.0	63.4	67.5	72.2
300	44.0	48.0	53.2	57.2	61.4	65.2	69.8
315	43.4	47.4	52.9	56.6	60.9	64.6	69.1
400	40.6	44.6	50.9	54.4	58.5	61.9	66.2
500	37.9	41.9	49.2	52.6	56.5	59.8	63.9
600	35.8	39.8	47.0	50.4	54.3	57.6	61.7
630	35.2	39.2	46.4	49.8	53.7	57.0	61.1
700	33.9	37.9	45.2	48.6	52.5	55.8	59.9
800	32.3	36.3	43.6	47.0	50.9	54.2	58.3
900	30.9	34.9	42.2	45.6	49.5	52.8	56.9
1000	29.7	33.7	40.9	44.3	48.2	51.5	55.6
1250	27.0	31.0	38.3	41.7	45.6	48.9	53.0
1500	24.8	28.8	36.1	39.5	43.4	46.7	50.8

Table 7.1 (Continued)

Frequency	SL 1	SL 2	SL 3	SL 4	SL 5	SL 6	SL 7
1600	24.1	28.1	35.3	38.7	42.6	45.9	50.0
2000	21.4	25.4	32.7	36.1	40.0	43.3	47.4
2500	18.8	22.8	30.0	33.4	37.3	40.6	44.7
3000	16.6	20.6	27.8	31.2	35.1	38.4	42.5
3150	16.0	20.0	27.3	30.7	34.6	37.9	42.0
4000	13.2	17.2	24.4	27.8	31.7	35.0	39.1
5000	10.5	14.5	21.8	25.2	29.1	32.4	36.5
6000	8.3	12.3	19.6	23.0	26.9	30.2	34.3
6300	7.7	11.7	19.0	22.4	26.3	29.6	33.7
7000	6.5	10.5	17.8	21.2	25.1	28.4	32.5
8000	4.9	8.9	16.2	19.6	23.5	26.8	30.9
9000	3.5	7.5	14.8	18.2	22.1	25.4	29.5
10 000			13.5	16.9	20.8	24.1	28.2

7.8 Temporal Variability of Ambient Noise

The ambient noise at a location varies with time due to the multiplicity of sources including wind, rain, ship traffic, and biologic activity. These time constants vary from fairly short, a few minutes, to quite long. Variability due to weather typically changes slowly whereas, in coastal regions, the variability due to local small boat activity can be extreme, with daytime peak and weekend surges. Measured shipping noise in the deep ocean has variability on the order of 6 dB, with a time constant of a few hours to several days. Sea state noise has a similar variability, but time constants tend to be longer (see Urick [4]). Figure 7.5 shows an example of measured deep water ambient noise taken six times over an 18 day period as compared to the expected value from historical data.

7.9 Depth Effects on Noise

Since most ambient noise originates from on or near the surface of the ocean, it is not surprising that the observed ambient noise is a function of the receiver depth. In deep water, at the lower frequencies where shipping noise dominates, the ambient noise level generally declines slowly with depth until the receiver reaches the critical depth, the depth below the thermocline where the sound speed starts to exceed the surface sound speed. At this depth, the quieting becomes more rapid, with the quietest conditions being on the bottom of the ocean. At higher frequencies, where wave action dominates, up to 10 kHz, ambient noise is relatively independent of depth because the noise is originating locally. Above 10 kHz, the ambient noise level will drop rapidly with depth as the noise is absorbed in the water, until the frequency region is reached where thermal noise dominates.

7.10 Directionality of Noise

The directionality of noise in the ocean is determined by multiple factors. Most of the noise sources listed in the 10 Hz to 10 kHz region arise from locations on or near the surface of the

Table 7.2 Wave noise level

Frequency	SS 0	SS1	SS2	SS3	SS4	SS5	SS6
1	50.9	57.7	62.2	65.0	66.7	68.4	70.9
1.25	50.9	57.7	62.2	65.0	66.7	68.4	70.9
1.6	50.9	57.7	62.2	65.0	66.7	68.4	70.9
2	50.9	57.7	62.2	65.0	66.7	68.4	70.9
2.5	50.9	57.7	62.2	65.0	66.7	68.4	70.9
3.2	50.9	57.7	62.2	65.0	66.7	68.4	70.9
4	50.9	57.7	62.2	65.0	66.7	68.4	70.9
5	50.9	57.7	62.2	65.0	66.7	68.4	70.9
6.4	50.9	57.7	62.2	65.0	66.7	68.4	70.9
8	50.9	57.7	62.2	65.0	66.7	68.4	70.9
10	50.9	57.7	62.2	65.0	66.7	68.4	70.9
12.5	50.9	57.7	62.2	65.0	66.7	68.4	70.9
16	50.9	57.7	62.2	65.0	66.7	68.4	70.9
20	50.9	57.7	62.2	65.0	66.7	68.4	70.9
25	50.9	57.7	62.2	65.0	66.7	68.4	70.9
32	50.9	57.7	62.2	65.0	66.7	68.4	70.9
40	50.9	57.7	62.2	65.0	66.7	68.4	70.9
50	50.9	57.7	62.2	65.0	66.7	68.4	70.9
64	50.9	57.7	62.2	65.0	66.7	68.4	70.9
80	50.9	57.7	62.2	65.0	66.7	68.4	70.9
100	50.9	57.7	62.2	65.0	66.7	68.4	70.9
125	50.9	57.6	62.2	65.0	66.7	68.4	70.9
160	50.8	57.5	62.2	65.0	66.7	68.4	70.9
200	50.6	57.3	62.2	65.0	66.7	68.4	70.9
250	50.2	57.0	62.1	65.0	66.7	68.4	70.9
320	49.7	56.6	61.9	64.9	66.7	68.4	70.9
400	49.0	56.0	61.5	64.7	66.6	68.4	70.9
500	48.2	55.4	61.1	64.4	66.4	68.2	70.8
640	47.3	54.7	60.5	63.9	66.0	67.9	70.5
800	46.1	53.8	59.7	63.2	65.4	67.4	70.0
1000	44.8	52.9	58.8	62.4	64.7	66.7	69.4
1250	43.4	51.8	57.8	61.5	63.8	65.8	68.5
1600	41.6	50.4	56.5	60.2	62.6	64.6	67.3
2000	39.8	49.0	55.2	58.9	61.3	63.3	66.0
2500	38.1	47.4	53.7	57.4	59.8	61.8	64.4
3200	36.2	45.6	51.9	55.7	58.0	60.0	62.6
4000	34.3	43.7	50.0	53.8	56.2	58.1	60.7
5000	32.6	42.0	48.3	52.1	54.4	56.4	59.0
6400	30.8	40.2	46.4	50.2	52.6	54.5	57.1
8000	28.9	38.3	44.5	48.3	50.7	52.7	55.3
10 000	27.1	36.5	42.8	46.6	48.9	50.9	53.5
12 500	25.4	34.8	41.0	44.8	47.2	49.1	51.7

Table 7.2 (Continued)

Frequency	SS 0	SS1	SS2	SS3	SS4	SS5	SS6
16 000	23.4	32.8	39.1	42.9	45.2	47.2	49.8
20 000	21.7	31.1	37.3	41.1	43.5	45.4	48.0
25 000	19.9	29.3	35.6	39.4	41.7	43.7	46.3
32 000	18.1	27.5	33.7	37.5	39.9	41.9	44.5
40 000	16.2	25.6	31.9	35.7	38.0	40.0	42.6
50 000	14.4	23.8	30.1	33.9	36.3	38.2	40.8
64 000	12.6	22.0	28.3	32.1	34.4	36.4	39.0
80 000	10.7	20.1	26.4	30.2	32.6	34.5	37.1
100 000	9.0	18.4	24.6	28.4	30.8	32.8	35.4

ocean. The energy from these sources is then propagated to the receiver. Therefore, it is not surprising that the noise may have a substantial vertical anisotropic nature. Figure 7.6 shows the distribution of ambient noise in the vertical [12].

In the deep ocean, at frequencies where shipping noise is dominant, it is common to see substantial variations in noise level versus the horizontal direction. There are two effects occurring: (1) discrete shipping, those ships that are close, are resolvable on sonars and (2) distant shipping may be clumped together because of the use of standard shipping lanes. The variability from discrete shipping is determined by how close ships come. This variability can be enormous. The underlying distribution from very distant shipping has a variability that is small in areas with little shipping (e.g., the middle of the South Pacific) to very large near major ports and shipping lanes. Figure 7.7 is an example of the

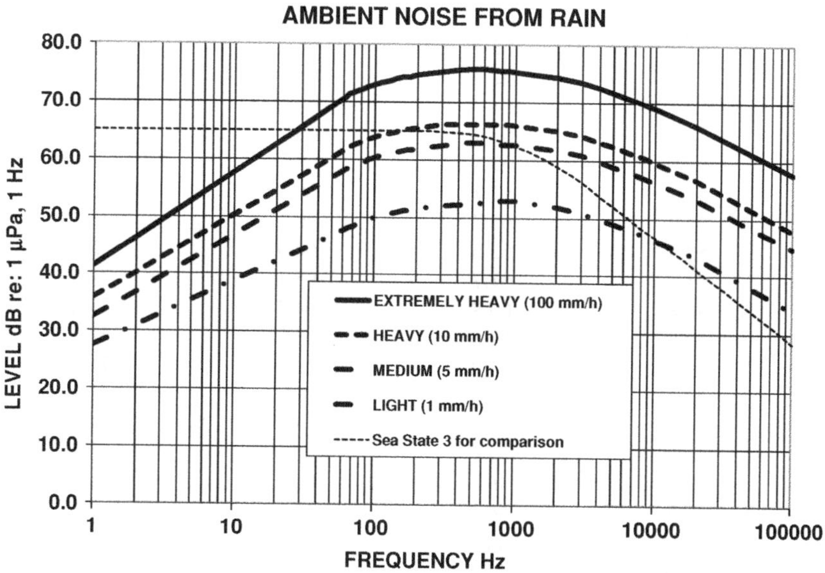

Figure 7.4 Ambient noise from rain for acoustic modeling and analysis [5]

Table 7.3 Rain generated noise level

Frequency	Light (1 mm/h)	Medium (5 mm/h)	Heavy (10 mm/h)	Extremely heavy (100 mm/h)
1	27.5	32.4	35.7	41.3
2	30.9	36.8	40.1	46.2
4	34.3	41.1	44.4	51.2
8	37.7	45.4	48.8	56.1
16	41.1	49.8	53.1	61.1
32	44.6	54.1	57.5	66.0
62	47.8	58.2	61.8	70.7
66	48.1	58.6	62.1	71.4
70	48.4	58.8	62.4	71.6
76	48.8	59.2	62.8	72.0
80	49.0	59.4	63.0	72.2
86	49.3	59.7	63.3	72.5
90	49.5	59.9	63.5	72.7
96	49.7	60.1	63.7	72.9
100	49.9	60.3	63.9	73.1
110	50.2	60.6	64.2	73.4
120	50.4	60.8	64.4	73.6
130	50.6	61.0	64.6	73.8
140	50.8	61.2	64.8	74.0
150	51.0	61.4	65.0	74.2
160	51.2	61.6	65.2	74.4
170	51.3	61.7	65.3	74.5
180	51.3	61.7	65.3	74.5
190	51.4	61.8	65.4	74.6
200	51.5	61.9	65.5	74.7
300	52.0	62.5	66.1	75.3
400	52.2	62.7	66.1	75.7
500	52.4	63.0	66.1	75.8
600	52.6	63.0	66.2	75.8
700	52.8	62.8	66.2	75.6
800	52.9	62.7	66.2	75.5
900	52.9	62.7	66.1	75.5
1000	52.9	62.6	66.0	75.4
2000	51.8	61.7	65.1	74.5
3000	50.8	60.8	64.2	73.6
4000	49.8	59.8	63.2	72.6
5000	49.0	59.0	62.4	71.8
6000	48.3	58.3	61.7	71.1
7000	47.7	57.7	61.1	70.5
8000	47.2	57.2	60.6	70.0
9000	46.7	56.7	60.1	69.5
10 000	46.3	56.3	59.7	69.1
20 000	43.2	53.2	56.6	66.0
50 200	38.0	48.4	51.6	61.2
100 000	34.1	44.8	47.9	57.6

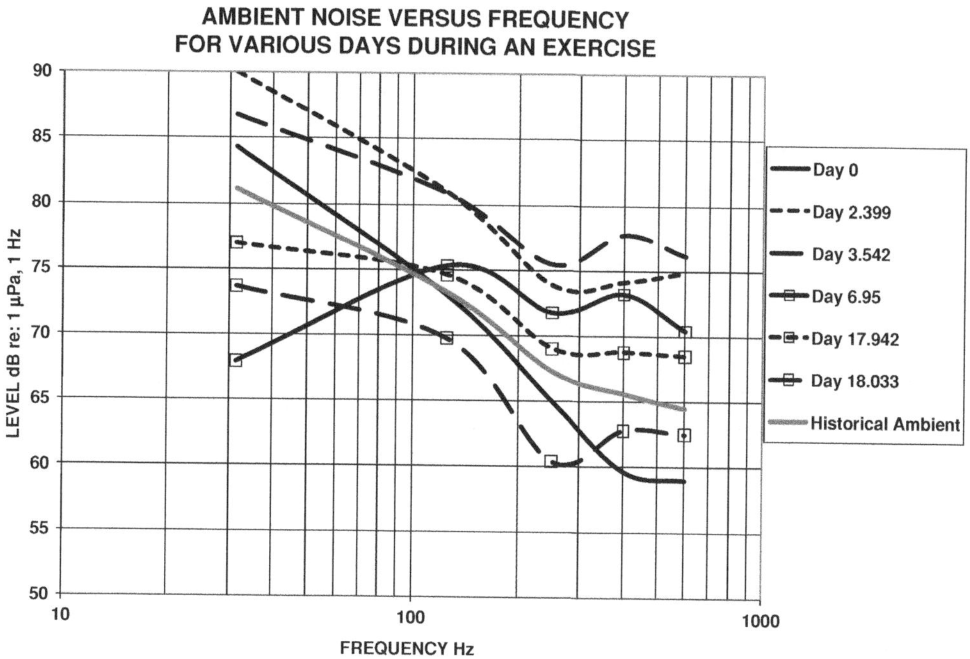

Figure 7.5 Variability in measured ambient noise in the Atlantic

variability of measured shipping noise in the horizontal, at 50 Hz, averaged over a one hour time period, in the Gulf of Oman. The peaks correspond to the directions of major shipping lanes; such curves are commonly referred to as Le Roses. In coastal areas, fishing and other small boat activity may be so great as to make the underlying background, long-term average, very smooth. However, these close, discrete contacts cause high variability in the short term.

7.11 Under Ice Noise

Arctic ambient noise can be divided into two areas: (1) deep under pack ice and (2) near the edge of the ice shelf. Due to the ice, there is no wave-induced noise deep under pack ice and very few ships anywhere nearby. As a consequence, the noise levels can be extremely quiet, well below sea state zero, reaching the limits of electronic noise for amplifiers. What noise there is comes from the ice cover. This noise is generally intermittent and characterized as booms, cracks, or crashes [13].

By contrast, the edge of the ice shelf, frequently referred to as the marginal ice zone (MIZ), is very noisy. Waves and ships are present, in addition to wind and waves, driving the ice to collide and crack almost continuously. As a receiver moves from 40 kyds under the pack ice to the MIZ and then 40 kyds out into open water, it is common to see all frequencies between 100 Hz and 1 kHz. First observed is a 20 dB rise at the MIZ, followed by a 10 dB drop in clear water.

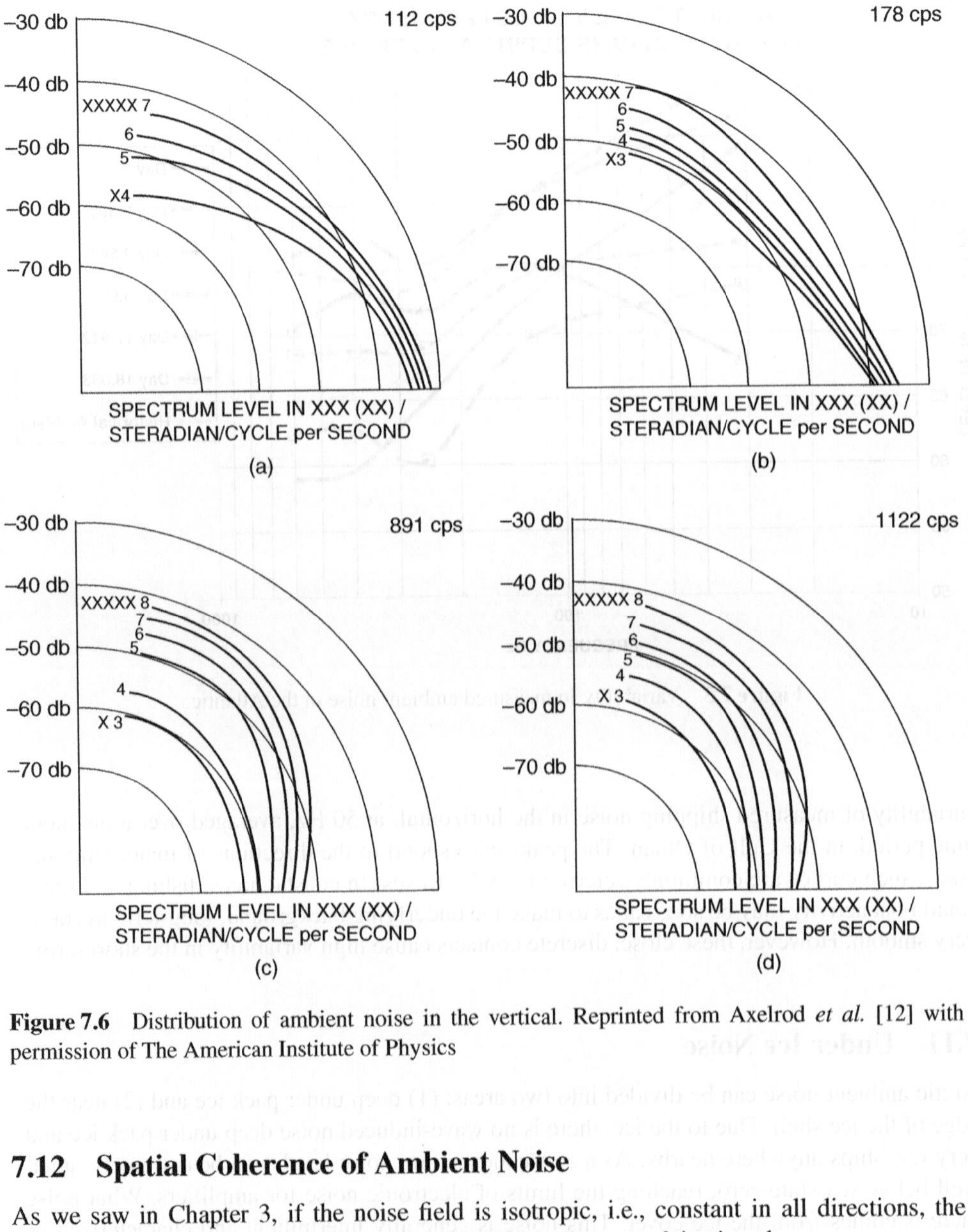

Figure 7.6 Distribution of ambient noise in the vertical. Reprinted from Axelrod *et al.* [12] with permission of The American Institute of Physics

7.12 Spatial Coherence of Ambient Noise

As we saw in Chapter 3, if the noise field is isotropic, i.e., constant in all directions, the correlation between two hydrophones separated by a distance, d, is given by

$$\rho(d) = \frac{\sin\left(\dfrac{2\pi d}{\lambda}\right)}{\left(\dfrac{2\pi d}{\lambda}\right)} \tag{7.5}$$

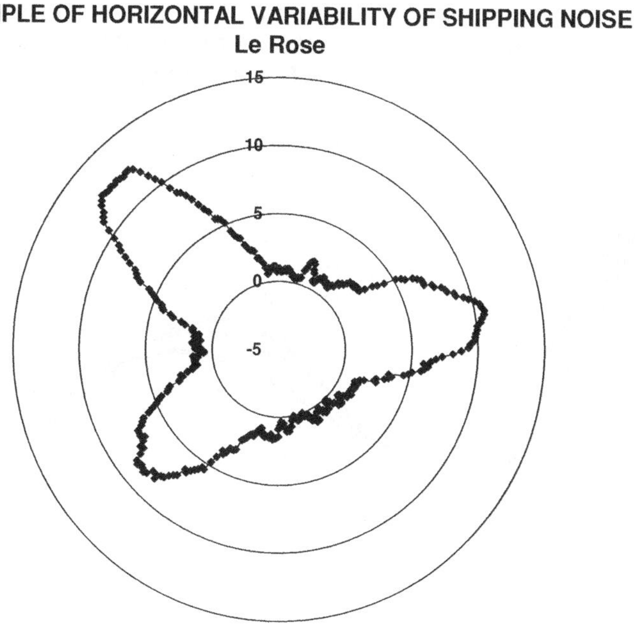

Figure 7.7 Variability of measured shipping noise in the horizontal

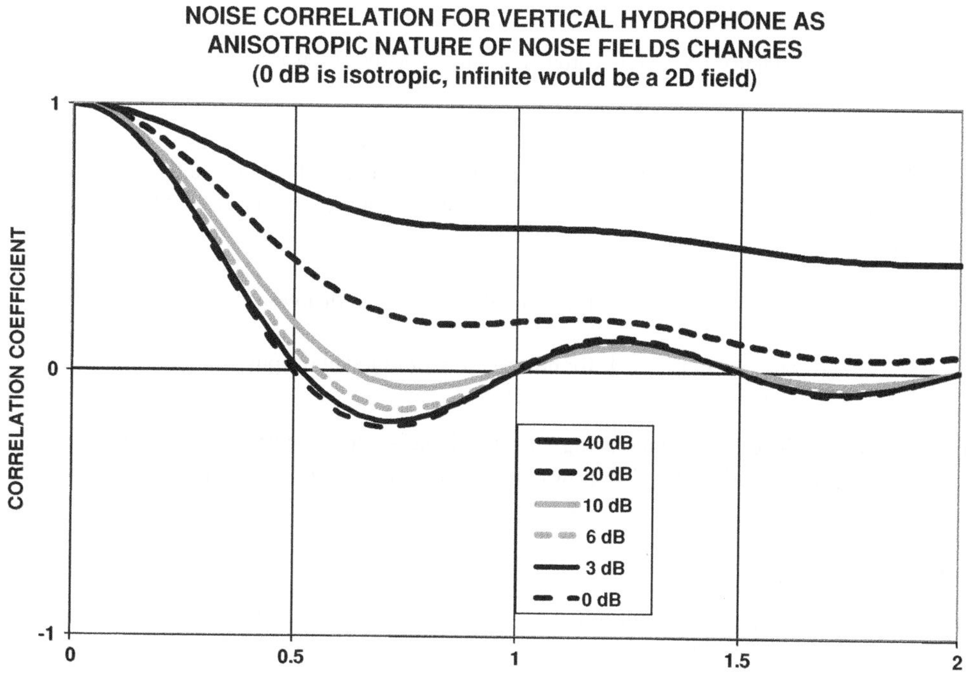

Figure 7.8 Noise correlation versus vertical separation

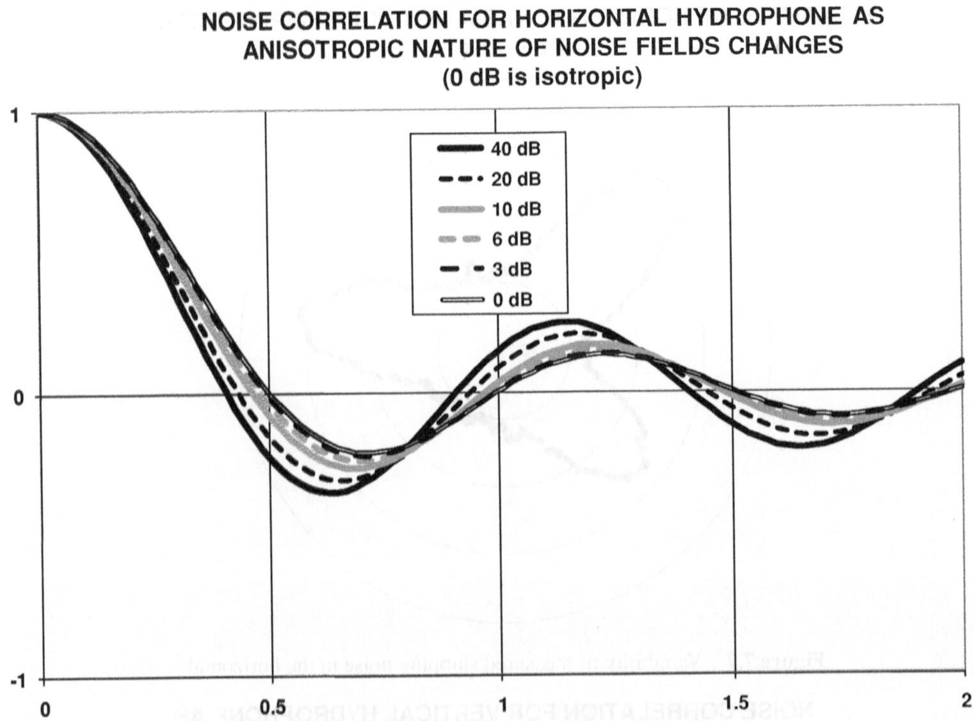

Figure 7.9 Noise correlation versus horizontal separation

The orientation of the hydrophones is irrelevant because the noise is isotropic. As has been shown by Talham [14], this is a special case not the general case. If the noise field were two-dimensionally isotropic, i.e., uniform in the horizontal, then the correlation for hydrophones would be

$$\begin{aligned} \rho(d) &= 1 && \text{for vertically oriented hydrophones} \\ \rho(d) &= J_0\left(\frac{2\pi d}{\lambda}\right) && \text{for horizontally oriented hydrophones} \end{aligned} \tag{7.6}$$

Figure 7.8 shows the effect of increasing anisotropic noise on the correlation coefficients for vertical hydrophones and Figure 7.9 shows the effect for horizontal hydrophones.

References

[1] Knudsen, V. O., Alford, R. S., and Emiling, J. W., "Survey of Underwater Sound, Report 3, Ambient Noise," NRDC 1848, September 1944.

[2] Wenz, G. M., "Acoustic Ambient Noise in the Ocean: Spectra and Sources," *Journal of the Acoustical Society of America*, **34**, 1962, 1936.

[3] Crouch, W. W., "Ambient Sea Noise: A Review of the Literature," Naval Underwater Systems Center, TR 4179, 1972 (available from the Defense Technical Information Center, AD893161).

[4] Urick, R. J., *Ambient Noise in the Sea*, Los Altos Hills, CA: Peninsula Publishing, 1986.

[5] Sadowski, V., Katz, R., and McFadden, K., "Ambient Noise Standards for Acoustic Modeling and Analysis," Naval Underwater Systems Center, TD 7265, 1984.

[6] McDonald, M. A., Hildebrand, J. A., and Wiggins, S. M., "Increase in Deep Ocean Ambient Noise in the Northeast Pacific West of San Nicolas Island, California, *Journal of the Acoustical Society of America*, **120**, 2006, 711–718.

[7] Mellan, R. H., "Thermal-Noise Limit in the Detection of Underwater Acoustic Signals," *Journal of the Acoustical Society of America*, **24**, 1952, 478.

[8] Ezrow, D. H., "Measurement of the Thermal-Noise Spectrum of Water," *Journal of the Acoustical Society of America* **34**, 1962, 550.

[9] Franz, G. J., "Splashes as Sources of Sound in Liquids," *Journal of the Acoustical Society of America*, **31**, 1959, 1080.

[10] Heindsmann, T. H., Smith, R. H., and Arnerson, A. D., "Effect of Rain upon Underwater Noise Levels," *Journal of the Acoustical Society of America*, **27**, 1955, 378.

[11] Bom, N., "Effect of Rain on Underwater Noise Level," *Journal of the Acoustical Society of America*, **45**, 1968, 150.

[12] Axelrod, E. H., Schoomer, B. A., and Von Winkle, W. A., "Vertical Directionality of Ambient Noise in the Deep Ocean at a Site Near Bermuda," *Journal of the Acoustical Society of America*, **37**, 1965, 77.

[13] Milne, A. R., and Ganton, J. H., "Ambient Noise under Arctic-Sea Ice," *Journal of the Acoustical Society of America*, **36**, 1964, 855.

[14] Talham, R. J., "Noise Correlation Functions for Anisotropic Noise Fields," *Journal of the Acoustical Society of America*, **69**, 1981, 213.

8

Reverberation

When active sonar transmits, the sound returns to the receiver from many other sources besides the target of interest. Any return that is not from a target of interest or from a discrete target-like object is referred to as reverberation. Sources of reverberation in the ocean include the surface, the bottom, and the volume of water. Volume reverberation sources include marine life, bubbles, and other inhomogeneities in the water. The sum of these sources is received after a ping as a long slowly decaying signal. For modern high-powered active sonars, this is frequently the dominant source of noise and a major concern during the design of such sonars.

8.1 Scattering, Backscattering Strength, and Target Strength

The discontinuities in acoustic characteristics results in a reradiation or scattering of a portion of the incident signal.

8.1.1 Surface and Bottom Scattering

In Chapter 6, Transmission Loss: Interaction with Boundaries, we discussed reflection or forward scattering loss for the surface of the ocean and noted that this is not actually a loss in the sense that energy is being absorbed, but rather that the lost signal is scattered in other (nonspecular) directions. Figure 8.1 shows an example of such redistribution of intensity.

Consider a plane acoustic wave insonifying an area of a rough interface, at an angle ϕ_{i}. The intensity of the incident wave is

$$I_{\mathrm{i}} = \frac{p_{\mathrm{i}}^2}{2\rho c} \tag{8.1}$$

The available power on an area, A, is

$$P_{\mathrm{i}} = A I_{\mathrm{i}} \cos(\phi_{\mathrm{i}}) \tag{8.2}$$

Underwater Acoustics: Analysis, Design and Performance of Sonar Richard P. Hodges

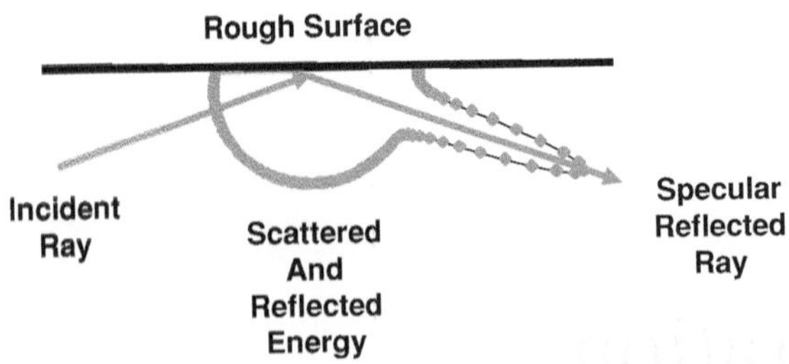

Figure 8.1 Scattering of sound from a rough boundary

The portion, ε, of this power that is not transmitted through or reflected is scattered into the half plane. The angular distribution, $S(\phi_i, \phi_s)$, of this power depends on the characteristics of the rough surface. It is generally normalized such that

$$\varepsilon(\phi_i) = \int S(\phi_i, \phi_s)\, d\phi_s \tag{8.3}$$

The intensity scattered in a given direction is

$$I_s(R, \phi_s) = \frac{P_i S(\phi_i, \phi_s)}{R^2} = \frac{A I_i \cos(\phi_i) S(\phi_i, \phi_s)}{R^2} \tag{8.4}$$

The ratio of the scattered intensity at the reference distance (usually 1 meter or 1 yard) to the incident intensities is called the scattering strength (σ_s):

$$\sigma_s(\phi_i, \phi_s) = A S(\phi_i, \phi_s) \cos(\phi_i) \tag{8.5}$$

If we let $\phi_i = \phi_s$ then the result, of particular interest to monostatic active sonar, is called the boundary backscattering strength:

$$\sigma_{bs}(\phi_i) = A S(\phi_i, \phi_s) \cos(\phi_i) \tag{8.6}$$

Expressed in decibels this is

$$\mathrm{BS_s} = 10 \log \left(\frac{\sigma_{bs}}{A} \right) \tag{8.7}$$

The boundary backscattering strength, $\mathrm{BS_s}$, is referenced to a unit area (1 m^2 or 1 yd^2).

The derivation of the scattering strength for a volume, σ_v, proceeds similarly:

$$\mathrm{BS_v} = 10 \log \left(\frac{\sigma_{bv}}{V} \right) \tag{8.8}$$

The volume backscatter strength is reference to a unit volume.

8.1.1.1 Lambert's Law

Consider a rough surface that is a perfect scatterer in the sense that neither ε nor $S(\phi_i, \phi_s)$ is a function of the angles. Given this, $\varepsilon(\phi_i) = \varepsilon_0$ and $S(\phi_i, \phi_s) = \varepsilon_0/(2\pi)$, which simply states that the scattered intensity is uniform over the half plane. This gives

$$\sigma_s(\phi_i, \phi_s) = \frac{A\varepsilon_0}{2\pi}\cos(\phi_i) \tag{8.9}$$

This is the same as the backscattering strength.

If the surface is no longer assumed to be isotropic and also depends on roughness, the scattering intensity will depend on the distribution of the surface. As an example, the average radiated intensity could be taken as proportional to the cosine of the scattering angle:

$$S(\phi_i, \phi_s) = \frac{\cos(\phi_s)}{\pi} \tag{8.10}$$

This leads to a scattering strength of

$$\sigma_s(\phi_i, \phi_s) = \frac{A\varepsilon_0\cos(\phi_s)\cos(\phi_i)}{\pi} \tag{8.11}$$

and a backscattering strength of

$$\sigma_{bs}(\phi_i) = A\varepsilon_0\,[\cos(\phi_i)]^2 \tag{8.12}$$

This result in decibels is known as Lambert's law for backscattering strength:

$$\mathrm{BS}_{\mathrm{Lambert}}(\phi_i) = \mathrm{BS}_0 + 20\log[\cos(\phi_i)] \tag{8.13}$$

If the surface is perfectly reflective, then BS_0 equals −5 dB relative to 1 m^2. Measured values are generally in the −10 to −40 dB range for bottom scattering. Figure 8.2 shows an example of Lambert's law for backscattering strength versus angle for a BS_0 of −37 dB.

It is common to see a generalized version of Lambert's law used to fit scattering data, where the assumption of uniform scattering is replaced by a cosine to a power law so that

$$\mathrm{BS}_{\mathrm{Gen\,Lambert}}(\phi_i) = \mathrm{BS}'_0 + 10n\log[\cos(\phi_i)] \tag{8.14}$$

8.1.1.2 Facet Scattering

Consider now a surface made up of a mosaic of facets, each with a random tilt. Each facet reflects an incident wave mainly around the specular (normal to facet surface) direction. As a wave moves away from the vertical, from normal to average surface, there will be fewer and fewer facets with the correct orientation; the backscattering strength will decrease rapidly. Let

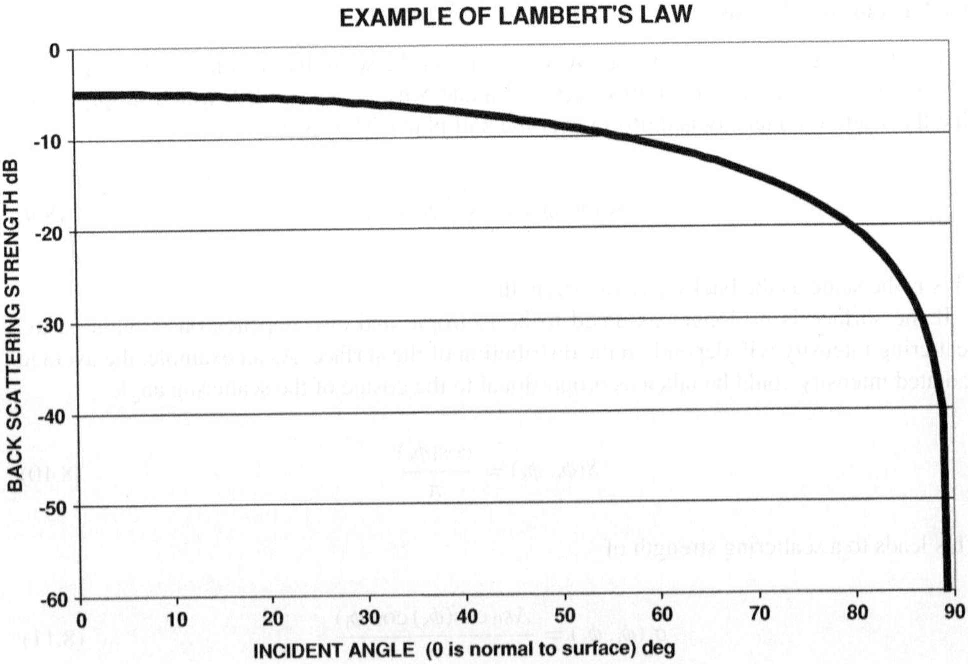

Figure 8.2 Example of backscattering strength versus incident angle using Lambert's law

ϕ be the angle of a facet with the horizontal and assume the slopes follow a normal distribution:

$$\rho\left[\tan(\phi)\right] = \frac{1}{2\pi}\exp\left\{-\left[\frac{\tan(\phi)^2}{2\gamma^2}\right]\right\} \tag{8.15}$$

where γ^2 is the variance of the slope. If the plane wave reflection coefficient is R then

$$\sigma_{\text{bs}}(\phi) = \frac{|R|^2}{8\gamma^2\left[\cos(\phi)\right]^4}\exp\left\{\frac{-\left[\tan(\phi)\right]^2}{2\gamma^2}\right\} \tag{8.16}$$

The derivation of this equation can be found in Brekhovskikh and Lysanov [1]. Note that neither of these methods results in a frequency dependence.

If we assume a Pierson–Moskowitz model for the sea surface (see Chapter 6, Transmission Loss: Interaction with Boundaries) then the standard deviation, h, of the wave height is given by

$$h = 0.00533w^2 \tag{8.17}$$

where h is in meters and w is in meters per second. Cox and Munk [2] gave a relationship between wind speed and the slope of swells as

$$\gamma^2 = 0.001\,(3 + 5.12w) \tag{8.18}$$

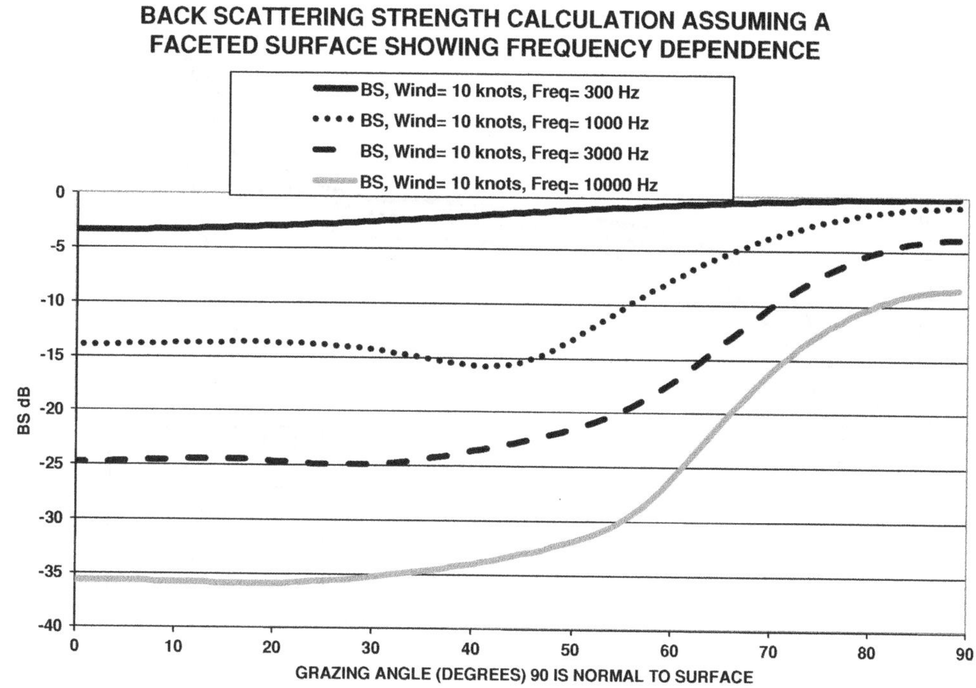

Figure 8.3 Theoretical backscattering strength versus grazing angle for a faceted sea surface at various frequencies; 90° is normal to the surface

with a correlation length, l, of

$$l = \sqrt{2}\frac{h}{\gamma} \tag{8.19}$$

If the size of the facets are assumed to equal the correlation length, the result for backscattering looks like Figure 8.3. Notice that under these assumptions the frequency dependence is quite pronounced.

Figure 8.4 shows the measured surface backscattering strength [3]. Note the similarity in the shapes of the curves.

8.1.1.3 Sea Surface as a Diffraction Grating

Marsh and colleagues [4, 5] derived another theoretical model (Figure 8.5) for sea surface backscattering. The model is independent of frequency and sea state and roughly matches data from sea state 0 to sea state 4. This model assumes that the sea surface acts like a diffraction grating. This assumption yields the following result:

$$\mathrm{BS_s} = 10\log\left\{\frac{[\tan(\phi)]^4}{32g^2}\omega^5 A(\omega)^2\right\} \tag{8.20}$$

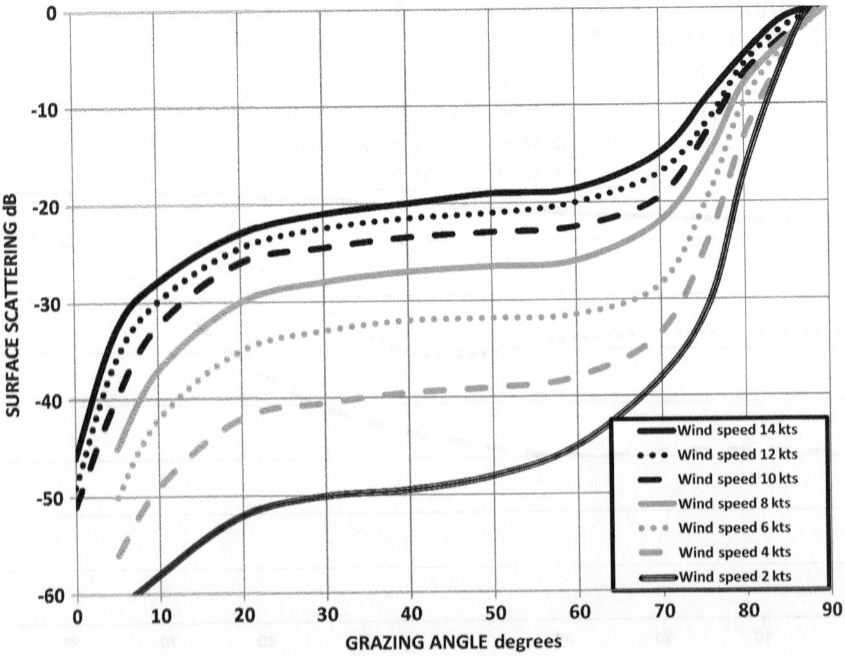

Figure 8.4 Curve fit to measured surface backscattering strength at 60 kHz [3]

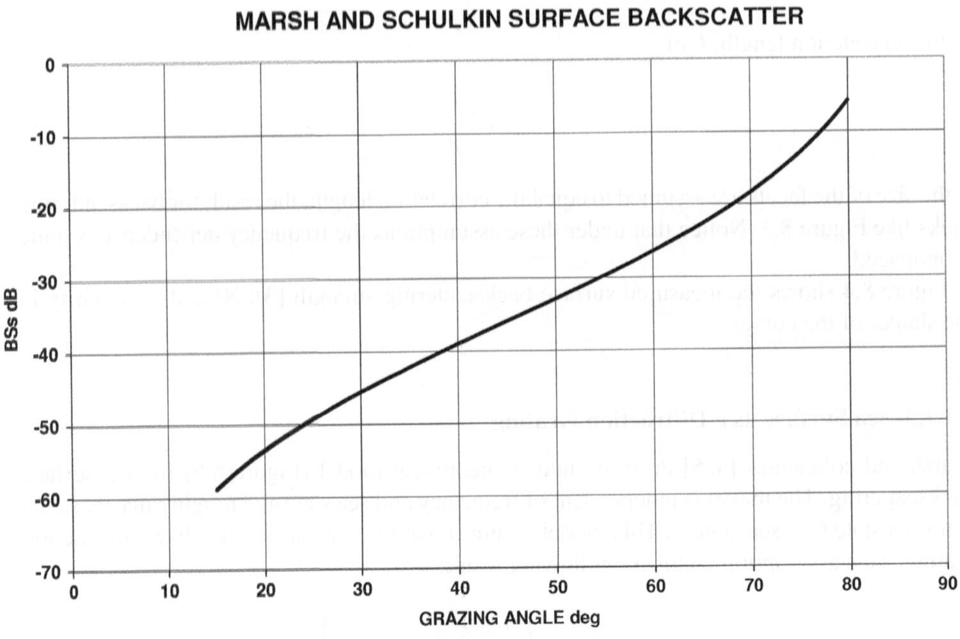

Figure 8.5 Marsh and Schulkin model for ocean surface backscattering strength as a function of grazing angle

If we assume that the energy spectrum of the sea surface is given by [6]

$$A(\omega)^2 = 0.0074g^2\omega^{-5} \tag{8.21}$$

then the result is

$$\mathrm{BS_s} = -36 + 40\log[\tan(\phi)] \tag{8.22}$$

8.1.1.4 Scattering from near Surface Bubbles

As discussed in Chapter 6, waves cause air bubbles to be entrained in the water near the surface. This is particularly evident at high frequencies (>10 kHz) and can be the dominate source of scattering. Publications by Crowther [7], McDaniel and Gorman [8], and more recently by McDaniel [9] have resulted in the following model:

$$\sigma_{\mathrm{bs}} = \frac{\beta_v \delta_r}{4\pi\delta}\left[\frac{1 + 8\beta\exp(-2\beta) - \exp(-4\beta)}{2\beta}\right] \tag{8.23}$$

where

$$\beta = \frac{\beta_v}{\sin(\phi)} \tag{8.24}$$

δ_r is the reradiation damping coefficient at resonance (0.0136), and

$$\begin{aligned} \beta_v &= 10^{(-5.2577+0.4701U)}\left(\frac{f}{25}\right)^{0.85} \quad U < 11\ \mathrm{m/s} \\ \beta_v &= \beta_v(11)\left(\frac{U}{11}\right)^{3.5} \quad U \geq 11\ \mathrm{m/s} \end{aligned} \tag{8.25}$$

$$\delta = 0.255\,(f)^{1/3} \tag{8.26}$$

8.1.1.5 Empirical Fits to Measured Data for Surface Reverberation

Explosive sources with an omnidirectional receiver were used by Chapman and Harris [10] to measure ocean surface backscattering strength versus angle and frequency. The fit to the data frequently used by models is

$$\begin{aligned} \mathrm{BS_s} &= 3.3\beta\log\left(\frac{\phi}{30}\right) - 42.4\log(\beta) + 2.6 \\ \beta &= 158\left(vf^{1/3}\right)^{-0.58} \end{aligned} \tag{8.27}$$

where

v = wind speed (kts)
f = frequency (Hz)

Figures 8.6 and 8.7 show plots of this model.

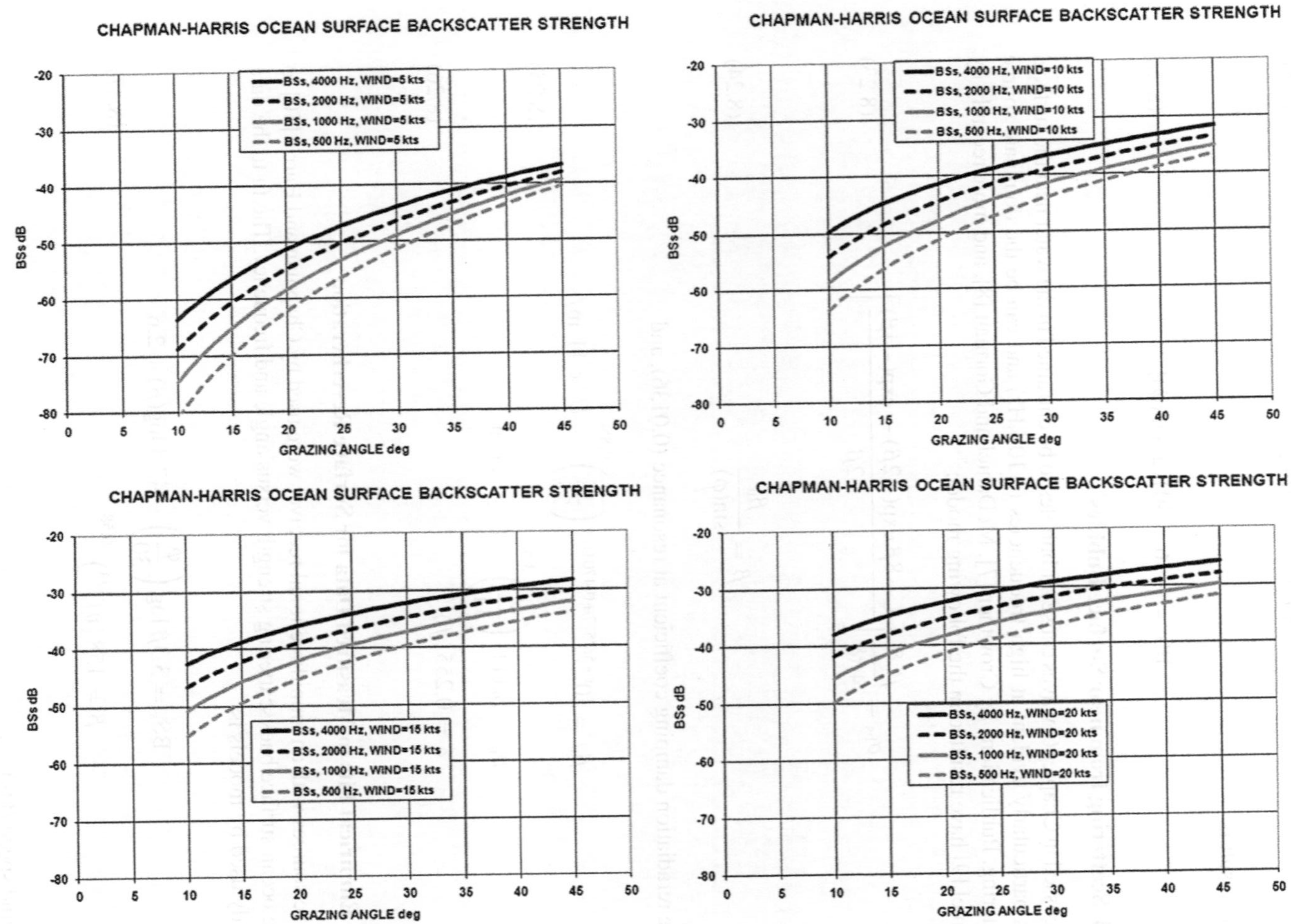

Figure 8.6 Chapman–Harris surface backscattering strength versus grazing angle, for wind speeds of 5, 10, 15, and 20 knots and frequencies of 500, 1000, 2000, and 4000 Hz

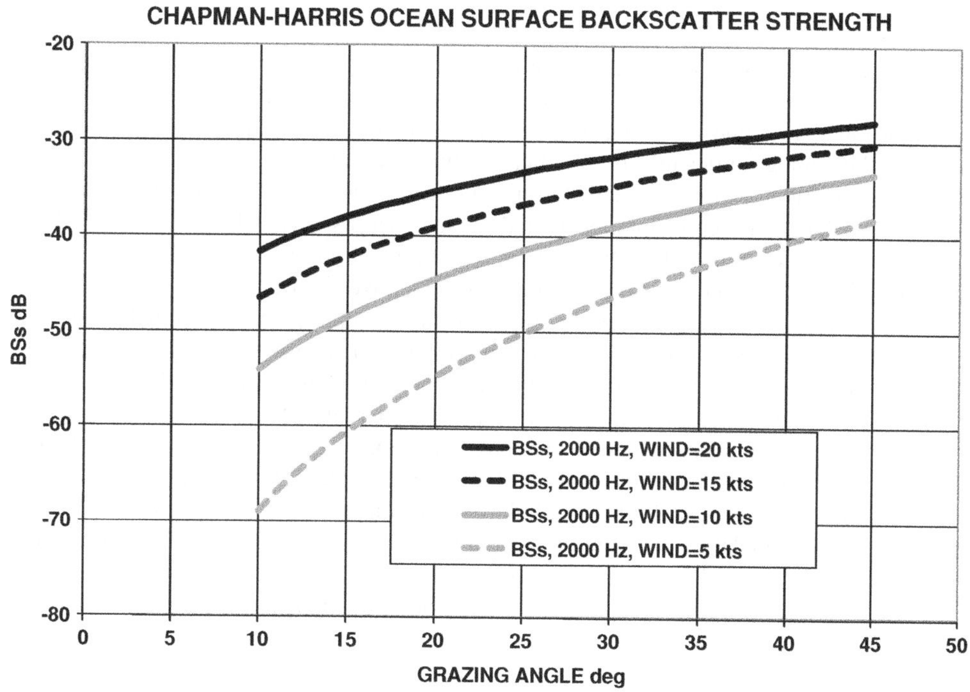

Figure 8.7 Chapman–Harris surface backscattering strength versus grazing angle, for wind speeds of 5, 10, 15, and 20 knots at 2000 Hz

Due to the wave nature of sound, the sea surface will look smooth and backscattering should vanish at long wavelengths and shallow grazing angles. Given this, the backscattering strength must depend on the ratio of $2h\sin(\theta)/\lambda$, where h is the mean peak-to-trough height, λ is the acoustic wavelength, and θ is the grazing angle. Schulkin and Shaffer [11] used available surface backscattering data and fit it using this ratio. The result is

$$\mathrm{BS_s} = 9.9\log[fh\sin(\phi)] - 45.3 \tag{8.28}$$

where

f = frequency (kHz)
ϕ = grazing angle (degrees)
$h = 0.0026v^{(5/2)}$ = height in feet
v = wind speed (kts)

Here, the acoustic wavelength has been replaced by the frequency, f, in kHz. This equation matches the measured data with a standard deviation of 5 dB for values of $fh\sin(\phi)$ between 5 and 200, and is not valid near normal incidence.

8.1.2 *Volume Scattering*

Sound scattered from a volume in the ocean is due primarily to marine life in the water column, although bubbles entrained near the surface can, particularly at high frequencies (>10 kHz), be a major source of sound scattering. While occurring less frequently, particulate matter and sound speed inhomogeneities may be significant sources of scattering.

Volume backscatter is highly variable throughout the oceans. The scattering strength varies not only with location, depth, frequency, and season, but also with the time of day. The diurnal variation is caused by the Diel vertical migration, a pattern of movement that some organisms living in the ocean's near-surface zone undertake each day. The organisms that exhibit this pattern of behavior range in size from microscopic plankton through to much larger organisms, such as fish. The Diel vertical migration is a response to ecological gradients, resources, prey, and predators. In the case of phytoplankton, while light for photosynthesis declines exponentially from the ocean's surface, the availability of nutrients typically rises down the water column. Consequently, some phytoplankton migrate down the water column at night to obtain nutrients, but return to the surface during daylight. By contrast, zooplankton and larger animals do not usually require light for growth. However, these animals may be vulnerable to larger predators that hunt by sight in well-lit surface waters, where their prey is most abundant. As a result, some of these animals undertake the opposite pattern of vertical migration, traveling to the surface at night to feed and descending to safer depths during the day.

8.1.2.1 Volume Backscattering Strength

Volume backscattering strength is referenced to a unit volume. Databases of volume backscattering strength, such as the one maintained by the U.S. Navy in its Ocean Atmospheric Master Library (OAML), are generally expressed as column scattering strength. This is defined as the equivalent in decibels of the scattering strength per unit volume integrated over the water column. Column strength is then dimensionless. Although this gives a single value for a location, it should not be interpreted as implying that the scattering strength is constant with depth or, as noted above, constant over time. Figure 8.8 shows an example of volume scattering strength versus depth for several times on the same day.

8.1.3 *Bottom Scattering*

Like bottom reflection loss discussed in Chapter 6, bottom scattering is much more complex than surface scattering because the impedance change is much less at the interface and because the bottom is a solid that can support transverse waves. Incident energy is partitioned into transverse and compressional waves and, particularly at low frequencies, will penetrate into the bottom and reemerge after interacting with structures deeper in the bottom. Therefore, it is not appropriate to classify bottoms simply by a single parameter like sea state. This said, all of the theoretical models discussed for surface scattering have been applied to bottom scattering. Many frequently inconsistent measurements have been reported in the literature; this is almost certainly associated with a lack of detailed knowledge of bottom conditions beyond the immediate top layer. Historically, empirical data were referenced to the top surface

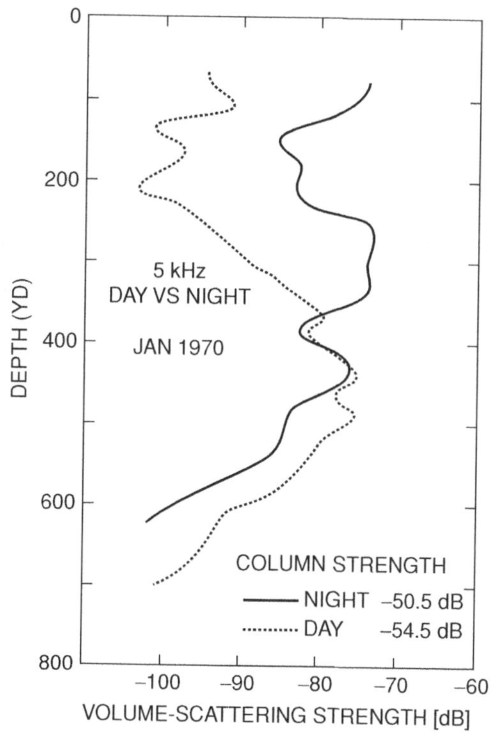

Figure 8.8 Volume scattering strength versus depth, Eastern Pacific at 5 kHz showing diurnal (Diel) variation. Reprinted from Vent [12] with permission of The American Institute of Physics

bottom characteristics (smooth or rough) and composition (mud, sand, pebbles, rock, etc.; see Table 6.6). Scattering is higher from a rough bottom and higher for a rock bottom than for softer bottoms. The actual scattering from the surface layer is a combination of roughness and scattering from the material comprising the bottom.

Bottom backscattering strengths are generally higher than ocean surface backscattering strengths. It is not uncommon for these strengths to follow a Lambert's law dependence on angle with a BS_0 varying from -45 dB (for mud) to -25 dB (for rock). For smoother bottoms, a definite trend of increasing backscattering strength is seen with frequency. For rough bottoms (shingle, pebble, rock), no frequency dependence is seen if the structure is large compared to a wavelength.

For frequencies above 10 kHz, the Applied Physics Laboratory of the University of Washington (APL-UW) [13] has developed a scattering model that depends both on bottom composition and frequency.

8.1.4 *Reverberation Target Strength*

The sound returned by a scattering source depends on the amount of area or volume illuminated by the transmitter and the amount seen by the receiver. Very much like the target strength, we will calculate in Chapter 9, Active Target Strength, a target strength that can be calculated in the form:

$$\mathrm{Nts_{rev}} = \mathrm{TS_{rev}} = \mathrm{BS_{b,v}} + 10\log(A \text{ or } V) \tag{8.29}$$

The reverberation received is given by

$$\begin{aligned} \mathrm{RL} &= \mathrm{Lp} + \mathrm{Nts_{rev}} - 2\mathrm{Nw_{rev}} \\ \mathrm{RL} &= \mathrm{SL} + \mathrm{TS_{rev}} - 2\mathrm{TL_{rev}} \end{aligned} \tag{8.30}$$

8.1.5 Calculation of Reverberation for Use in the Sonar Equation

8.1.5.1 Surface, Bottom, or Layer Reverberation Level

Today, there are sophisticated computer programs (see Chapter 6) that are used to perform this calculation for all but the simplest analysis cases. However, it is valuable to understand the nature of what is being calculated. Consider a simple case of r-squared spreading with the surface, the bottom, or a strong volume scattering layer. The reverberation level (RL) is given by

$$\mathrm{RL} = 10\log\left[\frac{I_0}{r^4} S_s \int \mathrm{Bt}(\theta, \phi)\mathrm{Br}(\theta, \phi)\, \mathrm{d}A\right] \tag{8.31}$$

where

Bt = beam pattern for the transmitter
Br = beam pattern for the receiver
I_0 = source level transmitted
r = range
θ = horizontal angle
ϕ = vertical angle
dA = area element

From Figure 8.9, a planar scatter, dA, is given by

$$\mathrm{d}A = \frac{c\tau}{2} r\, \mathrm{d}\theta \tag{8.32}$$

As discussed above, the vertical angle affects the value of S_s, the intensity of the signal reaching the scattering plane and the amount returning to the receiver. If we assume that the transmit and receive beam pattern are the same and that φ is close to the maximum response axis, then the integral is independent of the vertical angle and represents a reduction in the reverberation level or a gain in performance over a nondirectional system:

$$\mathrm{RL} = 10\log\left(\frac{I_0}{r^4} S_s \frac{c\tau}{2} r\right) + 10\log\left[\int_0^{2\pi} \mathrm{Bt}(\theta, \phi)\mathrm{Br}(\theta, \phi)\, \mathrm{d}\theta\right] \tag{8.33}$$

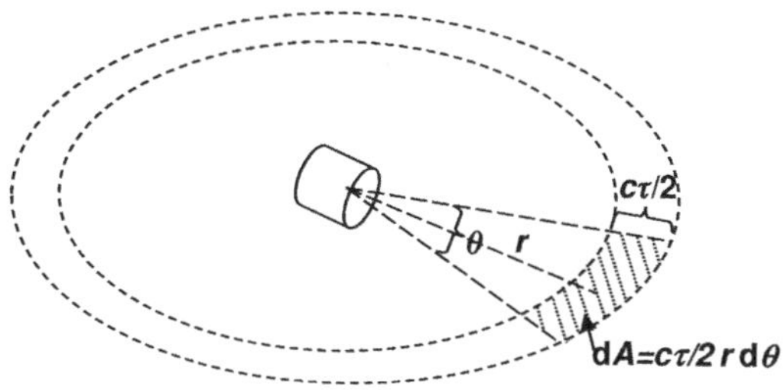

Figure 8.9 Scattering area for surface reverberation. The transducer array is at the center

The second term in Equation 8.33 is usually referred to as the equivalent two-way horizontal beamwidth (Figure 8.10). If the sensor is omnidirectional, the result would be 2π. In older texts, the ratio of this term to 2π expressed in decibels is referred to as the directivity for surface (planar) reverberation and is usually written as J_s or NRI:

$$J_s = \text{NRI} = 10\log\left(\frac{2\pi}{\theta}\right) \tag{8.34}$$

Table 8.1 gives the equivalent two-way horizontal beamwidth expressed in decibels for some simple array types. All dimensions are assumed to be greater than a wavelength (λ).

The surface or bottom (planar) reverberation level is given by

$$\begin{aligned} \text{RLs} &= \text{Lp} - 40\log(r) + 10\log\left(S_s\frac{c\tau}{2}r\theta\right)) \\ \text{RLs} &= \text{SL} - 40\log(r) + 10\log\left(S_s\frac{c\tau}{2}r\theta\right) \end{aligned} \tag{8.35}$$

or defining m_s as $2\pi S_s$ gives

$$\begin{aligned} \text{RLs} &= \text{Lp} - 30\log(r) + 10\log(m_s) + 10\log\left(\frac{c\tau}{2}\right) - J_s \\ \text{RLs} &= \text{SL} - 30\log(r) + 10\log(m_s) + 10\log\left(\frac{c\tau}{2}\right) - J_s \end{aligned} \tag{8.36}$$

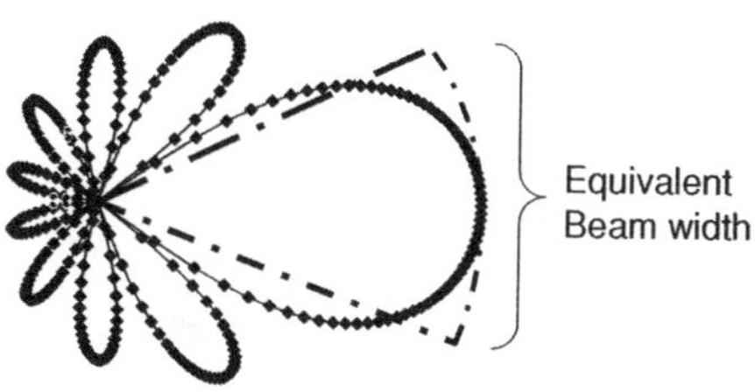

Figure 8.10 Actual two-way beam pattern and equivalent ideal beam pattern

Table 8.1 Equivalent two-way beam widths

Array type	Surface reverberation 10 log(Θ)	Volume reverberation 10 log(Ψ)
	$10\log\left(\int_0^{2\pi} \mathrm{Bt}(\theta,0)\mathrm{Br}(\theta,0)\mathrm{d}\theta\right)$	$10\log\left(\int_0^{2\pi}\int_{-\pi/2}^{\pi/2} \mathrm{Bt}(\theta,\phi)\mathrm{Br}(\theta,\phi)\cos(\phi)\mathrm{d}\phi\mathrm{d}\theta\right)$
Omnidirectional	$10\log(2\pi) = 7.98$	$10\log(4\pi) = 10.99$
Horizontal line	$10\log[\lambda/(2\pi L)] + 9.2$	$10\log[\lambda/(2\pi L)] + 14.2$
Rectangular array width W and height H	$10\log[\lambda/(2\pi W)] + 9.2$	$10\log[\lambda/(2\pi WH)] + 7.4$
Circular plane array radius R	$10\log[\lambda/(2\pi R)] + 6.9$	$20\log[\lambda/(2\pi R)] + 7.7$

8.1.6 Volume Reverberation Level

Consider the simple case of r-squared spreading to a scattering volume. The reverberation level is given by

$$\mathrm{RLv} = 10\log\left[\frac{I_0}{r^4}S_v\int \mathrm{Bt}(\theta,\phi)\mathrm{Br}(\theta,\phi)\,\mathrm{d}V\right] \quad (8.37)$$

where

- Bt = beam pattern for the transmitter
- Br = beam pattern for the receiver
- I_0 = source level transmitted
- r = range
- θ = horizontal angle
- ϕ = vertical angle
- $\mathrm{d}V$ = volume element

From Figure 8.11 a volume scatter, $\mathrm{d}V$, is given by

$$\mathrm{d}V = \frac{c\tau}{2}r^2\,\mathrm{d}\Psi \quad (8.38)$$

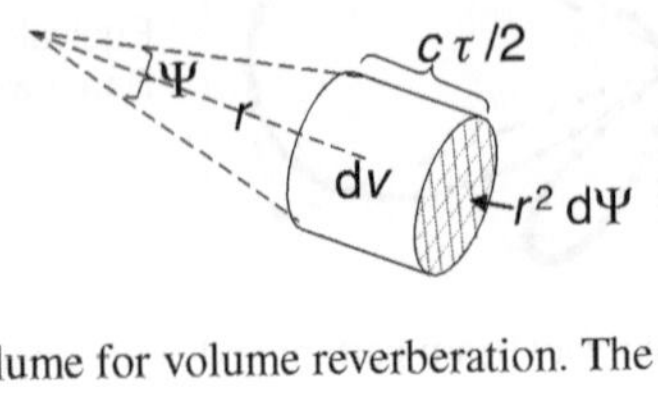

Figure 8.11 Scattering volume for volume reverberation. The transducer array is at the left

where

$d\Psi$ = solid angle element

$$\mathrm{RLv} = 10\log\left[\frac{I_0}{r^4}S_v\frac{c\tau}{2}r^2\right] + 10\log\left[\int_0^{2\pi}\int_{-\pi/2}^{\pi/2}\mathrm{Bt}(\theta,\phi)\mathrm{Br}(\theta,\phi)\cos(\phi)\,d\phi\,d\theta\right] \quad (8.39)$$

The volume reverberation level is given by

$$\begin{aligned}\mathrm{RLv} &= \mathrm{Lp} - 40\log(r) + 10\log\left(S_v\frac{c\tau}{2}r^2\Psi\right)\\ \mathrm{RLv} &= \mathrm{SL} - 40\log(r) + 10\log\left(S_v\frac{c\tau}{2}r^2\Psi\right)\end{aligned} \quad (8.40)$$

where

Ψ = equivalent solid angle for the two-way beam pattern

In older texts, the ratio of this term to 4π expressed in decibels is sometimes referred to as the directivity for surface (planar) reverberation and is usually denoted as J_v or NRI. If m_v is defined as $4\pi\, S_v$, then

$$\begin{aligned}\mathrm{RLv} &= \mathrm{Lp} - 20\log(r) + 10\log(m_v) + 10\log\left(\frac{c\tau}{2}\right) - J_v\\ \mathrm{RLv} &= \mathrm{SL} - 20\log(r) + 10\log(m_v) + 10\log\left(\frac{c\tau}{2}\right) - J_v\end{aligned} \quad (8.41)$$

Figure 8.12 shows the behavior of signal, surface reverberation, and volume reverberation versus range for the r-squared assumption. Notice that at short range the surface reverberation is the dominant source. As the range increases, volume reverberation and then ambient noise begin to dominate.

Figure 8.13 shows estimates for reverberation and signal versus time for an area in the Philippine Sea in the summer from the CASS-GRAB model. The calculations include a hard edged beam pattern that drops to minus infinity outside the main lobe. Notice that unlike the r-squared case in Figure 8.12, the reverberation and signal do not vary smoothly. This is caused by two major factors. First, the sound speed profile (SSP), which is bending the rays, is creating shadow zones. Second, the vertical beam pattern of the transmitter and receiver changes how the surface, bottom, and volume scattering layer are illuminated. The fathometer returns present in the plot are direct reflections (near normal incidence) off the bottom and occur at travel times equal to multiples of twice the water depth for a shallow source/receiver.

8.2 Reverberation Frequency Spread and Doppler Gain Potential

The spectral density of received reverberation depends on three factors: the transmitted spectral shape (Fourier transform of a pulse, see Appendix A), motion of the transmitter and receiver, and environmental spreading. As is shown in Section 8.2.1, except at very short ranges or at exceptional speeds, the target simply shifts the entire signal because of its speed in the line of

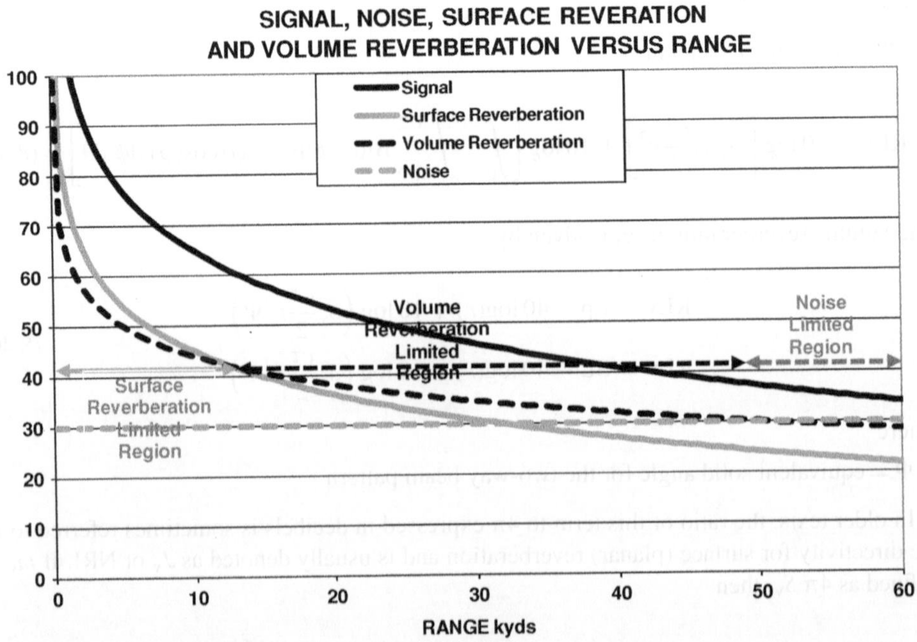

Figure 8.12 Range dependence of signal, surface reverberation, and volume reverberation, assuming an r-squared loss environment

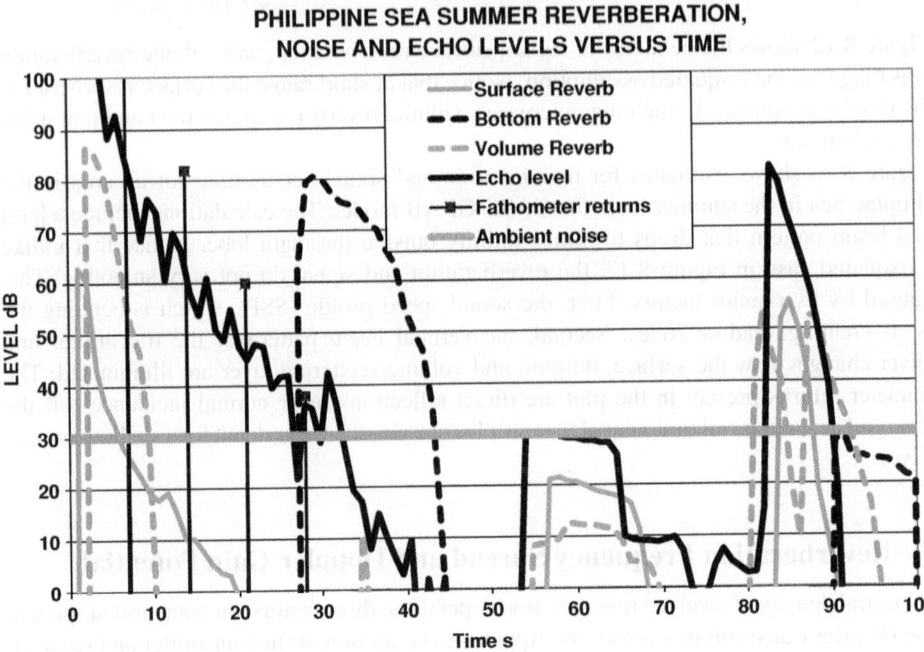

Figure 8.13 Reverberation and signal versus time for an area of the Philippine Sea in summer

sight (LOS), but does not spread the signal out in frequency. For simplicity, the discussion here will be limited to continuous wave (CW) pulses. This is not because coded pulses (CP) like frequency modulated (FM) transmissions do not exhibit spectral spreading, but rather because their ambiguity functions do not usually allow for an accurate discrimination of Doppler and because their effective pulse lengths are usually so short (1/BW) that they are less affected by reverberation.

8.2.1 Power Spectral Density of a CW Pulse

The pulse spectral shape can be determined by taking the Fourier transform of the transmitted pulse. If the pulse is a rectangular CW with a pulse length (τ) centered at a frequency, f_c, then the signal is

$$\begin{aligned} S &= A\sin(2\pi f_c t) \quad 0 \le t \le \tau \\ S &= 0 \qquad\qquad\qquad \text{otherwise} \end{aligned} \tag{8.42}$$

The spectral shape is given by

$$F(f) = \frac{\{\sin[\pi(f - f_c)\tau]\}^2}{[\pi(f - f_c)\tau]^2} \tag{8.43}$$

The bandwidth of a rectangular pulse is given by

$$\mathrm{BW} = \frac{\left\{\int_0^\infty \left[\frac{\sin(\pi f \tau)}{\pi f \tau}\right]^2 \mathrm{d}f\right\}^2}{\left\{\int_0^\infty \left[\frac{\sin(\pi f \tau)}{\pi f \tau}\right]^4 \mathrm{d}f\right\}} \tag{8.44}$$

$$\mathrm{BW} = \frac{3}{4\tau} \tag{8.45}$$

Notice that the effective bandwidth (BW) is inversely proportional to the pulse length and just slightly less than one over the pulse length. Figure 8.14 shows the power density versus frequency for two different pulse lengths.

Amplitude shading can be used to reduce the side lobes in an analogous way to reducing the side lobes on a beam pattern (see Section 3.5). Many schemes have been used for modern sonars. Figure 8.15 shows a comparison of the power density of a rectangular CW pulse with a simple cosine shaded pulse where the signal envelop starts at zero and rises to a maximum at $+\tau/2$, and then declines to zero at the end following a cosine shape. Here, the horizontal axis could be changed to the equivalent Doppler shift in knots by using Equation 1.25. Notice that the shaded pulse has a wider main lobe, but much smaller side lobes.

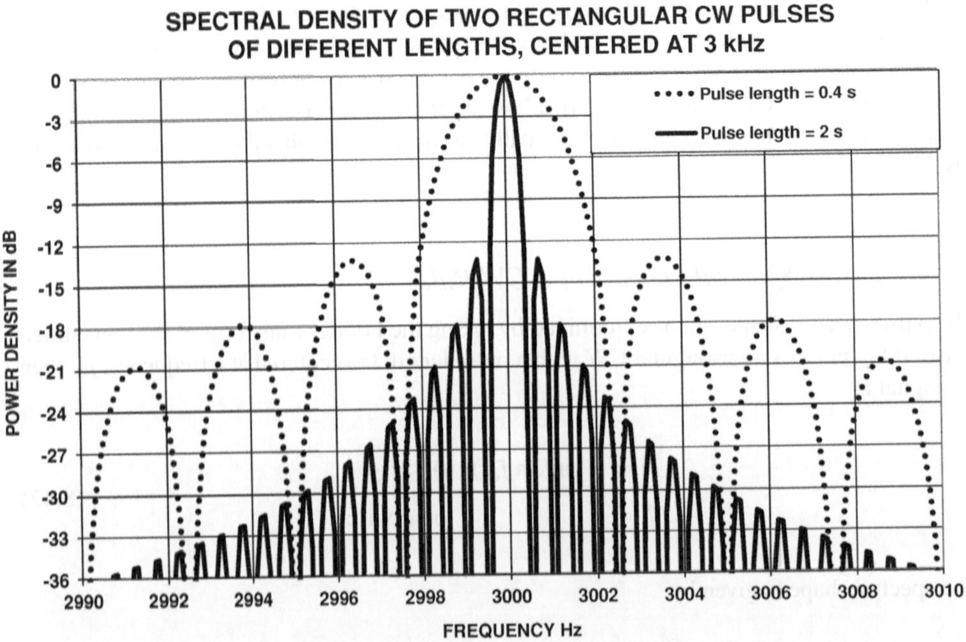

Figure 8.14 Spectral power density of two rectangular CW pulses of different lengths

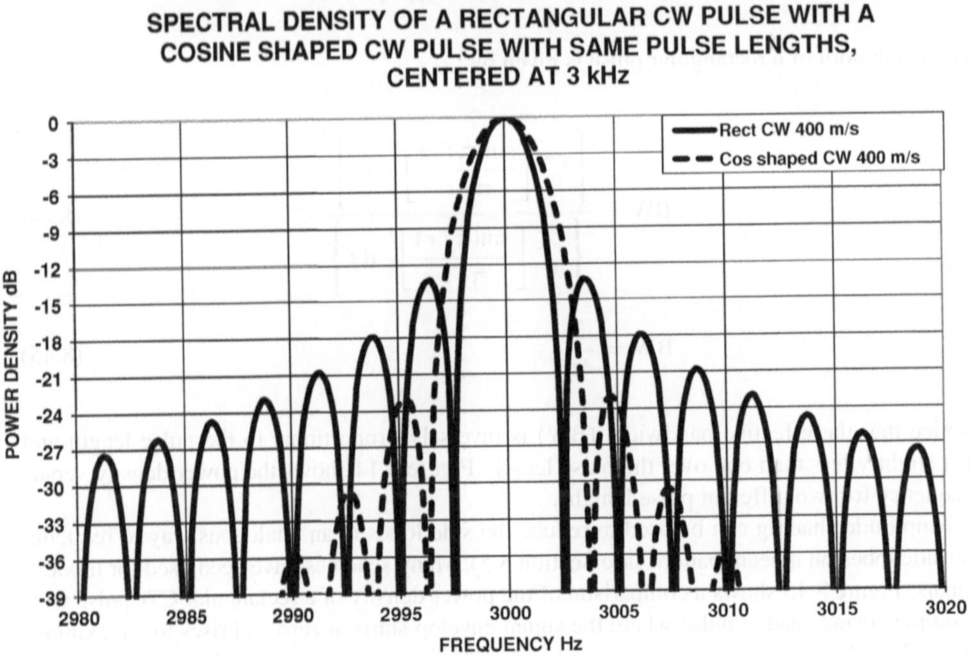

Figure 8.15 Comparison of spectral power density for a rectangular CW and a cosine amplitude shaped CW pulse

8.2.2 Environmental Frequency Spreading

The sources of surface and volume reverberation are moving, which introduces Doppler spreading. The Applied Physics Laboratory of the University of Washington (APL/UW) [13] has developed a model for environmental spreading at high frequencies. The spreading is assumed to be Gaussian. The standard deviation for surface scattering/reflection is given by

$$\sigma_{\text{surf}}(\text{m/s}) = 0.894\left(0.005\,33u^2\right)^{0.44} \tag{8.46}$$

where

u = wind speed (m/s)

As examples:

SS3 is 13.8 kts wind or 7.1 m/s, which gives $\sigma_{\text{surf}} = 0.5$ (m/s) = 0.97 kts
SS4 is 19.3 kts wind or 9.9 m/s, which gives $\sigma_{\text{surf}} = 0.67$ (m/s) = 1.31 kts

Volume reverberation spreading is assumed to be driven by surface circulation and changes with depth exponentially, $\exp(-kz)$, where k is the wave number of the peak surface–wave spectrum and z is depth. The volume spreading starts with the surface value at zero depth and has a lower limit of 0.13 m/s (0.25 kts) at deep depths. No frequency spreading is assumed for bottom scattering.

8.2.3 Frequency Spreading Due to Transmitter and Receiver Motion

For a monostatic active sonar on a moving platform, the transmitted frequency is shifted depending on the direction under consideration:

$$f_{\text{shifted}} = f_{\text{c}} + 0.000\,69 f_{\text{c}} u \cos(\text{AOB}) \tag{8.47}$$

where

f_{c} = transmitter frequency (Hz)
u = sonar's speed through the water (kts)
AOB = angle of the bow being considered (0 is straight ahead and 180 is directly astern)

This formula assumes a speed of sound of 4890 fps (ft/s) and a monostatic active sonar, so that only one angle is required.

Modern active sonars usually ensonify all, or a large fraction, of the azimuth bearings simultaneously and then use receivers to form many narrow beams covering the sector being searched. In general, to compensate for own platform Doppler shift, each beam is processed with what is called own Doppler nullification (ODN). ODN means that the beam processes a frequency band centered on the expected Doppler shifted frequency. However, receive beams do not have responses in only one direction. Therefore, there is a spread in reverberation due to own ship motion and real beam patterns. For a forward looking beam, where the ODN has adjusted for the full own ship speed, the reverberation coming in from side lobes 90° out centered at minus own ship speed relative to the ODN adjustment. The shape of the frequency spread due to own ship motion is a function of the beam pattern response, the look direction, and own ship speed. If the beam is pointed perpendicular to own ship motion and if the beam

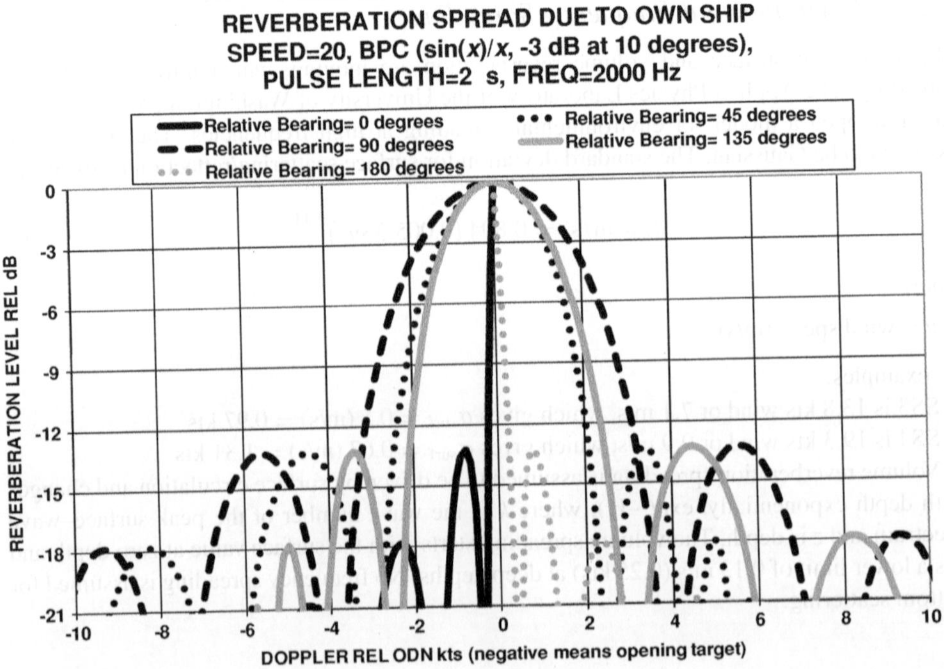

Figure 8.16 Spread in reverberation due to own ship motion, at relative bearings of 0, 45, 90, 135, and 180

pattern response is symmetric, then the spread in reverberation will be symmetric. This is not true in all other directions. For example, looking directly ahead, the spread only occurs in the direction corresponding to targets that are trying to move away and vice versa in the stern direction.

Figure 8.16 shows an example of own ship motion induced reverberation spread for several look directions. Each curve has been own Doppler nullified and adjusted of own ship speed in the line of sight (ODN adjusted) so that a return from a stationary or broadside target corresponds to 0 kts. Negative values correspond to a target whose motion contributes to an opening speed in the line of sight (SILOS). The beam pattern response used here is a sin(x)/x pattern.

Figure 8.17 shows all three sources of spreading combined. The curve has been adjusted for own ship speed in the line of sight (ODN adjusted) so that a return from a stationary or broadside target corresponds to 0 kts. Negative values correspond to a target whose motion contributes to an opening speed in the line of sight (SILOS). The beam pattern response used here is a sin(x)/x pattern.

8.2.4 *Frequency Spreading Due to Target*

If the signals were reflected off a target that is closing with a speed in the line of sight (SILOS) of 5 kts, the reflected signal would be shifted up in frequency as is shown in Figure 8.18 by

$$\Delta f = 0.000\,69 f v \cos(\alpha) = 0.000\,69 \times 10 \cos(60) = 10.35 \text{ Hz} \tag{8.48}$$

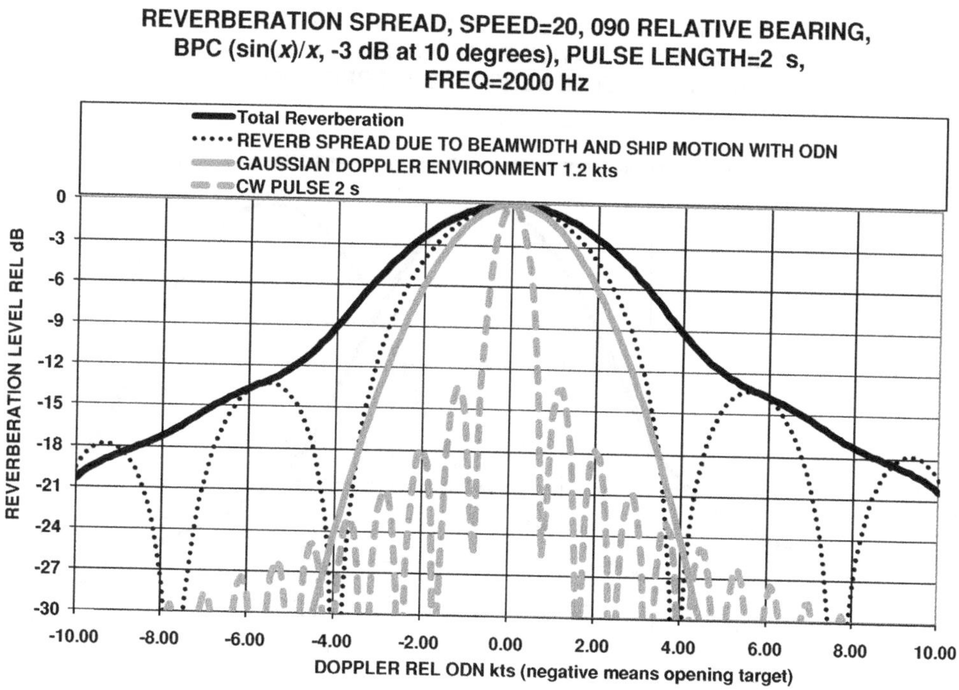

Figure 8.17 Spread in reverberation due to own ship motion, the environment, and pulse shape

where

f = frequency (Hz)
α = target aspect in degrees (0 is bow on, 90 or 270 beam, and 180/360 is a stern view)
v = target speed (kts)

Note that because the Doppler shift is proportional to frequency, there will be some bandwidth change, given by

$$\Delta BW = 0.000\,69 BW \times SILOS \tag{8.49}$$

The change in bandwidth is only 0.33 % in this case. For a 100 knot speed in the line of sight (SILOS), the change in bandwidth would be only 7 %. It is therefore a good assumption that targets do not change the signal bandwidth. An exception to this would occur if the target were so close that different portions of the target were at very different bearings. In this case, the difference is caused by different values of α and the change in shift would be given by the sine of the difference in angle. In the case above, if the bow was at 59° and the stern at 61°, the spreading would be about ±0.3 Hz and a submarine sized object would need to be at about 1 nautical mile away to produce this angular spread.

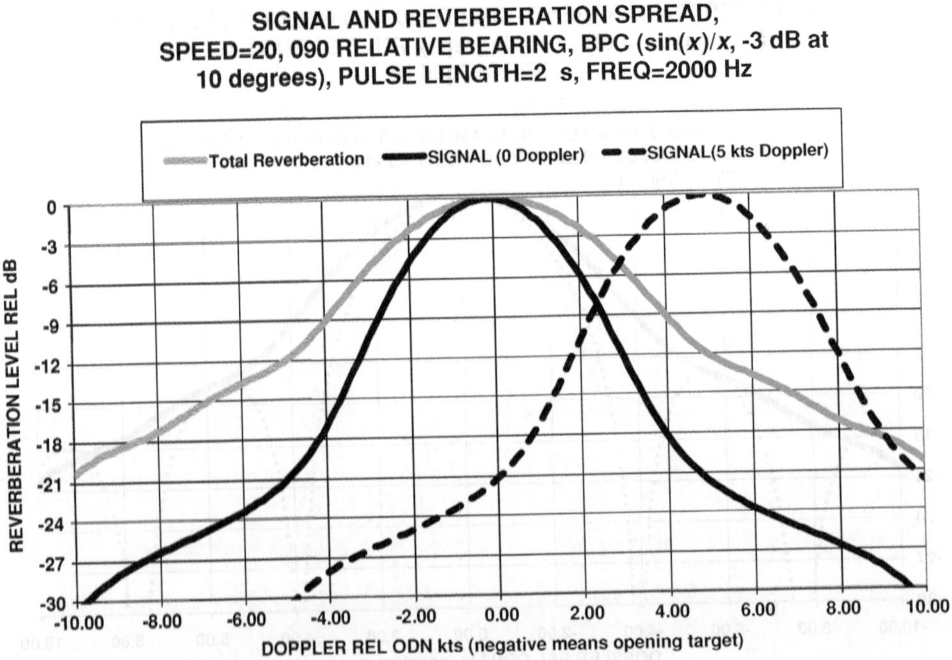

Figure 8.18 Example of frequency spread (Doppler equivalent) of signal and reverberation

8.3 Important Observation with Respect to Reverberation

It might be tempting to reduce the reverberation by reducing the power transmitted, but the level of reverberation from any source is proportional to the source level, as is the signal received from the target. Therefore, increasing or decreasing the power transmitted does not affect the signal-to-reverberation ratio whereas it does directly affect the signal-to-self or ambient noise ratio. Unless the reverberation is so high as to drive the receiver into a nonlinear region, reducing the source level will either not effect performance or will decease it proportionally. In detecting a target, the reverberation returning has a maximum value corresponding to a target Doppler of zero knots speed in the line of sight. This, for a CW pulse, gives the worst signal-to-reverberation ratio. If the target contributes Doppler, the signal returned shifts away from the peak reverberation and detectability improves. While this discussion has only considered the frequency spread of reverberation, the signal returned has a spread determined by the pulse shape, target spreading and the environmental spreading components. Figure 8.18 shows an example of signal and reverberation spread. If a fast Fourier transform (FFT) were to be applied to Figure 8.18, centered on the signal, the signal-to-reverberation ratio would be much better for the 5 knot Doppler target than for a 0 knot Doppler target.

References

[1] Brekhovskikh, L. M., and Lysanov, Yu, P., *Fundamentals of Ocean Acoustics*, 2nd edition, Berlin: Springer-Verlag, 1992.

[2] Cox, C., and Monk, W. "Measurement of the Roughness of the Sea Surface from Photographs of the Sun Glint," *Journal of the Optical Society of America*, **44**, 1954, 838.

[3] Garrison, G. R., Murphy, S. R., and Potter, D. S., "Measurements of the Backscattering of Underwater Sound from the Sea Surface," *Journal of the Acoustical Society of America*, **32**, 1960, 104.

[4] Marsh, H. W., "Exact Solution of Wave Scattering by Irregular Surfaces," *Journal of the Acoustical Society of America*, **33**, 1961, 330.

[5] Marsh, H. W., Schulkin, M., and Kneale, S. G., "Scattering of Underwater Sound by the Sea Surface," *Journal of the Acoustical Society of America*, **33**, 1961, 334.

[6] Marsh, H. W., "Sound Reflection and Scattering from the Sea Surface," *Journal of the Acoustical Society of America*, **35**, 1963, 240.

[7] Crowther, D. B., "Acoustic Scattering from Near-Surface Bubble Layers," in *Incavitation and Inhomogeneities in Underwater Acoustics* (Ed. W. Lauterborn), New York: Springer-Verlag, 1980, pp. 195–204.

[8] McDaniel, S. T., and Gorman, A. D., "Acoustic and Radar Sea-Surface Backscatter," *Journal of Geophysical Research*, **87**, 1982, 4127–4136.

[9] McDaniel, S. T., "Sea Surface Reverberation: A Review," *Journal of the Acoustical Society of America*, **94**, 1993, 1905–1922.

[10] Chapman, R. P., and Harris, J. H., "Surface Backscattering Strengths Measured with Explosive Sound Sources," *Journal of the Acoustical Society of America*, **34**, 1962, 1592.

[11] Schulkin, M., and Shaffer, R., "Backscattering of Sound from the Sea Surface," *Journal of the Acoustical Society of America*, **36**, 1964, 1699.

[12] Vent, R. J., "Acoustic Volume-Scattering Measurements at 3.5, 5.0 and 12 kHz in the Eastern Pacific Ocean: Diurnal and Seasonal Variations," *Journal of the Acoustical Society of America*, **52** (Part 2), 1972, 1.

[13] "APL-UW High-Frequency Ocean Environmental Acoustic Models Handbook," Applied Physics Laboratory of the University of Washington (APL/UW), Technical Report TR 9407, October 1994.

9

Active Target Strength

In active sonar, target strength refers to the ability of a target to return an echo. Depending on the interest of the observer, the target may be a submarine, mine, whale, or sunken ship. The ability of the surface, bottom, or various scattering layers to return sound is generally called reverberation because their indefinite extent produces a continuum of returns as opposed to a discrete echo.

9.1 Target Strength Definition

In the sonar equation, target strength, Nts or TS, is 10 log of the ratio of the incident acoustic intensity, I_i, to the reflected acoustic intensity, I_r, referenced to a specified distance (usually 1 yard or 1 meter) from the acoustic center of the target. The measurement of the return ping intensity is made in the far field and adjusted for the propagation loss experienced in the medium back to the reference distance (which is frequently physically inside the actual target):

$$\mathrm{Nts} = \mathrm{TS} = 10\log\left(\frac{I_r}{I_i}\right) \tag{9.1}$$

where

I_r = reflected acoustic intensity at the reference distance
I_i = incident acoustic intensity

Figure 9.1 shows the reflection of sound from an object with a plane wave incident and a spreading wave being reflected. The target strength for this reflection is given by Equation 9.1. As drawn, this represents a bistatic reflection, i.e., the reflection direction of interest is not opposite to the incident direction, as it would be for a monostatic reflection. Where monostatic target strength is a function of the aspect of such an object as it is rotated in the plane, bistatic target strength is a function of this angle and the angle between the incident and reflected angles.

Figure 9.2 illustrates the two nomenclatures used when discussing bistatic reflection. The first form specifies the source or incident angle, θ_i, and the receiver or reflected angle, θ_r,

Underwater Acoustics: Analysis, Design and Performance of Sonar Richard P. Hodges

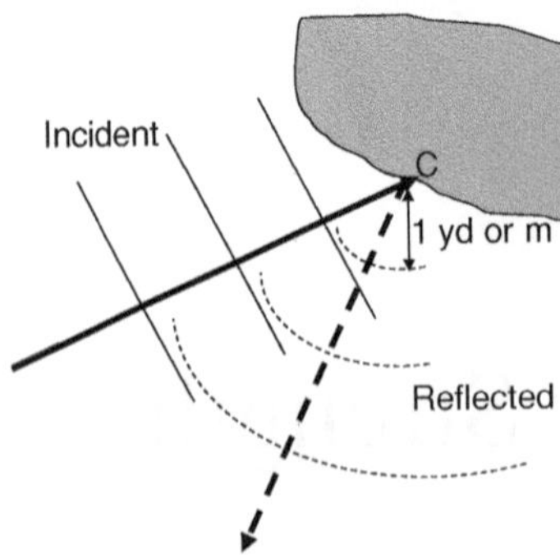

Figure 9.1 Plane wave reflecting from a target

relative to some axis of the target. The second form specifies the bistatic target aspect, θ_{asp}, which is the center of the incident and reflected angles relative to some axis of the target, and the bistatic angle, θ_{bis}, which is the angle between the incident and reflected angles. For comparison, the monostatic target strength only requires θ_{asp}.

The choice of 1 yard or 1 meter for a reference distance is arbitrary and frequently results in positive values for target strength for all but small objects (e.g., mines, fish). This should not be interpreted as meaning that more sound is being reflected than is incident; the conservation of energy would not allow this.

Consider a perfect sphere. (For this example, perfect means not only in shape, but also immobile, rigid, and very large compared to a wavelength.) Assuming a sphere is an isotropic reflector, one can compute the target strength. The incident power is given by $I_i \pi R^2$. The reflected power will spread out uniformly with a distance, r, covering an area of $4\pi r^2$. The reflected intensity, I_r, is

$$I_r = \frac{I_i \pi R^2}{4\pi r^2} \tag{9.2}$$

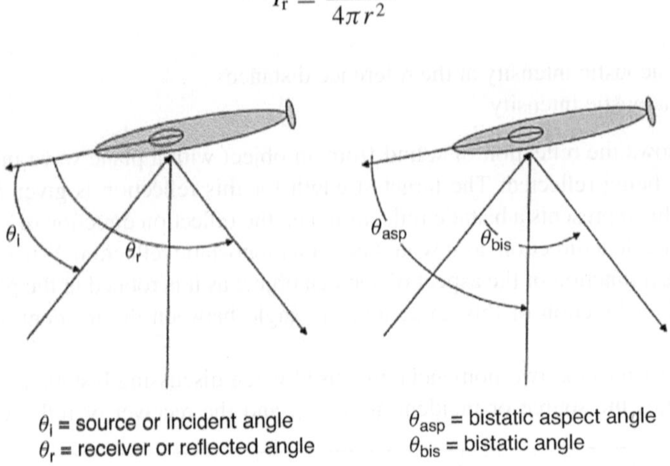

Figure 9.2 Angles for bistatic target strength

Referenced to 1 yard or meter:

$$\frac{I_r}{I_i} = \frac{R^2}{4} \tag{9.3}$$

where R is the radius of the sphere and the target strength of a sphere is

$$\text{Nts} = \text{TS} = 10\log\left(\frac{R^2}{4}\right) \tag{9.4}$$

Given this, a perfect sphere with a radius of 2 yards or meters, depending on the reference distance, will have a target strength of 0 dB and be independent of orientation. In testing, a sphere can be used to provide a reference level for use in comparisons with actual targets of interest. However, a hollow metal sphere rarely meets the required criterion because it may distort with depth and is not sufficiently rigid.

9.2 Active Target Strength of a Large Sphere

The above is an example of specular reflection where the wave nature of sound is ignored. When an object is not large compared to the wavelength or has features that are not large, the wave nature of sound complicates the calculation of its target strength. The above derivation of target strength assumes that the sphere isotropically redistributed the incident sound. The same conclusion can be found using simple geometry. Figure 9.3 shows the monostatic reflection of

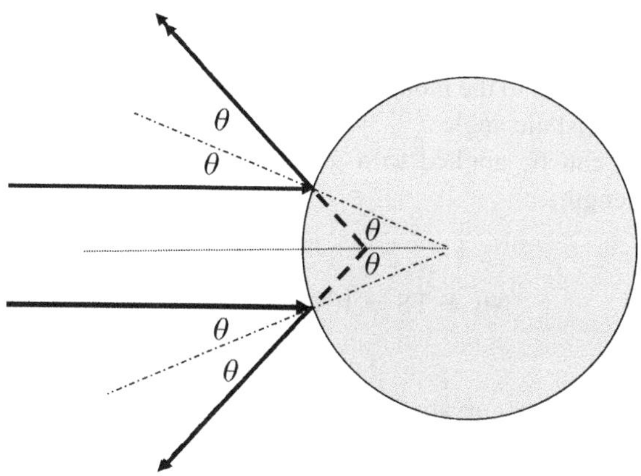

Incident Power = Reflected Power at 1 yard or meter

$I_i\,(R \times 2\theta)^2 = I_r\,(1 \times 4\theta)^2$

$I_r / I_i = R^2 / 4$

$\text{Nts} = \text{TS} = 10\log(R^2 / 4)$

Figure 9.3 Monostatic active target strength of a perfect large sphere

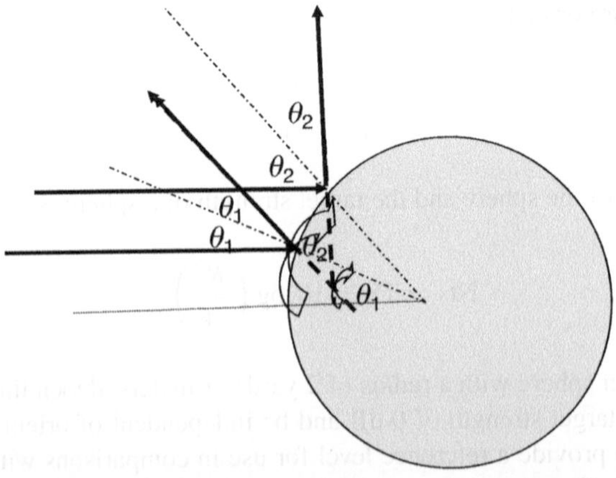

Incident Power = Reflected Power at 1 yard or meter

$$I_i\,[R\times(\theta_2-\theta_1)]^2 = I_r\,(1\times 2\,(\theta_2-\theta_1))^2$$

$$I_r / I_i = R^2/4$$

$$\text{Nts} = \text{TS} = 10\log(R^2/4)$$

Figure 9.4 Bistatic active target strength for a perfect large sphere

sound by a large sphere. The pencil beam hitting the sphere is spread by the curvature of the sphere. The incident power (intensity × area) is reflected into angular sectors twice as wide. Target strength is then calculated as the intensity at the unit reference distance.

The same approach is used in Figure 9.4 to compute the bistatic target strength of a sphere. This target strength is identical to the monostatic target strength and does not depend on either the bistatic aspect or the bistatic angle.

This same approach can be applied to a general convex surface with curvatures large compared to the wavelength:

$$\text{Nts} = \text{TS} = 10\log\left(\frac{R_1 R_2}{4}\right) \tag{9.5}$$

where R_1 and R_2 are the radii of curvature in orthogonal directions.

9.3 Active Target Strength of a Very Small Sphere

Rayleigh [1] developed the theory of sound scattering by a small, rigid, immobile sphere. Very small in size, the circumference of this sphere is much smaller than the wavelength. The rigidity of the sphere refers to its compressibility, being much less than that of the medium. Immobility suggests that the sphere does not take part in the acoustic motion of waves, i.e., the sphere is very dense compared to the medium. This result, with suitable changes in variables, is given by

$$\frac{I_r}{I_i} = \frac{16\pi^4 R^6}{9r^2\lambda^4}\left[1 + \frac{3}{2}\cos(\theta_{bis})\right]^2 \tag{9.6}$$

where

$\lambda =$ wavelength
$R =$ radius of the sphere
$r =$ range
$\theta_{bis} =$ bistatic angle

Target strength, referenced to a unit distance, is given by

$$\text{Nts} = \text{TS} = 10\log\left\{\frac{16\pi^4 R^6}{9\lambda^4}\left[1 + \frac{3}{2}\cos(\theta_{bis})\right]^2\right\} \tag{9.7}$$

Monostatic target strength is given by

$$\text{Nts} = \text{TS} = 10\log\left\{\frac{16\pi^4 R^6}{9\lambda^4}\left[1 + \frac{3}{2}\right]^2\right\} = 10\log\left(\frac{1082R^6}{\lambda^4}\right) \tag{9.8}$$

Figure 9.5 shows the target strength for 1 kHz and 3 kHz of a perfect sphere versus the size of the sphere. The target strength is only frequency dependent in the small region ($2\pi R \ll \lambda$). While the curves here are shown as smoothly transitioning from the small radius region to the large radius region, a ringing actually occurs during this transition that is not shown in the figure.

Rayleigh [1] also derived the target strength for small spheres that possessed a compressibility, κ' (reciprocal of bulk modulus), and density, ρ', that were different from the acoustic medium (κ, ρ). This theory shows that the

$$\left[1 + \frac{3}{2}\cos(\theta_{bis})\right]^2$$

term becomes

$$\left[1 - \frac{\kappa'}{\kappa} + 3\frac{\left(\frac{\rho'}{\rho} - 1\right)}{\left(1 + 2\frac{\rho'}{\rho}\right)}\cos(\theta_{bis})\right]^2 \tag{9.9}$$

Figure 9.6 shows the resulting correction to target strength to account for differences in compressibility and density. Notice that a plastic sphere would have a target strength 10 dB less than expected for its geometric size and a glass sphere about 4 dB less.

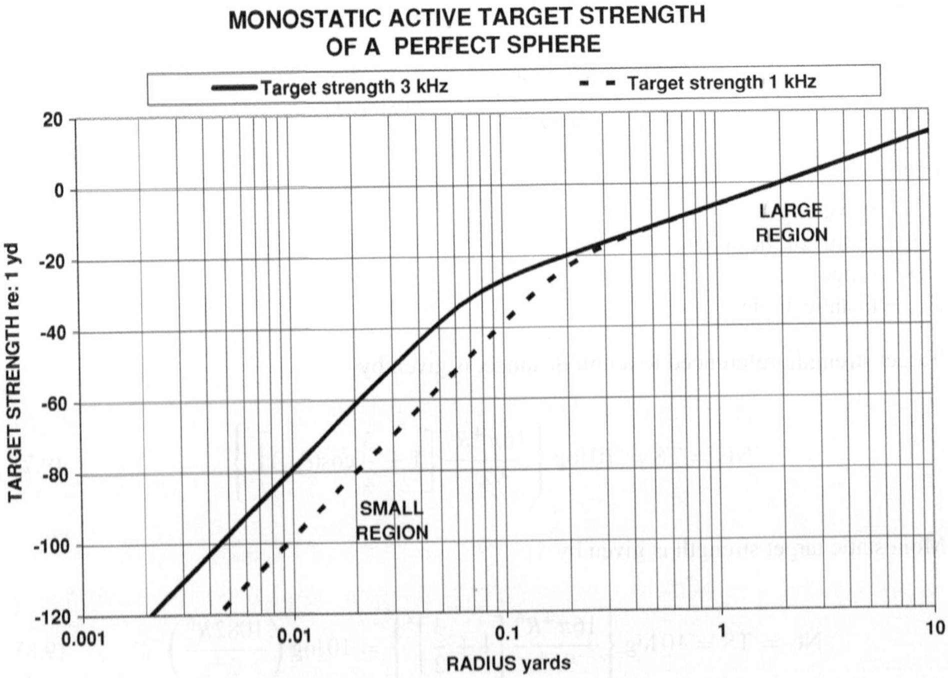

Figure 9.5 Target strength of a perfect sphere

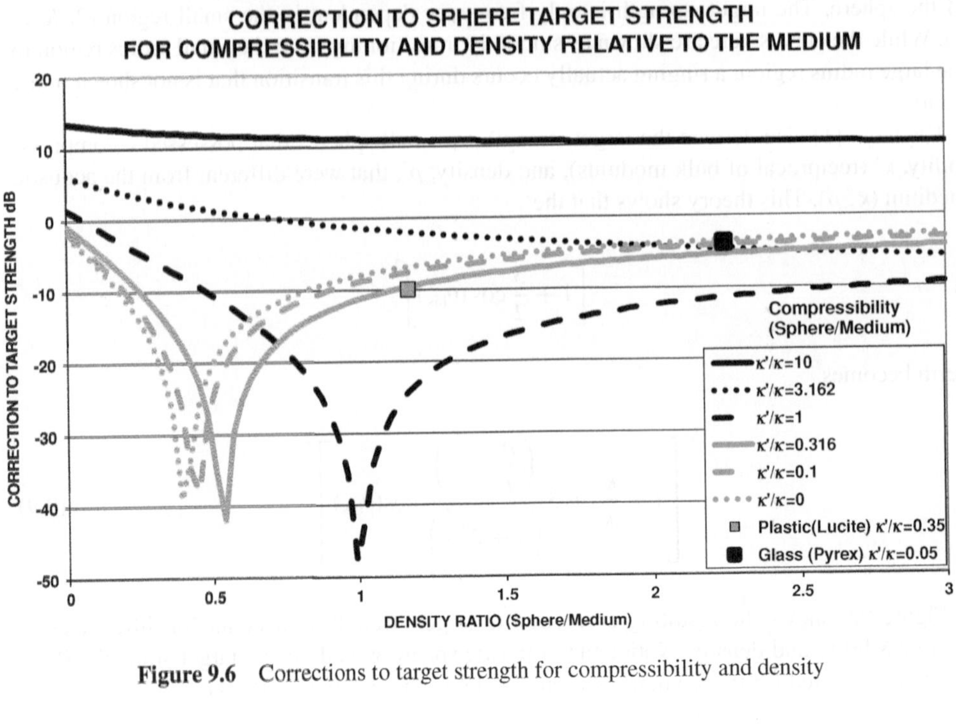

Figure 9.6 Corrections to target strength for compressibility and density

Table 9.1 Target strength estimates for simple shapes

Geometric form	Target strength = 10 log(A)	Direction	Conditions
Sphere (large)	$R^2/4$	All	$kR \gg 1$ $r \gg R$
Sphere (small)	$1082R^6/\lambda^4$	Normal	$kR \ll 1$ $kr \gg 1$
Ellipsoid	$[bc/(2a)]^2$	Parallel to a, b, c are semimajor axis	$ka, kb, kc \gg 1$ $r \gg a, b, c$
Cylinder	$RL^2/(2\lambda)[\sin(\beta)/\beta]^2 \cos^2(\theta)$	At θ with normal, $\beta = kL \sin(\theta)$	$kR \gg 1$ $r > L^2/\lambda$
Convex surface	$R_1R_2/4$	Normal	$kR_1, kR_2 \gg 1$ $kr \gg 1$
Smooth convex object	$A/(16\pi)$	Average in all directions, A = total surface area	All dimensions and curvature large compared to λ
Rectangular plate	$L_1L_2/\lambda^2[\sin(\beta)/\beta]^2 \cos^2(\theta)$	At angle θ to normal in plane of side L_1, $\beta = kL_1 \sin(\theta)$	$r > L_1{}^2/\lambda$ $kL_2 \gg 1$ $L_1 > L_2$
Circular plate	$(\pi R^2/\lambda)^2 [2J_1(\beta)/\beta]^2 \cos^2(\theta)$	At θ with normal, $\beta = 2kR \sin(\theta)$	$r > R^2/\lambda$ $kR >> 1$
Triangular corner reflector	$L^4/(3\lambda^2)(1 - 0.000\,76\theta^2)$	At θ to axis of symmetry, L = length of edge	Dimensions large compared to λ

where
$k = 2\pi/\lambda$
λ = wavelength
R = radius or radii of curvature
r = range from object
J_1 = first-order Bessel function

9.4 Target Strengths of Simple Geometric Forms

Table 9.1 presents the target strength for a variety of shapes. These values are idealized and therefore, under the best conditions, a guess for real objects with compressibility, densities, internal structures, and imperfections in construction. These values should only be used as a crude estimate for objects when actual measurements are not available. Computer programs exist that are able to calculate respectable estimates of target strength for complex objects for which the construction details are known. However, most of these programs are available only at a classified level.

9.5 Target Strength of Submarines

Measured target strengths for current submarines are classified and therefore not available here.

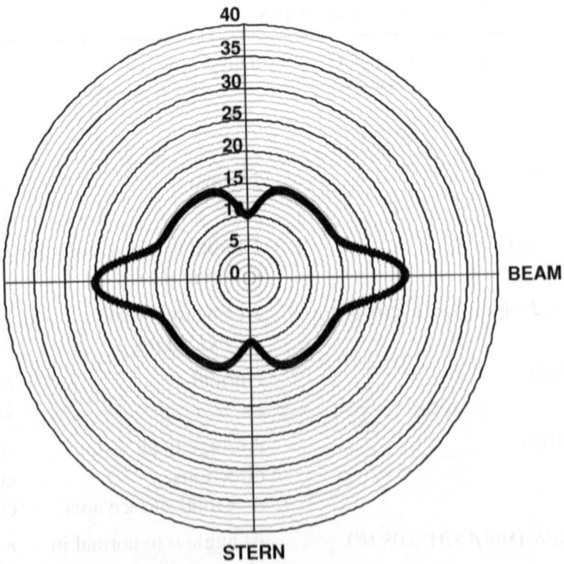

Figure 9.7 Classic active target strength curve

A "classic" target strength curve presented in many texts is called the butterfly shape (Figure 9.7). This shape shows a large increase in target strength near the beam, as expected for a cylindrical shape like a submarine. Smaller values in the bow and stern regions are due to the small cross-section of a submarine viewed from that aspect.

9.6 The TAP Model

The Technical Advisory Panel (TAP) model, available online (http://www.dtic.mil/srch/doc?collection=t3&id=ADA329259), is a target strength model for the bistatic response of a submarine due to four scattering mechanisms: (1) backscatter for a cylinder, (2) forward scatter from a cylinder, (3) scattering of elastic waves in a cylinder, and (4) scattering from hemispherical end caps. The model assumes that a target or submarine is a cylinder with hemispherical end caps and therefore ignores the actual construction of submarines, which may include double hulls, propellers, control surfaces, a sail or bridge structure, and ballast tanks. The Naval Research Laboratory has published a report, describing the equations in the TAP model and comparing its results to a more detailed model (SARA-2d) [2].

The left side of Figure 9.2 shows the geometry used in this model. The target is a finite cylinder of length, L, and radius, r, with hemispherical end caps. The backscatter contribution to the intensity, σ_{cbs}, is given by:

$$\sigma_{\mathrm{cbs}} = \frac{rL^2}{\lambda} \frac{\sin(\theta_{\mathrm{i}})\sin(\theta_{\mathrm{r}})}{\sin(\theta_{\mathrm{i}}) + \sin(\theta_{\mathrm{r}})} [j_0(\alpha)]^2 \tag{9.10}$$

$$\alpha = kL\frac{\cos(\theta_{\mathrm{i}}) + \cos(\theta_{\mathrm{r}})}{2} \tag{9.11}$$

where

j_0 = zeroth-order Bessel function
λ = wavelength
k = angular wave number ($2\pi/\lambda$)

The contribution to the intensity from the end cap, which is half that of a hemispherical end cap because submarines tend to be tapered cylinders and thus have smaller end caps, is given by

$$\begin{aligned} \sigma_{\text{ec bs}} &= \frac{0.69}{2}(kr)^4\frac{r^2}{4} \quad \text{if } kr < 1.1 \\ \sigma_{\text{ec bs}} &= \frac{1}{2}\frac{r^2}{4} \quad \text{if } kr \geq 1.1 \end{aligned} \tag{9.12}$$

The TAP model uses a modified version of Babinet's principle, with imposed reciprocity, for the contribution to intensity of forward scattering given by

$$\sigma_{\text{fs}} = \left(\frac{2rL}{\lambda}\right)^2 |\sin(\theta_{\text{i}})\sin(\theta_{\text{r}})|\,[j_0(\alpha)]^2 \tag{9.13}$$

The final term in the TAP model is the contribution to the intensity from elastic waves, σ_{ew}, given by

$$\sigma_{\text{ew}} = P(\theta_{\text{i}})P(\theta_{\text{r}}) \tag{9.14}$$

$$\begin{aligned} P(\theta) &= \frac{BL}{\pi}\cos\left[\frac{\pi}{2\nu}\frac{\delta\theta}{\left(\frac{\pi}{2}-\theta_{\text{lim}}\right)}\right] \quad \text{if } |\delta\theta| < \nu\left|\frac{\pi}{2}-\theta_{\text{lim}}\right| \\ P(\theta) &= 0 \quad \text{otherwise} \end{aligned} \tag{9.15}$$

$$\begin{aligned} \delta\theta &= |\theta-\pi| - \frac{\pi}{2} \\ \cos(\theta_{\text{lim}}) &= \frac{c}{c_\nu} \end{aligned} \tag{9.16}$$

where

c = speed of sound in water
c_ν = compressional or shear speed in the shell
ν = a scaling factor having a typical value of 2
B = an efficiency factor having a typical value of 0.2

Figure 9.8 shows the target strength generated by the TAP model for 300 Hz against a 360 foot long submarine with a 33 foot diameter, approximately the size of a Los Angeles Class SSN. This plot is for the case where the incident and reflected angles are on the same side of

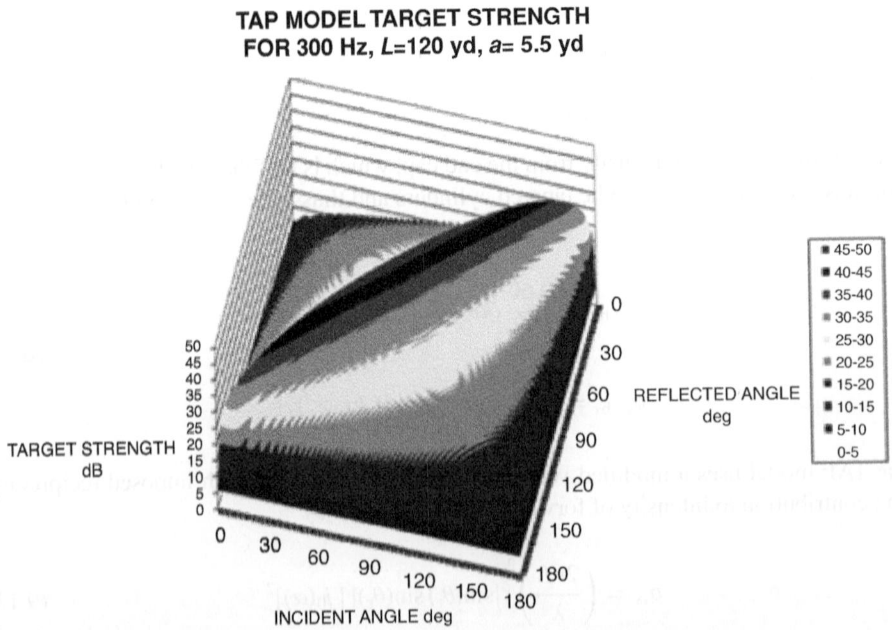

Figure 9.8 Example of the TAP model bistatic target strength as a function of incident and reflected angles

the target. The equations can be applied beyond this region. While the axes are labeled incident angle and reflected angle, the equations are symmetric so these labels can be interchanged without affecting the plot.

Figure 9.9 shows the target strength generated by the TAP model for 300 Hz against a 360 foot long submarine with a 33 foot diameter, approximately the size of a Los Angeles Class SSN. The axes are labeled bistatic angle and bistatic aspect (see the right side of Figure 9.2) and are not interchangeable. Figure 9.10 shows the monostatic target strength for this case, which is simply the diagonal of Figure 9.8 or the bistatic angle of zero in Figure 9.9.

9.7 Target Strength of Surface Ships

Measured target strengths for current warships are classified and therefore not available here.

9.8 Target Strength of Mines and Torpedoes

Measured target strengths for current mines and torpedoes are classified and therefore not available here. However, we can estimate the target strength by using the equations in Table 9.1.

For example, the Federation of American Scientists web site [3] gives the dimensions of a U.S. submarine-launched mobile mine as 161 inches long and 19 inches in diameter. Figure 9.11 is an estimate of target strength for a cylinder and a sphere, assuming that a cylinder with

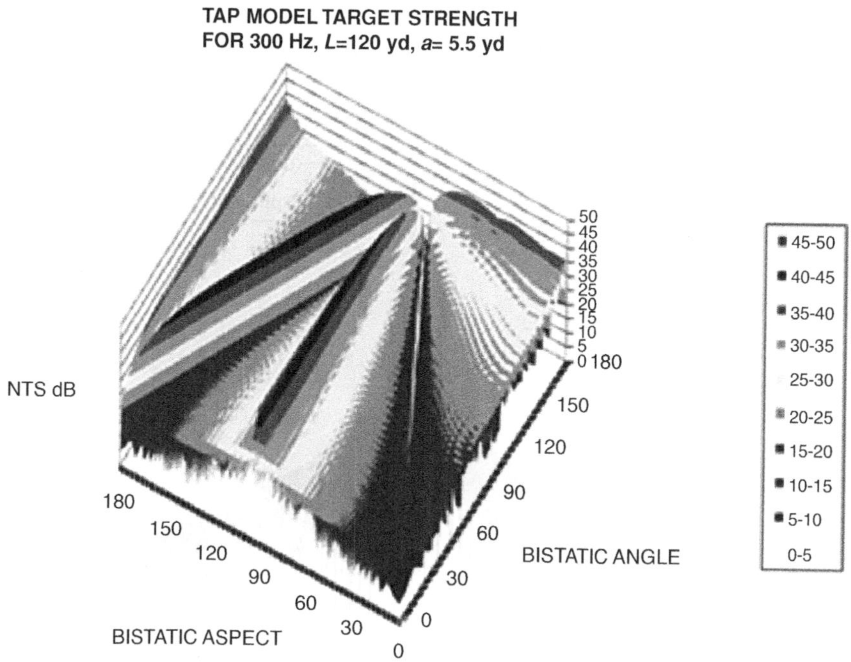

Figure 9.9 Example of the TAP model bistatic target strength as a function of bistatic aspect and bistatic angle

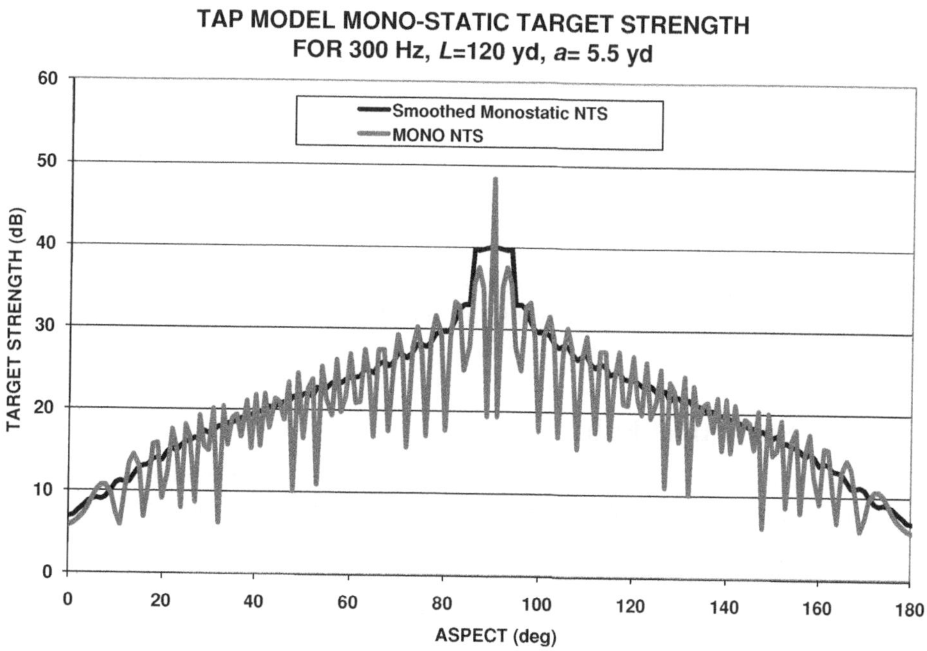

Figure 9.10 Example of the TAP model monostatic target strength as a function target aspect

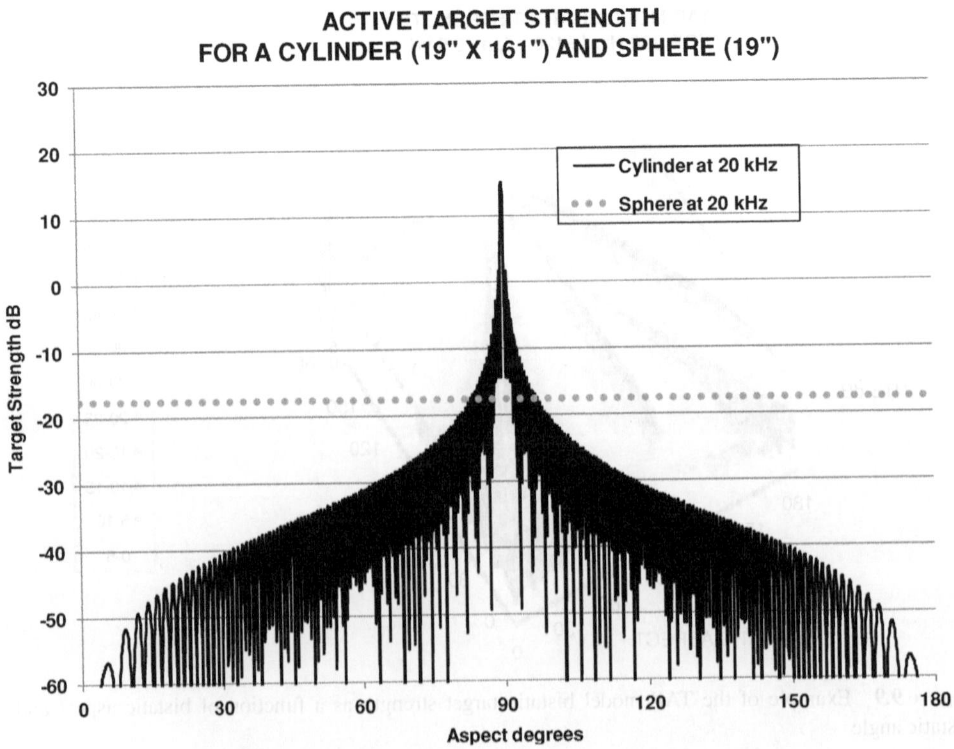

Figure 9.11 Estimated target strength for a cylinder and sphere

spherical end caps is a fair representation of this mobile mine. Figure 9.12 is the power sum of the cylinder and sphere at 20 kHz. As expected for a cylinder, the mine has a very large target strength directly on the beam. However, this peak value of 15 dB drops rapidly to 3 dB down at 0.23° off the beam. The median value for the cylinder is −41 dB and the median for the power sum of the cylinder and sphere is −17.6 dB.

The U.S. Mk 48 torpedo's dimensions are given as 19 feet long and 21 inches in diameter [3]. Using these values, the torpedo's target strength is shown in Figures 9.13 and 9.14.

9.9 Target Strength of Fish

The target strength of fish is not considered classified and is covered in considerable amounts of available literature. Marine biologists and the fishing industry have an interest in the target strength of fish. There are good summary articles on fish target strength available in the *Journal of the Acoustical Society of America* by Love [4] and Foote [5]. Love gives a simplified equation for a fish's target strength that equates to

$$\text{Nts(fish)} = \text{TS(fish)} = 20\log(L) - 29.2 \tag{9.17}$$

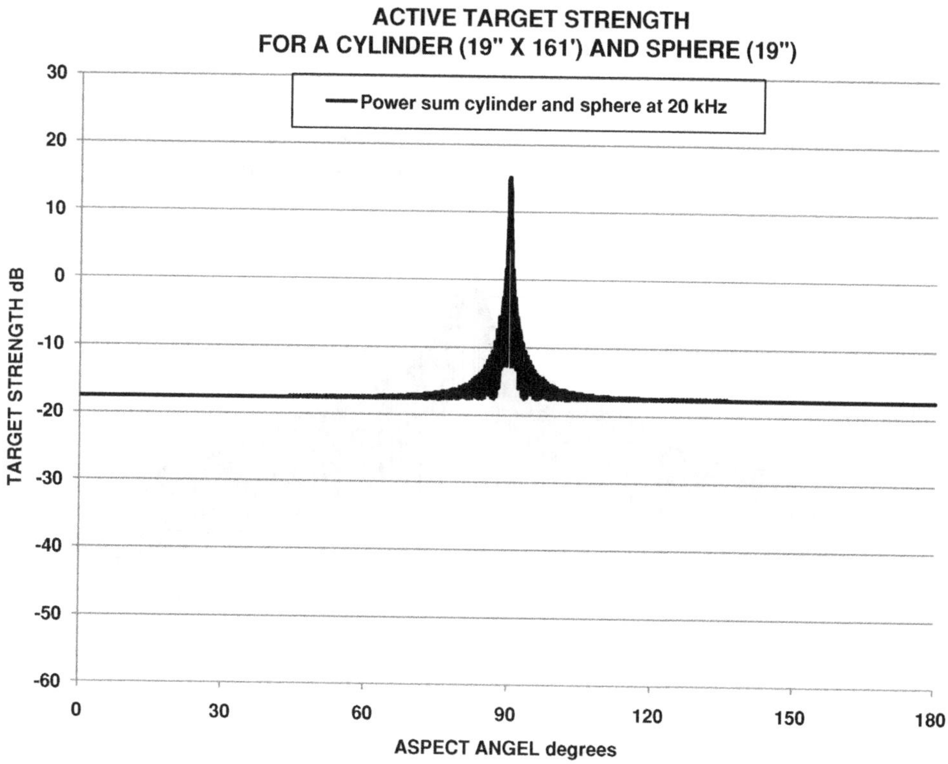

Figure 9.12 Estimated target strength for a submarine-launched mine at 20 kHz

The target strength of a fish can be estimated as follows. Most fish have swim bladders to maintain buoyancy. These bladders are a major contributor to the fish's target strength. Assuming the bladders are roughly ellipsoids, we can use the ellipsoid equation from Table 9.1 to estimate target strength:

$$\mathrm{Nts(ellipse)} = \mathrm{TS(ellipse)} = 10\log\left\{\left[\frac{bc}{2a}\right]^2\right\} \tag{9.18}$$

Let's assume that the variables a, b, and c are proportional to the fish's length, L. Then, set b to be about one-twelfth the length of the fish and let the height-to-width ratio (c/a) equal one. Adjusting for the difference in density and compressibility of the bladder by 2 dB we get

$$\mathrm{Nts(fish)} = \mathrm{TS(fish)} = 20\log\left(\frac{L}{24}\right) - 2 = 20\log(L) - 29.6 \tag{9.19}$$

This calculated target strength is remarkably close to the measured value. The assumption here is that the frequency is sufficiently high that the bladder can be considered large as compared to the wavelength.

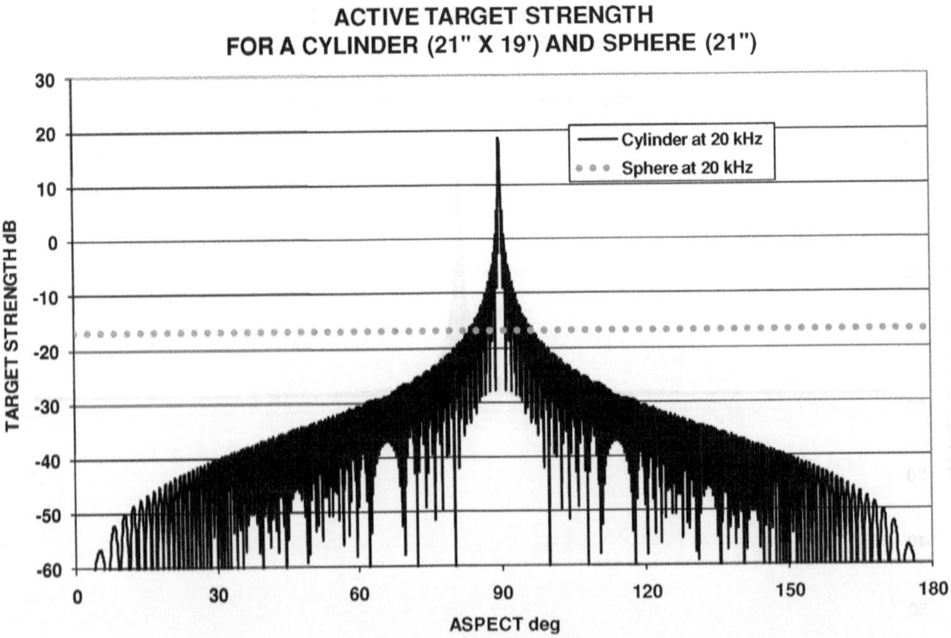

Figure 9.13 Estimated target strength for a cylinder and sphere

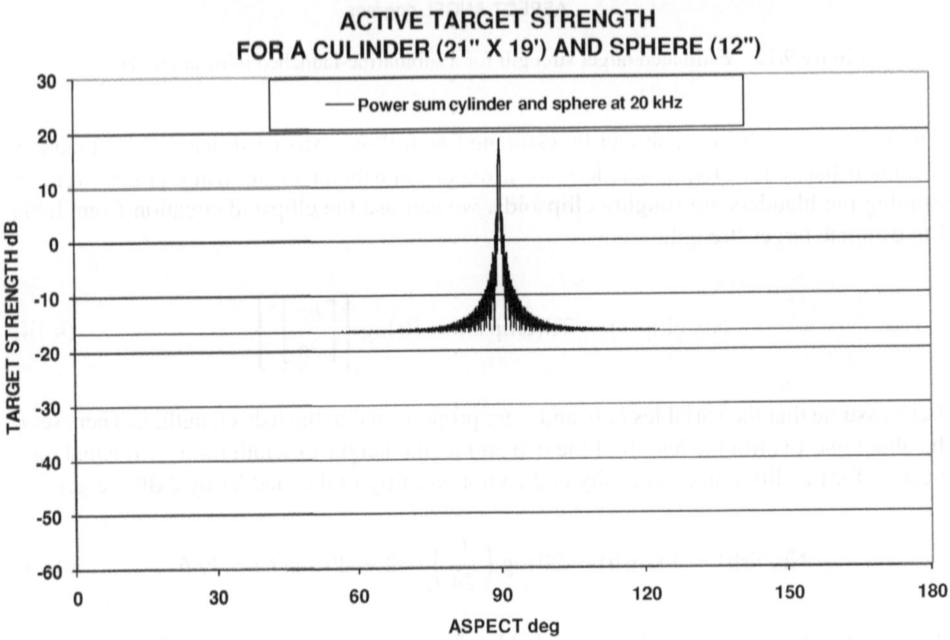

Figure 9.14 Estimated target strength for a U.S. Mk 48 torpedo at 20 kHz

For schools of fish, it is common to assume that the target strengths of each fish power add such that

$$\mathrm{Nts(school)} = \mathrm{TS(school)} = 20\log(L) - 29.2 + 10\log(n) \tag{9.20}$$

where n is the number of fish. This approximation is only valid if the resolution of the sonar pulse is such that it includes the entire number of fish, n, and assumes that the fish can be represented by a representative length, L. Short continuous wave (CW) or wide FM pulses will resolve a school of fish into many pieces, resulting in a lower average target strength.

References

[1] Rayleigh, J. W. S., *The Theory of Sound*, Vols I and II, New York: Dover Publishing, reprint 1945.

[2] Sarkissian, A., "Comparison of TAP Model and SARA-2d Results," Naval Research Laboratory, Report NRL/MR/7130–97-7974, DTIC 19970918110, 1997.

[3] Federation of American Scientists, "Mk 67 Submarine Launched Mobile Mine (SLMM)," Available at: http://www.fas.org/man/dod-101/sys/ship/weaps/mk-67.htm, updated 12 December 1998.

[4] Love, R. H., "Target Strength of an Individual Fish at Any Aspect," *Journal of the Acoustical Society of America*, **62** (6), December 1977.

[5] Foote, K. G., "Averaging of Fish Target Strength Function," *Journal of the Acoustical Society of America*, **67** (2), 1980, 504–516.

10

Radiated Noise

Radiated noise is the signal (Ls or SL) being used in the passive sonar equation and as part of the interference component of the active sonar equation. (Signals radiated by active sonar sources are discussed in Chapter 4, Active Sonar Sources. An excellent classic reference for further investigation of this topic is *Mechanics of Underwater Noise* [1].)

The source of radiated noise depends on what we are interested in detecting. Ships, submarines, torpedoes, and marine mammals are examples of sources of noise. The signals emitted by these sources can be extremely complex. Man-made platforms can have numerous rotational and reciprocating machinery components, as well as sounds associated with personnel on board. Typically, the propulsion components (engines, turbines, gear trains, pumps, and propellers) are the principal sources of noise. Passive sonars are designed with the purpose of exploiting radiated noise and distinguishing it from interfering noises, e.g., array self noise, ambient noise, and noise from platforms not of interest.

10.1 General Characteristics of Ship Radiated Noise

Radiated noise is observed as intensity measured in decibels. For most applications, the noise of interest is the far-field noise, which is the intensity measured at a large arbitrary range adjusted to a standard reference range, usually 1 yard or 1 meter from the center of the source, by numerically adding the propagation loss between the measurement range and 1 yard. The fact that this reference distance is inside most platforms does not matter, because we are interested in the far-field noise. As with other parameters in the sonar equation, it is imperative that the rest of the relevant units be specified. Specifically, the reference for intensity (1 μPa, 1 μbar, or 1 dyne/cm^2) and the bandwidth included, usually 1 Hz or "spectrum level," 1/3 octave, or 1/10 decade.

Radiated noise types are divided into two general categories, broadband and narrowband. A broadband signal covers a wide range of frequencies whereas a narrowband signal is composed of discrete tones. The transition between broad and narrow is fairly arbitrary. Many platforms have both forms of signal combined. The signal processing designed to detect these two general types is very different.

Signals can be further classified by their temporal characteristics: continuous, intermittent, and transient, with the primary difference between the last two being duration. An intermittent

Underwater Acoustics: Analysis, Design and Performance of Sonar Richard P. Hodges

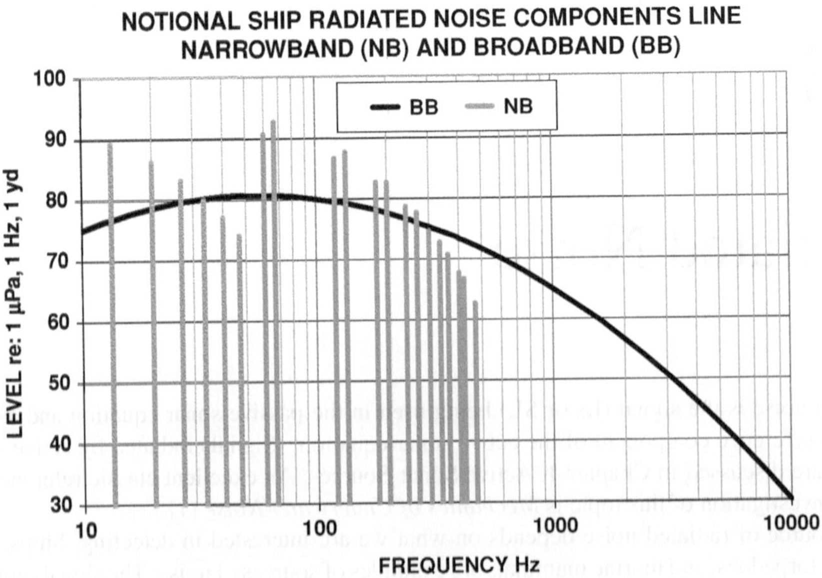

Figure 10.1 Notional ship radiated noise components

source might be a sump pump that comes on every hour and runs for five or ten minutes, whereas a transient source is usually something like a wrench being dropped, a door being slammed, or a rudder moving.

Not all narrowband signals have the same percentage bandwidth. For example, propeller speed (in rpm) may vary with time, causing the frequency of blade-related signals to change over a range of values, whereas electrical systems are generally highly regulated, resulting in extremely stable frequencies. These differences indicate that a sonar will perform best if it has very narrow filters for stable frequencies and wider filters for less stable frequencies.

Figure 10.1 shows a "typical" radiated noise curve for a ship with broadband and narrowband signals separated. There are three series of narrowband components corresponding to blade rate (6.67 Hz fundamental), diesel firing rate (66.7 Hz fundamental), and electrical (60 Hz fundamental). Figure 10.2 shows the resulting signature with both broadband and narrowband components combined. Note that some narrowband components have disappeared below broadband levels.

Figure 10.3 shows the effect on the shape and level of radiated noise if the reference bandwidth is 1/3 octave. Above about 4 Hz, the radiated noise levels are higher and the curves roll off less rapidly with frequency. The level at 10 kHz represents the intensity in a 2300 Hz band, whereas the level at 10 Hz represents the intensity in a 2.3 Hz band.

10.2 Propeller Radiated Noise

The noise radiated from propellers is different from internal machinery noise in that it is outside the ship and in direct contact with the medium. There are three principal types of noise

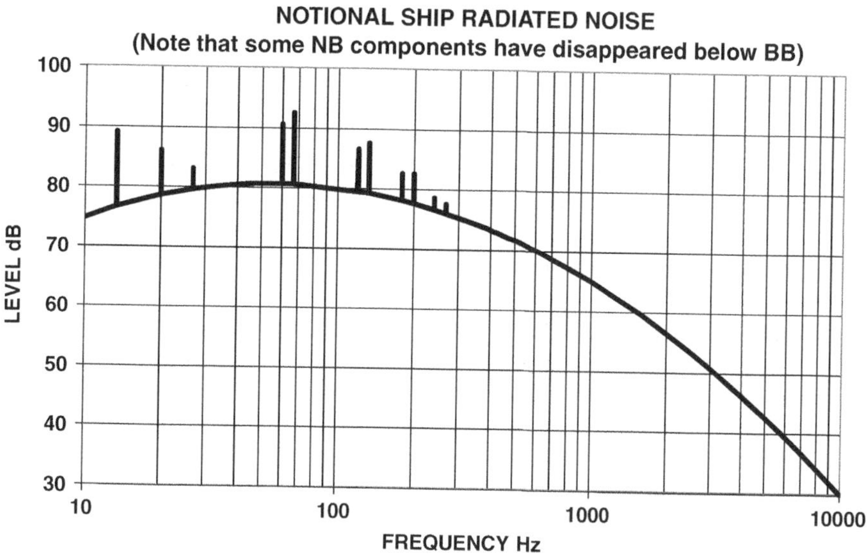

Figure 10.2 Notional ship radiated noise with broadband and narrowband components combined

generated by propellers: blade rate, cavitation, and resonance. Blade rate noise is caused by a change in thrust of the blades as they pass through flow irregularities. Blade rate signals radiate as a harmonic series of frequencies given by

$$f_{\mathrm{m}} = mnf_{\mathrm{r}} \tag{10.1}$$

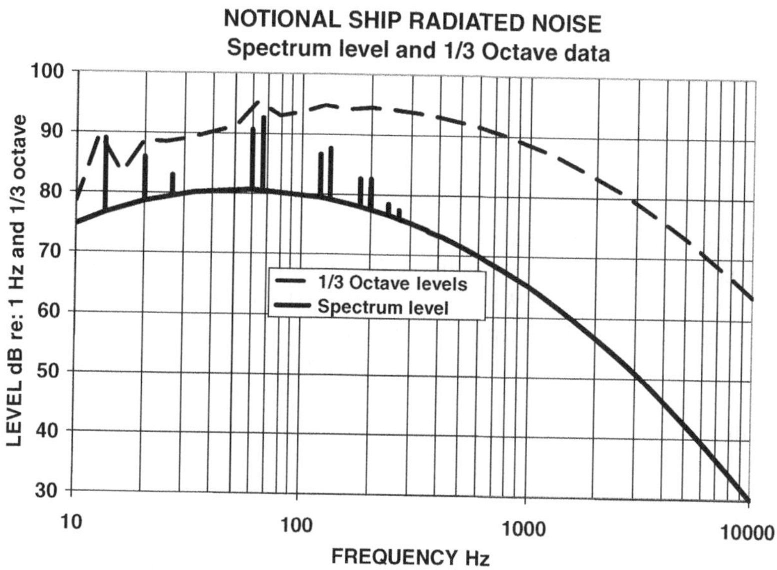

Figure 10.3 Notional ship radiated noise 1/3 octave level and spectrum level

where

m = harmonic number
n = number of blades
f_r = rotational frequency of the propeller

For example, a six-bladed propeller rotating at 50 rpm can be expected to have a series of tones at 5 Hz, 10 Hz, 15 Hz, etc. Over much of a ship's operating range, speed is proportional to the rpm of the propeller, usually referred to as turns per knot, TPK. If the blade rate fundamental can be determined and the number of blades is known, the rpm of the shaft, and consequently the speed of the ship, may be estimated.

Cavitation occurs when the flow velocity over portions of the propeller (or in some cases the hull or control surfaces) becomes high enough for Bernoulli forces to cause the local pressure to become negative, essentially exceeding the tensile strength of the water. This change in pressure forms gas bubbles made up of water vapor and dissolved air. A short time after formation, these bubbles collapse as they migrate away from the low-pressure area, and in doing so generate short transient pulses of sound that result in broadband noise. Since there are a large number of bubbles, the result sounds like a loud hiss.

In general, cavitation is the dominant noise type for surface ships at high frequencies. This noise type follows Bernoulli's law, which states that the dynamic pressure is proportional to the velocity squared. As the rpm of a propeller increases, the tip of the blades begin to show signs of cavitation first, followed by more and more blade surface area as the speed increases. This results in a large change in radiated noise (typically 10–20 dB) over a fairly narrow change in speed. To account for this, propellers are assigned a cavitation index (K_T) to reflect the likelihood of cavitating at various speeds:

$$K_T = \frac{P_0 - P_v}{0.5\rho v_t^2} \tag{10.2}$$

where

P_0 = static pressure at the propeller
P_v = vapor pressure of water
ρ = density of water
v_t = tip velocity of the propeller

The exact value of K_T at which cavitation occurs depends on the specifics of the design of the propeller and its current material condition. For most propellers, tip cavitation begins at values of K_T between 0.6 and 2.0. When K_T is above 6.0, cavitation is very unlikely and, if it occurs, is an indication of a damaged propeller. Cavitation is virtually certain for K_T values below 0.3. Given this, the depth at which cavitation occurs for a submarine is proportional to the square root of the speed, i.e., deeper submarines can go faster without incurring the noise associated with cavitation. Figure 10.4 shows the typical rapid increase in radiated noise as a propeller goes from no cavitation to full cavitation.

10.3 Machinery Noise

Machinery noise is generated by all moving parts of a ship. Vibrations originate from reciprocating parts such as the periodic explosions in the cylinders of a diesel engine; repetitive

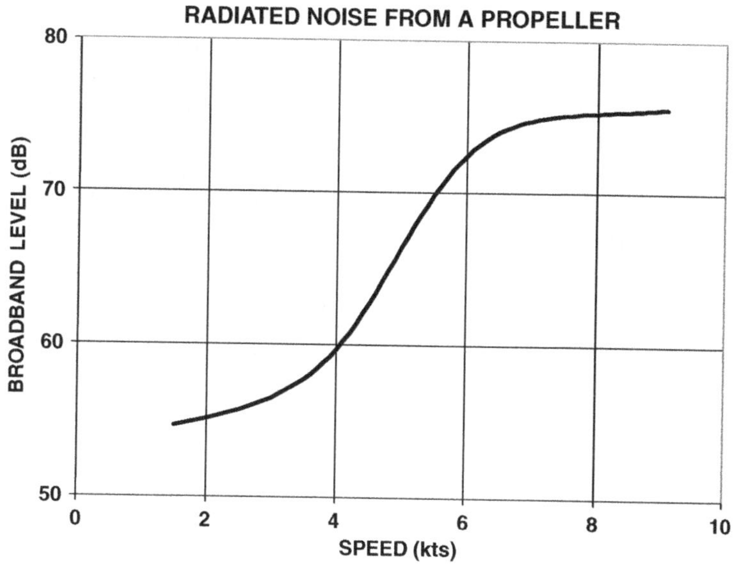

Figure 10.4 Radiated noise effects due to cavitation

discontinuities from turbine blade, gear teeth, and armature slots on electric motors; imbalances in rotating machinery; and turbulence and cavitation in fluid flow in the hydraulic system, pumps, and valves. The first three origins of vibrations give rise to narrowband components, while the last produces more broadband-like signals. If the first three noise sources are associated with the main propulsion of the vessel, they are likely to increase in both frequency and amplitude with an increase in speed. At certain speeds, these noise components may match structural vibrations of the same frequency, causing a significant increase in the radiated signal.

10.4 Resonance Noise

Resonance noise occurs when vibrations drive the fundamental modes of structural components, resulting in large amplitudes. The structures involved could be single plates, struts, propeller blades, or the entire hull structure. When propellers exhibit this noise, it is referred to as a "singing prop." Frequently, the resultant signals generated are signals with much larger bandwidths than normal machinery narrowband signals. These occupy the region between what is referred to as broadband and narrowband. The frequencies excited are a function of the structure affected, but the amplitude is strongly dependent on the drive frequency (blade rate and its harmonics) and amplitude. Therefore, one would expect hull resonances to be at much lower frequencies than propeller blade resonances.

10.5 Hydrodynamic Noise

Hydrodynamic noise, also known as flow noise, is caused by fluctuation and irregularities in the flow of fluid over a structure. This turbulence in the boundary layer creates a pressure field that is not an acoustic field. Although of primary importance when discussing sonar self noise (see Chapter 11, Self Noise), flow noise is typically a near-field phenomenon whose influence decreases very rapidly with distance. These vibrations can excite elements of structure, particularly at their resonance frequencies, producing noise that radiates efficiently. Typical resonance sources are from the hull, cavities, plates, and the propeller.

Cavities may be excited by flow moving across their openings (Helmholtz resonators), the equivalent of blowing air across the mouth of a soda bottle. The amplitude of such resonances is a strong function of speed and has in the past been sufficient to break welds on ballast tanks. The frequency of resonance is given by

$$f_{\mathrm{H}} = \frac{c}{2\pi}\sqrt{\frac{A}{VL}} \tag{10.3}$$

where

c = speed of sound
A = area of the opening
V = volume of the cavity
L = length of the opening

A simple solution to reducing this noise would be to close the opening of the cavity. Other approaches would involve inserting baffles or putting a slotted grill across the opening to shift the resonance frequency.

In fluid dynamics, turbulence or turbulent flow is a flow regime characterized by chaotic, stochastic property changes. This includes low momentum diffusion, high momentum convection, and rapid variation of pressure and velocity in space and time. Flow that is not turbulent is called laminar flow. The dimensionless Reynolds number characterizes whether flow conditions lead to laminar or turbulent flow. Laminar flow occurs at low Reynolds numbers where viscous forces are dominant and is characterized by smooth, constant fluid motion. The consideration of what is "low" for a Reynolds number depends on the shape and smoothness of the object in question. This could be as low as 0.1 for a sphere to as high as 2300 for a smooth pipe. However, turbulent flow occurs at high Reynolds numbers where inertial forces are dominant, producing random eddies, vortices, and other flow fluctuations. Similarly, the consideration for what is "high" can vary from 4000 to hundreds of thousands depending on the case of interest. The Reynolds number (Re) is given by

$$Re = v_{\mathrm{s}}\frac{L}{\nu} \tag{10.4}$$

where

v_{s} = mean fluid velocity
L = characteristic length
ν = kinematic fluid viscosity: $\nu = \mu/\rho$

As an example, consider a ship (L = 100 m) moving at 15 knots (7.7 m/s). The Reynolds number would be about 8×10^8, well into the turbulent flow region.

The transition between laminar and turbulent flow for specific bodies is indicated by a critical Reynolds number (Re_{crit}). This number is based on the pipe diameter and the mean velocity, v_{s}, within the pipe. For circular pipes, it is generally accepted to be 2.3×10^3. Because

Re_{crit} depends on the exact flow configuration it must be determined experimentally. There is a region near Re_{crit} of gradual transition where the flow is neither fully laminar nor fully turbulent and predictions of fluid behavior can be difficult. Consequently, designers will avoid any pipe configuration that falls near this value to ensure that the flow is either laminar or turbulent.

In turbulent flow, unsteady vortices appear on many scales and interact with each other. Turbulence causes the formation of eddies which are defined by the Kolmogorov length scale and a turbulent diffusion coefficient. Drag due to boundary layer skin friction increases with turbulence. The structure and location of boundary layer separation often changes, sometimes resulting in a reduction of overall drag. Because this laminar–turbulent transition is governed by the Reynolds number, it can be used as a scaling parameter predicting that if the size, but not shape, of a body is increased the transition will occur at a proportionally lower speed.

10.6 Platform Quieting

Platform quieting is a major consideration for modern day warships. Ship silencing begins with a design involving mounting potential noise sources on sound isolation mounts, which may then be mounted on sound isolated rafts. These rafts may then be mounted on additional sound isolated rafts until the desired level of silencing is achieved. Careful attention must be paid to potential "sound shorts" which bypass the isolation. For example, attaching a diesel engine sea water cooling pipe directly to the hull would bypass the isolation mounts of a sound isolated raft, providing a path for the sound to travel.

Quieting propellers can be accomplished in several ways. First, large-diameter, low-rpm propellers can shift energy to lower frequencies where detecting sonar may not perform as well. Next, careful attention to the manufacture and maintenance of the propeller can prevent premature cavitation. Finally, a skewed propeller can reduce the blade rate because the blade would enter the flow irregularities gradually. Large tankers and submarines pioneered the use of highly skewed propellers to reduce blade rate, in their case it was a consideration of structural vibration interfering with operation, thus reducing sonar detectability.

In addition, some submarines apply coatings to the outside of the hull to reduce radiated noise, as well as target strength. Surface warships may also employ bubble screens, called "Agouti" by the British and "Prairie Masker" by the U.S., to mask the radiated noise with a low impedance wall.

10.7 Total Radiated Noise

Donald Ross, in his book *Mechanics of Underwater Noise* [1], gives an equation to approximate the noise radiated by a merchant ship:

$$\text{Ls} = \text{SL} = 154 + 60\log\left(\frac{U}{25}\right) + 10\log\left(\frac{B}{4}\right) - 20\log(f) \quad \text{dB re : } 1\,\mu\text{Pa, 1 Hz, 1 yd.} \tag{10.5}$$

where

U = propeller tip speed in m^2/s
B = number of blades on the propeller
f = frequency in Hz

An alternative form of the equation is

$$Ls = SL = 154 + 60\log\left(\frac{V}{10}\right) + 9\log(DT) - 20\log(f) \quad \text{dB re : 1 μPa, 1 Hz, 1 yd.} \tag{10.6}$$

where

V = speed in knots
DT = ship displacement (tons)
f = frequency in Hz

Equations 10.5 and 10.6 assume that the propeller is the dominant source of radiated noise. Given this is probably true above 100 Hz, these equations should not be used for lower frequencies.

Figure 10.5 shows an example of the above equations for a ship of 100 000 deadweight tons (dwt) at 14 knots. Figure 10.6 shows the cumulative power versus frequency for the same case. Note that the power is concentrated at the lower frequencies. Figure 10.7 shows the approximate total power radiated by various platforms. The levels vary from hundreds of watts for a very large tanker/bulk transport ship to one hundred milliwatts for a World War II U.S. Navy submarine. World War II submarines did not incorporate the quieting features discussed in Section 10.6. The engines and machinery were not sound isolated, the propeller

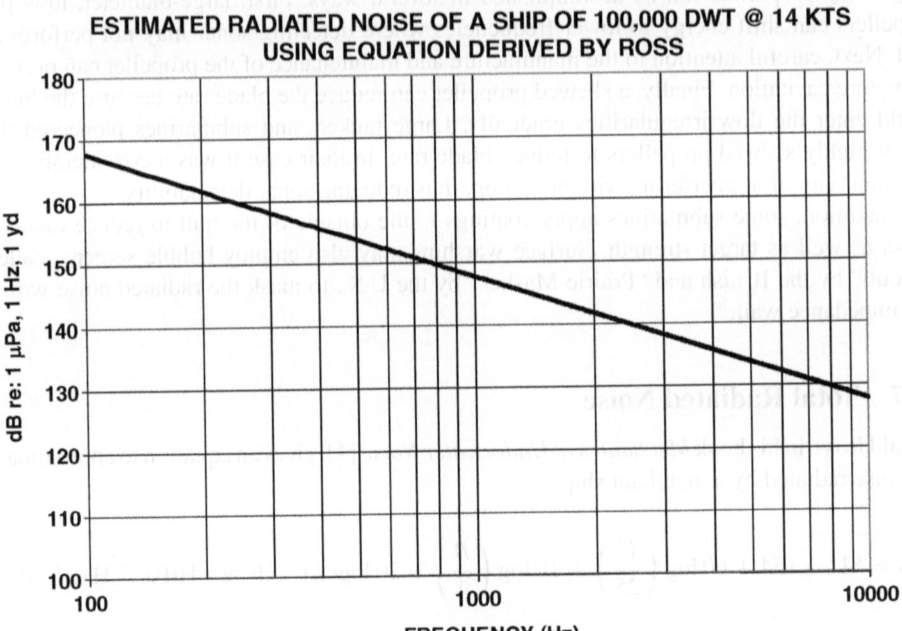

Figure 10.5 Estimated radiated noise

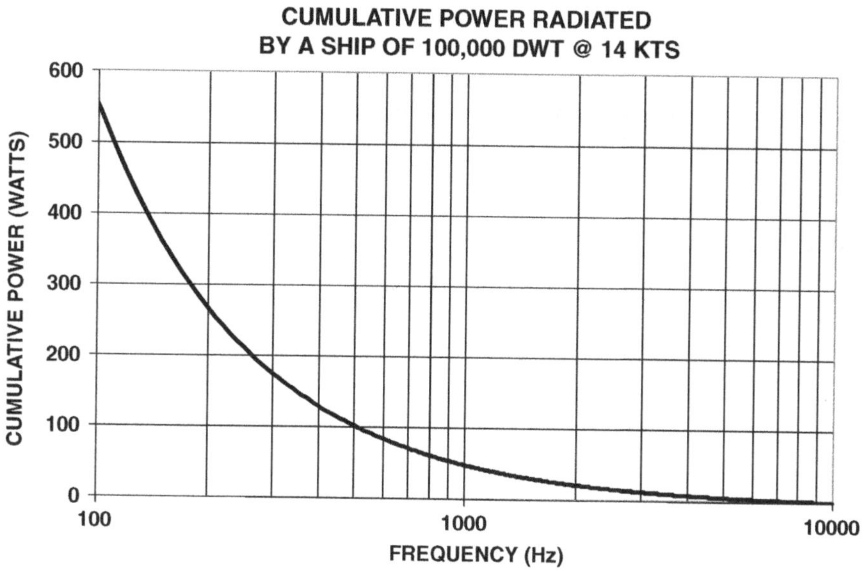

Figure 10.6 Estimated cumulative power radiated

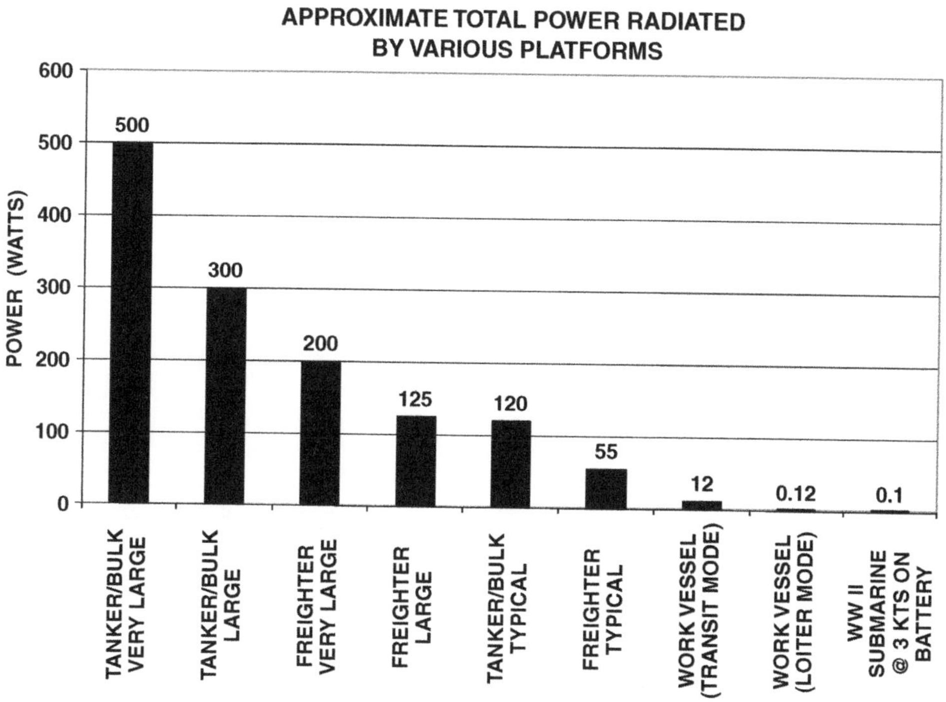

Figure 10.7 Approximate total power radiated

was not skewed, and sound reducing hull coatings were not applied. As a result, it should come as no surprise that modern submarines are much quieter than World War II submarines at comparable speeds. Consequently, in order to be successful, passive ASW sonar systems must be able to find targets generating milliwatt levels of noise among merchant traffic that is thousands of times noisier and thousands of times more numerous.

Reference

[1] Ross, D., *Mechanics of Underwater Noise*, Oxford: Pergamon Press, 1976, pp. 275–277.

11

Self Noise

One component of noise in the sonar equations, self noise (Ln or NL), is caused by sources in the immediate vicinity of the hydrophones. At first thought, we might think this noise is just radiated noise viewed up close, but this is not the case. Much of the components that make up self noise are not present in the far field. Potential self noise sources include radiated noise from the platform propagating to the hydrophones, structure-borne platform noise permeating the hydrophone supports, and the flow of water past the hydrophone. Electronic noise, although a potential source of self noise, is usually not a problem with modern amplifiers, except under the quietest conditions (e.g., under arctic pack ice or when electromagnetic interference (EMI) or ground loop problems exist).

Self noise will have a minimum value which is present regardless of the speed the array is traveling through the water. Increases to this noise will be due to flow noise and increases in noise from the host platform. Self noise is specified as a sound pressure level (SPL) and has a frequency and speed dependence.

Figure 11.1 shows the variety of paths noise generated by a platform can reach in the sonar array: (1) hull-borne, (2) direct water path (frequently the case for propeller noise), (3) other water paths involving volume scatters, and scattering or reflection off (4) the bottom or (5) surface of the ocean. While directional hydrophones or arrays can help reduce noise from these sources, it is more common to put isolating barriers called baffles between the source and the array. Baffles are structures made up of materials with a very different ρc from water so they reflect sound or absorb sound. Submarines may have bow-mounted arrays with "eyebrow" baffles installed above the array to shield them from surface reflected own ship noise. Similarly, destroyers may have sonars mounted in fairings protruding from the bow. A baffle is frequently included to prevent propeller noise from going directly to the array. In both cases, the bow was chosen because it is the furthest distance from dominant noise sources, e.g. the propeller and the engine room. Hydrophones or arrays mounted on ships have elaborate isolation devices to prevent hull-borne vibrations from entering the array. This includes sound mounts, as well as baffles.

11.1 Flow Noise

Flow noise is a form of hydrodynamic noise associated with flow conditions. It consists of the fluctuating pressure field created by the turbulent flow in the boundary layer about the

Underwater Acoustics: Analysis, Design and Performance of Sonar Richard P. Hodges

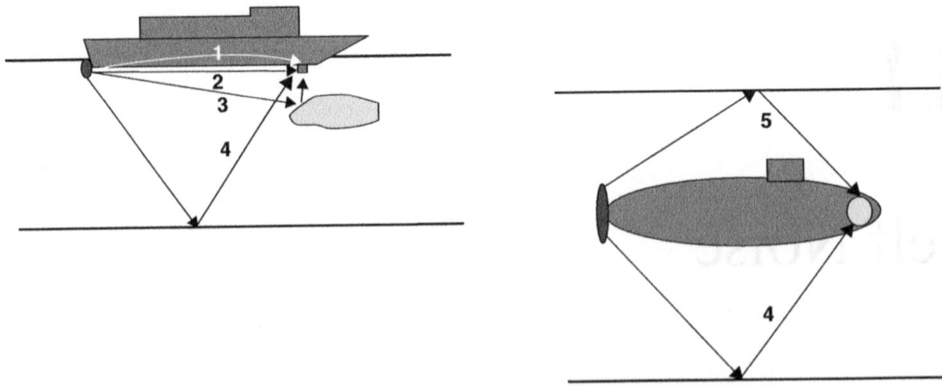

Figure 11.1 Self noise paths from a source to sonar array

array. This pressure field is not an acoustic field and its influence decreases very rapidly with distance. Arrays are generally enclosed in smooth domes or fairing, which have a standoff from the flow so that the influence of flow noise on self noise can decline. These domes are designed not to have structural vibration modes in the frequency region that the sonar will operate in. The shape of these domes is chosen to maintain laminar flow for as great a speed as possible.

As an object moves through water, the flow pattern around it changes. At low speeds, the flow is laminar, meaning the flow lines are parallel to the shape of the body. A transition to turbulent flow occurs when the speed exceeds a value determined by the Reynolds number, *Re*. The Reynolds number, named after Osborne Reynolds, is the ratio of inertial forces to viscous forces and quantifies the relative importance of these two types of forces for given flow conditions. This number is probably the most important dimensionless number in fluid dynamics and is used, usually with other dimensionless numbers, to provide a criterion for determining dynamic similitude. Typically, the Reynolds number is given as follows:

$$Re = \frac{\rho v_s L}{\mu} \tag{11.1}$$

$$Re = \frac{v_s L}{\nu} \tag{11.2}$$

where

v_s = mean fluid velocity
L = characteristic length
ν = kinematic fluid viscosity: $\nu = \mu/\rho$
μ = dynamic fluid viscosity
ρ = fluid density

For flow in pipes with a circular cross-section, the characteristic length, L, is the pipe diameter. For flow over a flat plate, the characteristic length, L, is the length of the plate and the characteristic velocity is the free stream velocity.

Laminar flow occurs at low Reynolds numbers ($Re < 2300$), where viscous forces are dominant, and is characterized by smooth, constant fluid motion. Turbulent flow also occurs

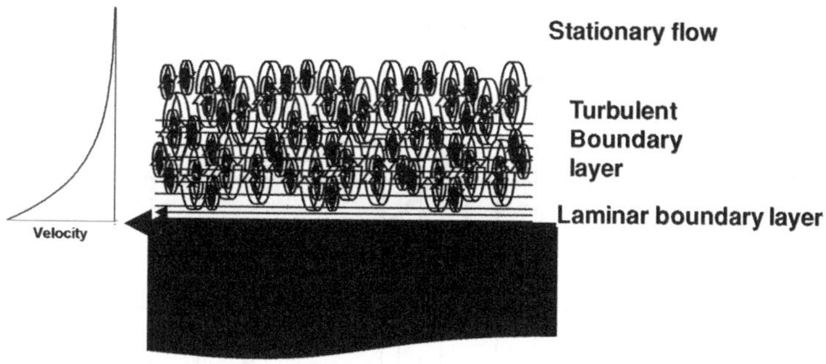

Figure 11.2 Flow structure for a body moving through stationary water. Note that the graph on the left shows the velocity versus distance from the surface

at high Reynolds numbers ($Re > 4000$) and is dominated by inertial forces, producing random eddies, vortices, and other flow fluctuations. The transition between laminar and turbulent flow occurs at a critical Reynolds number (Re_{crit}). Because this number depends on the exact flow configuration, it must be determined experimentally. There is a region near this value of gradual transition where the flow is neither fully laminar nor fully turbulent and predictions of fluid behavior can be difficult.

For circular pipes, the critical Reynolds number is generally accepted to be about 2300. However, engineers will avoid any pipe configuration that falls within 500 of this value to ensure that the flow is either laminar or turbulent; otherwise the flow will be very sensitive to slight changes and may not meet the requirements of the system.

In a boundary layer over a flat plate, the local regime of the flow is determined by the Reynolds number based on the distance measured from the leading edge of the plate. In this case, the transition to turbulent flow occurs at a Reynolds number of the order of 10^6. In this state, as is shown in Figure 11.2, a laminar region in the boundary layer is essentially at the same speed as the body. Here, the flow becomes turbulent and the relative velocity decreases with distance until it reaches zero, at which point there is no turbulence. The adherence to the Reynolds number assumes a smooth body. The requirement is not absolute smoothness, but that the roughness must be less than the thickness of the laminar boundary layer. It has been found [1] that the roughness height, h, requirement is given by

$$h = \frac{0.003}{u} \tag{11.3}$$

where

h = roughness height (m)
u = flow velocity in m/s; in engineering units the constant would be 6 milli-inches per knot

The total root-mean-square (rms) pressure, from the same reference, is given by

$$P_{rms} = \alpha\rho\frac{u^2}{2} \tag{11.4}$$

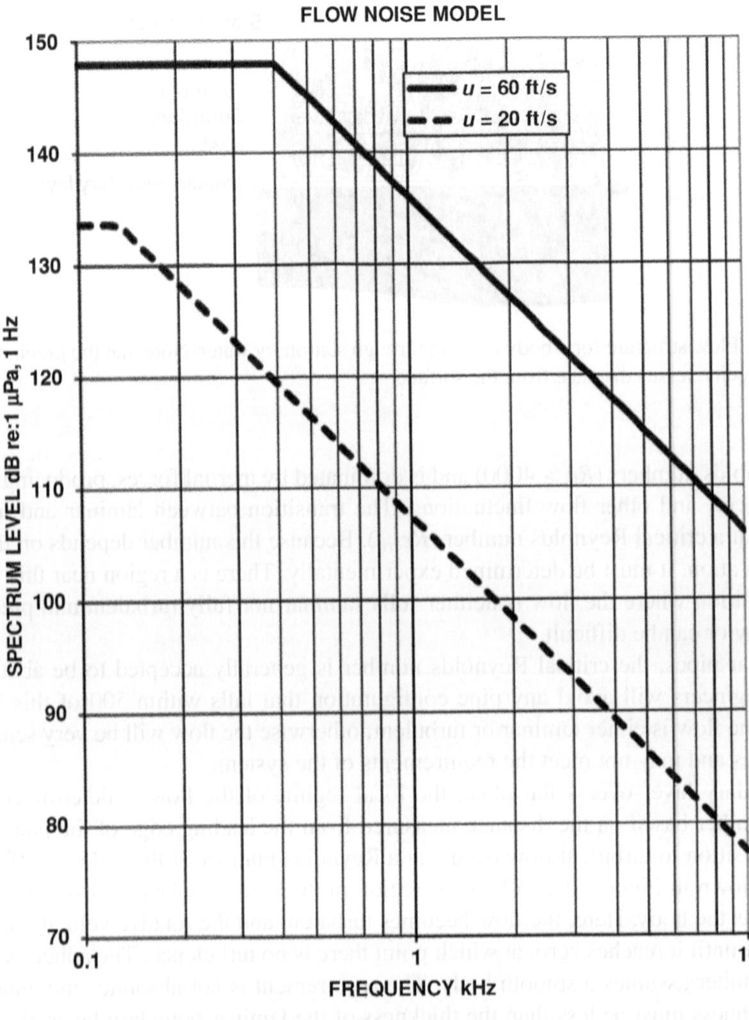

Figure 11.3 Flow noise model

where

α = dimensionless Kraichman constant with a value between 0.0018 and 0.012 centered roughly at 0.003

ρ = water density

u = free flow velocity

A model for the spectral shape of flow noise is shown in Figure 11.3.

The shape of the flow noise curve is flat until a transition frequency, given approximately by [1]

$$F_{\mathrm{t}} = \frac{u}{12\delta} \tag{11.5}$$

where

F_t = transition frequency (Hz)
u = free field velocity
δ = thickness of the boundary layer

Since u has units of distance per second and δ has units of distance, any set of consistent units can be used. Since the flow velocity is a continuous function of distance from the surface, the effective or displacement thickness is used. This is given by

$$\delta = \frac{\int u(h)\,\mathrm{d}h}{u_0} \tag{11.6}$$

The range of δ given by Skudrzyk and Haddle [1] is 0.153 in (20 ft/s) to 0.135 in (60 ft/s) and is relatively insensitive to speed.

Above the transition frequency, the spectrum falls off as $1/f^3$, so the model used is

$$\begin{aligned} &\text{Noise power} = C && f < F_t \\ &\text{Noise power} = C\left(\frac{F_t}{f}\right)^3 && f > F_t \end{aligned} \tag{11.7}$$

The intensity over all frequencies, which is proportional to the root-mean-square pressure (P_{rms}^2), is given by

$$P_{\mathrm{rms}}^2 = \int_0^{F_t} C\,\mathrm{d}f + \int_{F_t}^{\infty} C\left(\frac{F_t}{f}\right)^3 \mathrm{d}f = \frac{2}{3} C F_t \tag{11.8}$$

This can be written in decibels, relative to 1 μPa:

$$\begin{aligned} \mathrm{Ln}(f) &= 20\log\left(\alpha\frac{1}{2}\rho\frac{u^2}{0.000001}\right) - 10\log(F_t) && \text{for } f < F_t \\ \mathrm{Ln}(f) &= 20\log\left(\alpha\frac{1}{2}\rho\frac{u^2}{0.000001\sqrt{F_t}}\right) + 30\log\left(\frac{F_t}{f}\right) && \text{for } f > F_t \end{aligned} \tag{11.9}$$

where u and ρ are in meter-kilogram-second (mks) units and f is in Hz.

Self noise increases as speed to the third power or 30 log(u) below the break point frequency. Above the break point frequency, self noise increases as speed to the sixth power or 60 log(u).

It is easy to see why flow noise is a major design consideration. From Figure 11.3, we see that at 20 ft/s (about 12 kts) the noise level at 1 kHz is about 108 dB. This is much higher than the typical ambient noise levels in the sea of about 62 dB (see Chapter 7, Ambient Noise). If we want to be limited not by flow noise but by ambient noise, we must gain more than 46 dB; such gains require a combination approach. One approach is to provide a fairing to separate the hydrophones from the flow. This works because the flow noise is a pressure variation not an acoustic wave and therefore attenuates rapidly with distance. All things being equal, a large diameter towed array will have better flow noise characteristics than a thin one. Similarly, physically large hydrophones will respond better to flow noise than very small hydrophones.

Another approach is to use multielement or large hydrophones or large numbers of hydrophones. One such way is to compose "hydrophones" by wiring together several separate hydrophones and beamforming this group as if it were a single hydrophone. In this way, an array with 50 "hydrophones," each comprised of 10 elements, can have the gain against flow noise of a 500 hydrophone array, without the requirement of a 500 hydrophone beamformer.

11.2 Turbulent Noise Coherence

The nature of flow noise is quite different from an acoustic noise field. The correlation coefficient is very different from that of ambient noise (see Chapter 3, Transducers, Directionality, and Arrays), with important implications for array construction. For a single frequency, f, the correlation coefficient for hydrophones separated by a distance, d, is given by

$$\rho(d) = \kappa(s)\cos(2\pi s) \tag{11.10}$$

where s is the nondimensional Strouhal number, defined as

$$s = f\frac{d}{u_c}$$
$$\rho(d) = \kappa(s)\cos\left(2\pi f\frac{d}{u_c}\right) \tag{11.11}$$

where

f = frequency

u_c = convection velocity, which is the velocity with which turbulent patches are carried past the hydrophone, generally between 60 and 100 % of the flow velocity

The correlation function $\kappa(s)$ has been found by Bakewell [2] and Corcos [3] to be

$$\kappa_L(s) = \exp(-0.7\,|s|) = \exp\left(-0.7f\frac{d}{u_c}\right) \quad \text{parallel to flow}$$
$$\kappa_L(s) = \exp(-5.0\,|s|) = \exp\left(-5.0f\frac{d}{u_c}\right) \quad \text{perpendicular to flow} \tag{11.12}$$

Since the cosine is cyclic for large values of s, the correlation function dominates. As we saw in Chapter 3, Transducers, Directionality, and Arrays, the array gain in signal over noise equals 10 log(N), where N is the number of hydrophones when the noise correlation hydrophone to hydrophone ratio becomes small. For acoustic signals, the cross-correlation coefficient is given by

$$\rho_{ac}(d) = \frac{\sin\left(2\pi f\frac{d}{c}\right)}{2\pi f\frac{d}{c}} \tag{11.13}$$

This function becomes small when $f\ d/c$ becomes large. The same statement can be made for $\rho(d)$, except that the condition is when $f\ d/u_c$ becomes large.

For all platforms operating in the ocean $u_c \ll c$, since c is slightly less than 3000 kts.

11.3 Strumming Noise

Another form of hydrodynamic noise is strumming noise or the vibration of cables or struts. This noise is caused by vortex shedding as a flow moves past the cable. This vibration occurs at a frequency given by

$$f_{\text{strum}} = s\frac{v}{d} \tag{11.14}$$

where

s = dimensionless Strouhal number whose value is 0.18 over a large range of speeds and diameters
v = flow velocity
d = diameter

Since v/d has units of one over time, any set of consistent distances can be used as long as the speed has units of distance per second. This noise problem can be improved by providing fairing around the strut or cable.

References

[1] Skudrzyk, F. J., and Haddle, G. P., "Noise Production in a Turbulent Boundary Layer by Smooth and Rough Surfaces," *Journal of the Acoustical Society of America*, **32**, 1960, 19.
[2] Bakewell, H. P., "Longitudinal Space Time Correlation Function in Turbulent Air Flow," *Journal of the Acoustical Society of America*, **35**, 1963, 936.
[3] Corcos, G. M., "Resolution of Pressure in Turbulence," *Journal of the Acoustical Society of America*, **35**, 1963, 192.

12

Statistical Detection Theory

Detection theory is applying statistics to the decision process. Take the problem of detecting if a coin is biased. To do this, we would perform an experiment and measure how often heads came up and compare that to what a fair coin could be expected to do. Suppose the coin was flipped three times and it came up tails all three times. We could not say with 90 % confidence that the coin is biased because one time in eight, three tails will come up from the flips of a fair coin. Suppose the coin came up tails 50 times in a row, any reasonable person would conclude that the coin or the tossing procedure was biased since this would only come up one time in two to the fiftieth (1 in 1120 trillion) with a fair process. With some forethought, an experiment can be designed to do this in a less arbitrary manner. By confining our decisions to two outcomes or hypotheses, biased or not (signal present or not), the problem becomes a binary decision. This same technique can be applied to an *m*-ary process as well, e.g., if we were trying to detect a set of bits that were to represent the alphabet.

12.1 Introduction

Historically, the decision that no signal is present is referred to as the null hypothesis (H_0) and the hypothesis that it is present is called the alternative hypothesis (H_1). The data may be scalar or multidimensional. Our purpose is to divide the data space into two regions: the acceptance region, where we will accept H_0, and the rejection or critical region, where we will reject H_0 and accept H_1. The assumption is that these two regions exhaust all possible data points and that there is no ambiguous region. The four possible decision outcomes are shown in Table 12.1. Figure 12.1 illustrates a simple decision process.

For this system, the signal may be absent, equal zero, or present. Noise, $n(t)$, is also present. After taking a single sample at time t_0, the problem is to decide, based on this one sample, $x(t_0)$, whether the signal is present and not zero. This decision may be made more reliably by taking more than one sample, as we will consider later. For this case, the null hypothesis (H_0), the event that no signal is present, and the alternative hypothesis (H_1), the event where the signal $s(t)$ is present, are given by

H_0: no signal present $[x(t) = n(t)]$

H_1: signal is present $[x(t) = s(t) + n(t)]$

Underwater Acoustics: Analysis, Design and Performance of Sonar Richard P. Hodges

Table 12.1 Possible outcomes of a decision process

Correct detection	Signal present; decision "yes"	
False rest	Signal present; decision "no"	Type II error
False alarm	Signal not present, decision "yes"	Type I error
Correct rest	Signal not present; decision "no"	

Based on the single observation, $x(t_0)$, we must choose one of the hypotheses. The first step is to choose a criterion. One reasonable criterion is to choose the most likely hypothesis given the single observation. Let $P(H_0/x(t_0))$ be the conditional probability of H_0 given $x(t_0)$ and $P(H_1/x(t_0))$ be the conditional probability of H_1 given $x(t_0)$. These are a posteriori probabilities, derived after the fact. The decision rule is to choose H_0 if

$$P(H_0/x(t_0)) > P(H_1/x(t_0))$$
$$\frac{P(H_0/x(t_0))}{P(H_1/x(t_0))} > 1 \tag{12.1}$$

and choose H_1 otherwise.

It is often more convenient to express the decision rule in terms of the probability density functions. In other words, choose H_0 if

$$P(H_0/x < x(t_0) < x + dx) > P(H_1/x < x(t_0) < x + dx) \tag{12.2}$$

and H_1 otherwise. Using the definition of conditional probability:

$$P(H_0/x < x(t_0), x + dx) = \frac{P(x < x(t_0) < x + dx/H_0)P(H_0)}{P(x < x(t_0) < x + dx)} \tag{12.3}$$

where $P(H_0)$ is the probability that H_0 is true. The probability that the hypothesis H_1 is true is $1 - P(H_0)$. These are a priori probabilities.

Assigning a probability density function, $p(x)$, to the random variable, x, gives $P(x < x(t_0) < x + dx) = p(x)\,dx$ and $P(x < x(t_0) < x + dx)/H_0) = p_0(x)\,dx$, with the subscript denoting the conditional on H_0:

$$P(H_0/x(t_0)) = \frac{p_0 dx\, P(H_0)}{p(x)dx} = \frac{p_0(x)P(H_0)}{p(x)} \tag{12.4}$$

$$P(H_1/x(t_0)) = \frac{p_1\,[1 - P(H_0)]}{p(x)} \tag{12.5}$$

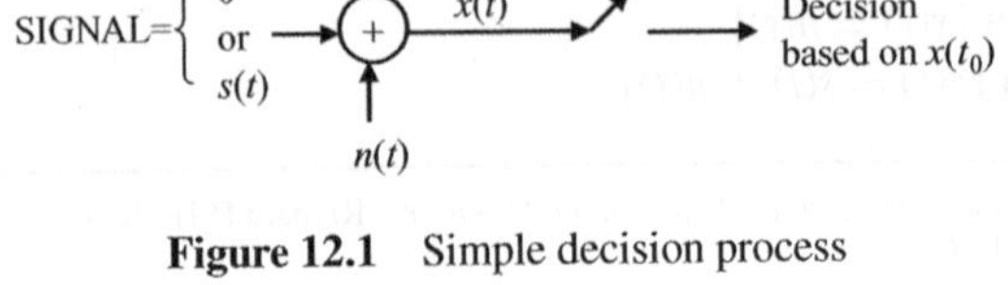

Figure 12.1 Simple decision process

The decision rule can then be written: choose H_0 if

$$\frac{p_0(x)P(H_0)}{p_1(x)[1 - P(H_0)]} > 1 \tag{12.6}$$

or

$$\frac{p_0(x)}{p_1(x)} > \frac{1 - P(H_0)}{P(H_0)} \tag{12.7}$$

Taking the reciprocal, the rule becomes: choose H_1 if

$$\lambda = \frac{p_1(x)}{p_0(x)} > \frac{P(H_0)}{1 - P(H_0)} \tag{12.8}$$

This ratio, λ, is of particular importance and is called the likelihood ratio. Since the log of x is monotonically increasing with x, it is frequently convenient to use the natural log of λ, which is called the log-likelihood ratio.

The most elegant mathematical method of approach to decision theory is through Bayesian analysis, where cost (positive and negative) would be assigned to the various responses of the receiver. The total cost would then be computed using the probabilities associated with each response and the detection strategy would be chosen to minimize the expected cost. In sonar and radar, the a priori probabilities and the costs of each kind of error are difficult to determine. For these cases, a criterion is needed that involves neither a priori probabilities nor estimated costs. This is the Neyman–Pearson criterion. The objective here is to maximize the probability of detection for a given probability of false alarm.

In order to make a decision, the receiver must measure some suitable property of the data. A typical input to a receiver might consist of random noise voltage to which is added a continuously varying signal voltage. Let us assume a time period, T, and a receiver bandwidth, W. Since noise is a random process, any discrete sample of the noise voltage taken at an instant is a random variable. If x_i denotes the ith sample, the probability density, $p(x_i)$, is a function where $p(x_i)\,\mathrm{d}x$ represents the probability that the voltage has a value between x_i and $x_i + \mathrm{d}x$.

According to the sampling theorem, if a wave is sampled at regular intervals of $1/(2W)$ seconds, as in Figure 12.2, it can be completely reconstructed from the $2WT$ samples, so that either the continuous function or the sequence of random variables ($x_1, x_2, x_3, \ldots, x_n$, where $n = 2WT$) can be used interchangeably without loss of information. Furthermore, the samples will be statistically independent, so that the joint probability density function of all n samples is the product of the n individual density functions:

$$p(x) = \prod_{i=1}^{n} p(x_i) \tag{12.9}$$

where x represents the combination of all n values: $x_1, x_2, x_3, \ldots, x_n$.

Assume that an input, $x(t)$, is presented to the receiver. It is not known if this input consists of noise only or signal plus noise. Let $P_N(x)$ be the conditional probability density function under the hypothesis that only noise is present and let $P_{S+N}(x)$ be the conditional probability

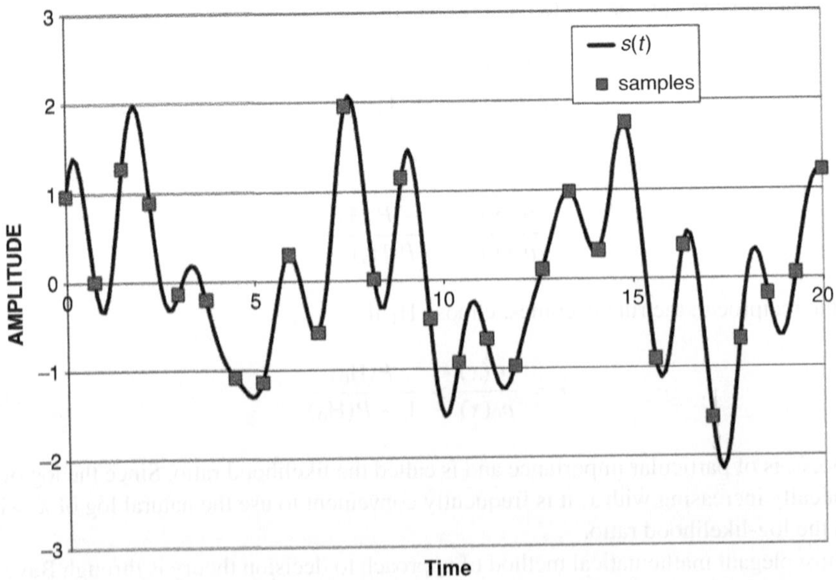

Figure 12.2 Sampling of the signal to meet the Shannon/Nyquist criterion [1]

density function under the hypothesis that signal plus noise is present. Each of these functions is a product of individual density functions, as in Equation 12.9. The likelihood ratio is

$$\lambda(x) = \frac{p_{S+N}(x)}{p_N(x)} \tag{12.10}$$

An optimum receiver must be able to compute the likelihood ratio corresponding to any input wave. The receiver is provided with a threshold, λ_0, such that the decision is "yes" or H_1 whenever $\lambda(x) \geq \lambda_0$ and "no" or H_0 otherwise.

Assuming that the signal is combined with white Gaussian noise (white indicating that the spectrum level is independent of frequency over the band W and Gaussian signifying that each individual sample has a Gaussian (or normal) distribution):

$$p_N(x_i) = \frac{1}{\sqrt{2\pi\sigma^2}} e^{-x_i^2/(2\sigma^2)} \tag{12.11}$$

where σ^2 is the variance of x_i, which is proportional to the noise power. This also assumes that there is no DC component to the noise (i.e. a mean of zero).

The joint probability density function of the noise only hypothesis, H_0, is

$$\begin{aligned} p_N(x) &= \prod_{i=1}^{n} \frac{1}{\sqrt{2\pi\sigma^2}} e^{-x^2/(2\sigma^2)} \\ p_N(x) &= \frac{1}{(2\pi\sigma^2)^{n/2}} e^{-1/(2\sigma^2)\sum_{i=1}^{n} x_i^2} \end{aligned} \tag{12.12}$$

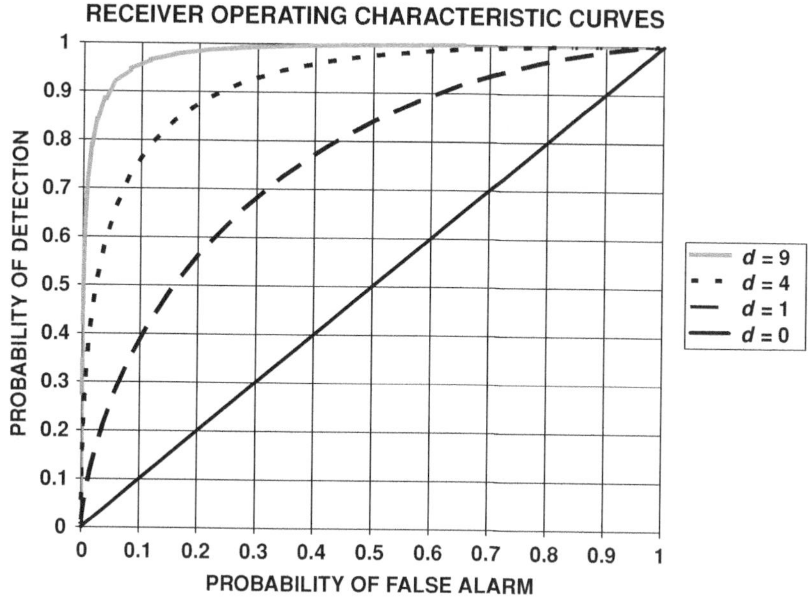

Figure 12.3 Receiver operating curve

Obtaining a closed-form solution for the likelihood ratio is very difficult except under the simplest assumptions. We will consider two simple, yet important, cases. The first case is when the signal is known exactly and the second case is where the signal consists of white Gaussian noise. To better understand these cases, we need to review some important terminology. Receiver operating characteristic curves, or ROC, were originally published by Peterson and Birdsall [2] to show the relationship between probability of detection and probability of false alarm as the detection threshold is decreased (moves from left to right) for various values of d (the detection index defined below). The original curves were published with normal or Gaussian probability on the horizontal and vertical scales. In the form presented in Figure 12.3, these curves are not very useful because the desired probability of false alarm is very low and the scale used does not allow the determination of operating points of interest.

12.2 Case 1: Signal Is Known Exactly

Let x_i be the voltage of the ith sample, s_i be the contribution of the signal, and n_i be the contribution of the noise:

$$x_i = s_i + n_i \tag{12.13}$$

If the signal is known exactly, then each value of s_i is known and only the noise contributes to the randomness of x_i. The probability density function for the hypothesis that the signal is present is

$$p_{S+N}(x) = \prod_{i=1}^{n} \frac{1}{\sqrt{2\pi\sigma^2}} e^{-n_i^2/(2\sigma^2)}$$
$$p_{S+N}(x) = \prod_{i=1}^{n} \frac{1}{\sqrt{2\pi\sigma^2}} e^{-(x_i-s_i)^2/(2\sigma^2)} \tag{12.14}$$
$$p_{S+N}(x) = \frac{1}{(2\pi\sigma^2)^{n/2}} e^{-1/(2\sigma^2)\sum_{i=1}^{n}(x_i-s_i)^2}$$

The likelihood ratio is

$$\lambda(x) = \frac{\left[\frac{1}{(2\pi\sigma^2)^{n/2}} e^{1/(2\sigma^2)\sum_{i=1}^{n}(x_i-s_i)^2}\right]}{\left[\frac{1}{(2\pi\sigma^2)^{n/2}} e^{1/(2\sigma^2)\sum_{i=1}^{n}(x_i)^2}\right]} \tag{12.15}$$
$$\lambda(x) = e^{-1/(2\sigma^2)\sum_{i=1}^{n} s_i^2} e^{-1/\sigma^2 \sum_{i=1}^{n} s_i x_i}$$

The first term in Equation 12.15 is a known constant since the values of s_i are known. The second term shows us what process an optimum processor should perform; it is the summation in the second exponential.

From sampling theory:

$$\frac{1}{\sigma^2}\sum_{i=1}^{n} s_i x_i = 2\frac{W}{\sigma^2}\int_0^T s(t)x(t)\,dt \tag{12.16}$$

This is the correlation function of the input $x(t)$ and the known signal, $s(t)$.

The s_i^2 terms in Equation 12.15 are the square of a sample of the signal, which is proportional to the signal power. With $m = 2WT$ samples, the average signal power S is

$$S = \frac{k}{2WT}\sum_{i=1}^{n} s_i^2 \tag{12.17}$$

where k is a proportionality constant, a reciprocal of impedance, and the average noise power, N, is

$$N = \frac{k}{2WT}\sum_{i=1}^{n} n_i^2 = k\sigma^2 \tag{12.18}$$

The signal energy is

$$E = ST = \frac{k}{2W}\sum_{i=1}^{n} s_i^2 \tag{12.19}$$

The exponent in the first term of Equation 12.15 can then be rewritten as

$$\frac{1}{2\sigma^2}\left[\sum_{i=1}^{n} s_i^2\right] = \frac{WE}{\sigma^2 k} = \frac{WE}{N} = \frac{E}{N_0} \tag{12.20}$$

where N_0 is proportional to the noise power per unit bandwidth.

We now will set a threshold value, β, for $\lambda(x)$. This is equivalent to setting a threshold α such that

$$\beta = \mathrm{e}^{-(E/N_0+\alpha)} \tag{12.21}$$

The probability of false alarm is then the probability that

$$\frac{1}{\sigma^2}\sum_{i=1}^{n} s_i x_i \geq \alpha \tag{12.22}$$

Using the hypothesis that x_i contains noise only, set $x_i = n_i$, which are by assumption independent, zero-mean Gaussian random variables with variance σ^2. Given that the values for s_i are known constants, the mean and variance of $s_i x_i$ in Equation 12.20 are

$$\begin{aligned} \mu(s_i x_i) &= E(s_i x_i) = 0 \\ \sigma^2(s_i x_i) &= E\left[(s_i n_i)^2\right] - \mu^2(s_i x_i) = E\left(s_i^2\right)\sigma^2 = s_i^2\sigma^2 \end{aligned} \tag{12.23}$$

Therefore, the sum in Equation 12.20 is a Gaussian random variable with a zero mean and a variance of

$$\frac{1}{\sigma^2}\sum_{i=1}^{n} s_i^2 = 2\frac{E}{N_0} \tag{12.24}$$

The probability of false alarm is

$$P(\mathrm{fa}) = \sqrt{\frac{N_0}{4\pi E}}\int_{\alpha}^{\infty} \mathrm{e}^{-N_0/(4E)\,y^2}\,\mathrm{d}y = \frac{1}{\sqrt{2\pi}}\int_{\sqrt{\frac{N_0}{2E}}\alpha}^{\infty} \mathrm{e}^{-z^2/2}\,\mathrm{d}z \tag{12.25}$$

where

$$y = \frac{1}{N}\sum_{i=1}^{n} x_i^2 \tag{12.26}$$

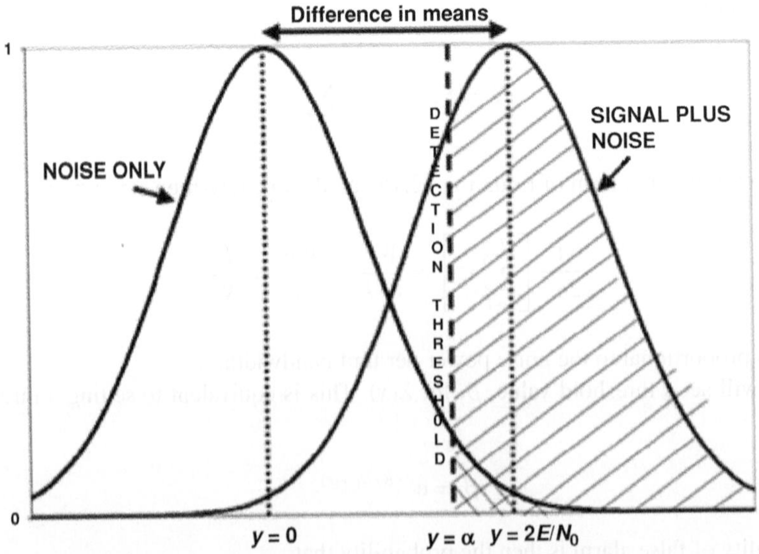

Figure 12.4 Plot of the probability density functions for H_0 and H_1

The probability of detection is the probability that

$$\frac{1}{N}\sum_{i=1}^{n} s_i x_i > \alpha \tag{12.27}$$

under the hypothesis, H_1, i.e., signal in addition to noise is present:

$$\mathrm{P(d)} = \sqrt{\frac{N_0}{4\pi E}}\int_{\alpha}^{\infty} \mathrm{e}^{-N_0/(4E)(y-(2E)/N_0)^2}\,\mathrm{d}y = \frac{1}{\sqrt{2\pi}}\int_{[\sqrt{(N_0/(2E))}\alpha-\sqrt{((2E)/N_0)}]}^{\infty} \mathrm{e}^{-z^2/2}\,\mathrm{d}z \tag{12.28}$$

Figure 12.4 shows the probability density functions for both distribution noise only (H_0) and for signal plus noise (H_1). These distributions are Gaussian with variance $2E/N_0$. The noise only distribution has a zero mean and the signal plus noise distribution has a mean of $2E/N_0$. Values of y to the right of α result in a decision of "yes, signal is present" and to the left "no, signal is not present." The shaded area is the probability of detection if the signal is present and the double shaded area is the probability of false alarm if only noise is present.

The difference in the means for these two distributions is key to evaluating the ability to detect. This difference is affected by the width of the distributions. The parameter d is defined as

$$d = \frac{(\text{Difference of means})^2}{\text{Variance}} \tag{12.29}$$

This value is called different names by different authors, including detection index, deflection signal to noise, and processor output signal to noise. In psychoacoustics, the square root of this value is frequently used:

$$d = \frac{\left(2\frac{E}{N_0} - 0\right)^2}{2\frac{E}{N_0}} = 2\frac{E}{N_0} \tag{12.30}$$

Using Equation 12.17, this can be rewritten as

$$d = 2WT\frac{S}{N} \tag{12.31}$$

Solving this equation for *S/N* gives the input signal to noise required to achieve a level of performance specified by the parameter, *d*, as

$$\frac{S}{N} = \frac{d}{2WT} \tag{12.32}$$

or in decibels as

$$\mathrm{SNR(dB)} = 10\log\left(\frac{S}{N}\right) = 10\log\left(\frac{d}{2WT}\right) \tag{12.33}$$

Rewriting Equations 12.21 and 12.22 and substituting $2E/N_0$ for d yields

$$\begin{aligned} P(\mathrm{d}) &= \frac{1}{\sqrt{2\pi}}\int_{(\alpha-d)/\sqrt{d}}^{\infty} \mathrm{e}^{-z^2/2}\,\mathrm{d}z \\ P(\mathrm{fa}) &= \frac{1}{\sqrt{2\pi}}\int_{\alpha/\sqrt{d}}^{\infty} \mathrm{e}^{-z^2/2}\,\mathrm{d}z \end{aligned} \tag{12.34}$$

The choice of $\alpha/\sqrt{d}$(use vinc) determines the probability of false alarm. Similarly, if $\alpha = d$ then the probability of detection is 50 %. This signal to noise is called the recognition differential (Nrd or RD) or the detection threshold (DT) (see the beginning of Chapter 13, Methodology for the Calculation of Recognition Differential, for a discussion of a stricter definition of these terms). Since α is a constant chosen to set the false alarm rate, the recognition differential can be written in dB as

$$\text{Nrd or RD} = \mathrm{DT} = \mathrm{SNR}(P(\mathrm{d}) = 0.5) = 10\log\left(\frac{\alpha}{2WT}\right) \tag{12.35}$$

This is the signal to noise input power to the processor required to achieve a 50 % probability of detection for a known signal of bandwidth W and duration T, given the specified probability of false alarm corresponding to the α chosen. In some literature, Nrd is defined, not in terms of power, but as 10 log (E/N_0) where $E = ST$. Here, the numerical value differs by 10 log(T)

or 20 dB for a 10 ms signal. Thus, as with any scientific equation, it is important to know the units in use and to balance them accordingly or else large errors will result.

12.2.1 *Observations on Case 1*

It is important to observe from the analysis above that false alarms are caused by random fluctuations in the noise field, not by other signals that are not the signal of interest. For example, when searching for submarines with an active sonar, the false alarm rate can be set by setting a threshold. A whale might be the source of a signal that looked enough like a submarine to cause a threshold crossing. Whales are not random noise; they are false contacts. This is an important distinction. The parameter α determines the probability of a false alarm, but the number of whales per square mile determines the probability of getting a whale false contact. Also, the probability of false alarm is being calculated for a single look direction and for a time interval equal to the length of the signal starting at precisely the moment the signal arrives. An active search sonar operator will have many ranges, bearings, and beam bins to search, so the false alarm rate of the system must be calculated as $1 - (1 - P(\text{fa}))^m$, where m is the number of choices needing to be made in some period of time.

For example, let us assume one false alarm per four hours could be tolerated from a system with 100 beams and a range resolution of 100 (200 yd intervals out to 20 000 yd). If the system pings once per round trip travel time (about once per 25 seconds), then in four hours there will be

$$\begin{aligned}&(100\text{ beams} \times 100\text{ range bins} \times 4\text{ h} \times 3600\text{ s/h})/(25\text{s/ping}) \\ &= 5.76 \times 10^6\text{ opportunities for a false alarm}\end{aligned} \tag{12.36}$$

A probability of false alarm of 1.7×10^{-7} is needed to achieve the desired system goal. In this example, the possibility of Doppler shift in the signal requiring searching an additional dimension of speed bins has been ignored.

The processor here *knows* what the signal without noise looks like. This is called coherent processing and is frequently implemented by assuming the signal received looks exactly like the signal transmitted. However, in active sonar, multiple paths of arrival are common, each with a different travel time and the possibility of a phase shift due to interaction with the interfaces. Such a received signal will not match the transmitted signal well, resulting in poorer than expected performance, and in extreme cases, performance even worse than incoherent processing where the assumption is that the signal consists of random noise.

12.3 Case 2: Signal Is White Gaussian Noise

In Case 1, it was assumed that the maximum amount of information was known about the signal. Case 2 is the other extreme. Here, it is assumed nothing is known about the signal except its statistical properties. The signal and the noise consist of band-limited white Gaussian noise (white indicating that the spectrum level is independent of frequency over the band W

and Gaussian signifying that each individual sample has a Gaussian (or normal) distribution). From Equation 12.11:

$$p_N(x_i) = \frac{1}{\sqrt{2\pi\sigma^2}} e^{-x_i^2/(2\sigma^2)} = \frac{1}{\sqrt{2\pi N}} e^{-x_i^2/(2N)} \tag{12.37}$$

where $\sigma^2 = N$ is the variance of x_i, which is proportional to the noise power. This also assumes no DC component to the noise (i.e. a mean of zero). This is the probability density distribution under the assumption of noise only and is identical to that in Case 1.

If the signal is present, each sample, x_i, will include a sample, s_i, of the signal and a sample, n_i, of the noise. Each s_i is a Gaussian random variable with zero mean and variance, S, where S is proportional to signal power. The sum of two Gaussian random variables is a Gaussian random variable with a mean equal to the sum of the means (here, zero). The variance is the sum of the variances (here, $N + S$). The joint probability density function is

$$p_{S+N} = \frac{1}{[2\pi(S+N)]^{n/2}} \mathrm{e}^{-1/[2(S+N)] \sum_{i=1}^{n} x_i^2} \tag{12.38}$$

The likelihood ratio is given by

$$l(x) = \frac{(2\pi N)^{n/2}}{[2\pi(S+N)]^{n/2}} \frac{\mathrm{e}^{-1/[2(S+N)] \sum_{i=1}^{n} x_i^2}}{\mathrm{e}^{-1/[2(N)] \sum_{i=1}^{n} x_i^2}}$$
$$l(x) = \left(\frac{N}{S+N}\right)^{n/2} \mathrm{e}^{S/(2N(S+n)) \sum_{i=1}^{n} x_i^2} \tag{12.39}$$

The optimum receiver must than calculate the sum of the squares of x_i. This is called a square-law integrator and measures the total energy of the input. All other components of the likelihood ratio are constants. Therefore, the decision criterion is

$$l(x) \geq \beta \tag{12.40}$$

which can also be expressed in terms of α such that

$$\frac{1}{N} \sum_{i=1}^{n} x_i^2 \geq \alpha^2 \tag{12.41}$$

where α is defined by

$$\beta = \left(\frac{N}{S+N}\right)^{n/2} \mathrm{e}^{S\alpha^2/[2(S+N)]} \tag{12.42}$$

As an aside, let us consider what results when we sum the squares of independent Gaussian random variables. First, the density function for x_i^2 is

$$y_i = x_i^2 \tag{12.43}$$

$$p(x) = \frac{1}{\sqrt{2\pi}} e^{-x_i^2/2}$$
$$p(y) = \frac{dx}{dy} 2p(x) \tag{12.44}$$

where the number 2 represents the multiplicity of x with respect to y, i.e., two x's to each y, one positive and one negative. This gives

$$y_2 = \sqrt{2\alpha} - \sqrt{2n-1} \tag{12.45}$$

$$p(y) = \frac{1}{2x} 2p(x) \tag{12.46}$$

or

$$p(y) = \frac{1}{\sqrt{2\pi y}} e^{-y/2} \tag{12.47}$$

This is a chi-square distribution with one degree of freedom. For the sum of x_i^2, the result is

$$p(s) = \frac{1}{\left(\sigma^2\right)^{n/2} 2^{n/2} \Gamma(n/2)} s^{n/2-1} e^{-s/(2\sigma^2)} \tag{12.48}$$

This is the gamma distribution, also known as the chi-square distribution with n degrees of freedom. For the special case of $n = 2$, this reduces to the exponential distribution.

Returning to the case at hand, the probability of detection and the probability of false alarm are given by

$$P(\mathrm{d}) = \frac{1}{\sqrt{2\pi}} \int_{y_1}^{\infty} e^{y^2/2}\, dy \tag{12.49}$$

where

$$y_1 = \sqrt{\frac{2N}{S+N}\alpha} - \sqrt{2n-1} \tag{12.50}$$

and

$$P(\mathrm{fa}) = \frac{1}{\sqrt{2\pi}} \int_{y_2}^{\infty} e^{-y^2/2}\, dy \tag{12.51}$$

where

$$y_2 = \sqrt{2}\alpha - \sqrt{2n-1} \quad (12.52)$$

The value of α can be determined if the bandwidth and desired probability of false alarm are known. For example, consider a 1000 Hz band with a processing time of 30 s ($n = 2 \times 1000 \times 30 = 60\,000$) and a desired probability of false alarm of 0.0001. To achieve a probability of false alarm of 0.0001, y_2 must be about 3.719 47 (see normal probability tables); solving for α yields 247.578. The definition of y_1 can then be used to determine the required signal to noise for a given P(d). For example, if $P(\mathrm{d}) = 0.5$, then $y_1 = 0$ and the required signal to noise is

$$\begin{aligned}\frac{S}{N} &= \frac{2\alpha^2}{2n-1} - 1 = \frac{y_2^2 + 2\sqrt{2n-1}y_2}{2n-1} \\ &= 0.0216 \text{ or } -16.66 \text{ dB}\end{aligned} \quad (12.53)$$

Let us assume that $\sqrt{2n-1} \gg y_2$, which is equivalent to assuming the signal to noise is small compared to one ($S/N \ll 1$). Under this assumption, the variance of the noise only case and the signal to noise case are nearly equal. (Even for a very low P(fa) of 1.28×10^{-12}, y_2 is only 7 and $n = 2WT$ would need to be $\gg 5$ [3]. For most cases this is an excellent assumption.) The required signal to noise for a 50 % probability of detection with a false alarm probability specified by y_2 is approximately given by

$$\frac{S}{N} = \frac{2y_2}{\sqrt{2n}} \quad (12.54)$$

This signal to noise is called the recognition differential (Nrd or RD) or the detection threshold (DT). The recognition differential can be written in dB as

$$\text{Nrd or RD} = \text{DT} = \text{SNR}(P(\mathrm{d}) = 0.5) = 5\log\left(\frac{d}{WT}\right) \quad (12.55)$$

where d must be $y_2{}^2$.

This type of processing is called incoherent processing because only the average properties of each component are known, not the detailed structure. If all other factors were equal (bandwidth, duration, and desired false alarm rate), coherent processing, if possible, will yield a better performance because the Nrd or RD for coherent processing improves as $-10\log(WT)$ or $-5\log(WT)$ for incoherent processing (provided that $WT > 1$). For the latter case, $WT < 1$ really means that the signal cannot be represented in any way other than statistically.

12.3.1 Observations on Case 2

It is important to observe from the analysis above that false alarms are caused by random fluctuations in the noise field, not by other signals that are not the signal of interest. For example, when searching for submarines with a passive sonar, the false alarm rate can be set by setting a threshold. A surface ship might be the source of a signal that looked enough like a submarine to cause a threshold crossing. Surface ships are not random noise; they are false contacts. This is an important distinction. The parameter a determines the probability of a false alarm, but the number of surface ships per square mile determines the probability of getting a surface ship false contact.

References

[1] Shannon, C. E., "The Mathematical Theory of Communication", *Bell Systems Technical Journal*, **26**, July and October, 1948, 379–424 and 623–637,

[2] Peterson, W. W., and Birdsall, T. G., "The Theory of Signal Detectability," University of Michigan Engineering Research Institute, Report 13, 1953.

[3] Abramowitz, M., and Stegun, I. A. (Eds), *Handbook of Mathematical Functions, with Formulas, Graphs, and Mathematical Tables*, New York: Dover Publications, 1 June 1965.

13

Methodology for Calculation of the Recognition Differential

In Chapter 12, Statistical Detection Theory, we computed the theoretical detection threshold (DT) or recognition differential (Nrd or RD) for two special cases. The assumptions made for these calculations may not hold true in real world systems and situations. To overcome this, it is necessary to develop a methodology for taking into account differences between these theoretical recognition differentials and what might be expected in the real world.

To facilitate this discussion, a quick review of key definitions is needed. Urick, in *Principles of Underwater Sound for Engineers* [1], defined the detection threshold (DT) as the ratio, in decibels, of the signal power in the receiver bandwidth to the noise power in a 1 Hz band, measured at the receiver input terminals required for detection at the assumed probability (normally 50 %) with the assumed probability of false alarm. Historically, the recognition differential (Nrd or RD) has been defined similarly to DT, except that it is the ratio, in decibels, of the signal power in the receiver bandwidth to the noise power in the receiver bandwidth, instead of referenced to a 1 Hz band. If the noise is flat over the band of interest, the two terms are related by $\mathrm{DT} = \mathrm{Nrd} + 10\log(W)$.

In complex systems, with many different bandwidths being processed, the specifications for recognition differentials are frequently written with respect to a reference bandwidth that may not be equal to any of the actual processor bands. This allows for easy comparison of the two terms without the need to correct for bandwidth.

In this chapter, we will consider four signal types:

1. Continuous broadband signals (passive broadband, or PBB).
2. Continuous narrowband signals (passive narrowband, or PNB), either relatively pure sine waves or signals whose bandwidth is small compared to its center frequency.
3. Target echoes resulting from active transmissions (active sonar).
4. Transient signals (aural detection).

Underwater Acoustics: Analysis, Design and Performance of Sonar Richard P. Hodges

For each type of signal, the procedure for calculating the recognition differential has five steps:

1. Computation of a theoretical recognition differential using idealized assumptions, e.g., Gaussian noise and a perfectly understood signal.
2. Correction for nonideal noise characteristics.
3. Correction for nonideal processor and display implementation.
4. Correction for nonideal signal characteristics.
5. Adjustment for additional at-sea losses.

13.1 Continuous Broadband Signals (PBB)

Broadband systems are designed to detect wideband signals in random noise. In particular, we mean signal and noise for which we believe only the power spectral densities are known. If both the signal and noise have amplitude statistics that are Gaussian, then the optimum detection system is a frequency filter, followed by a square law detector, a power averager, and a threshold detector (Figure 13.1). (See Chapter 12, Statistical Detection Theory, Case 2 for the special case of white Gaussian noise.)

If the spectra are known, the optimum shaping filter is

$$|H(f)^2| = S(f)/N(f)^2 \tag{13.1}$$

where

$S(f)$ = input power density for the signal at frequency f
$N(f)$ = input power density for noise at frequency f

This filter weighting is known as an Eckart filter. Since the spectra are generally not known with precision, the filter used is usually an approximation that results in less than optimum performance.

The output of the square law detector device is equal to the sum of the powers of the signal plus noise in the band of the input filter. Due to the randomness associated with the inputs, there are large fluctuations in time. The power averager device, also referred to as an integrator, provides a running average of the power for a time, T, to the threshold detector. The threshold detector tests the output of the power averager with a fixed threshold, declaring a signal when the threshold is exceeded. This is a statistical test with the value chosen to reflect the desired false alarm rate when no signal is present. As we saw in Chapter 12, this threshold is a function of the expected output when only noise and the desired false alarm rate are present.

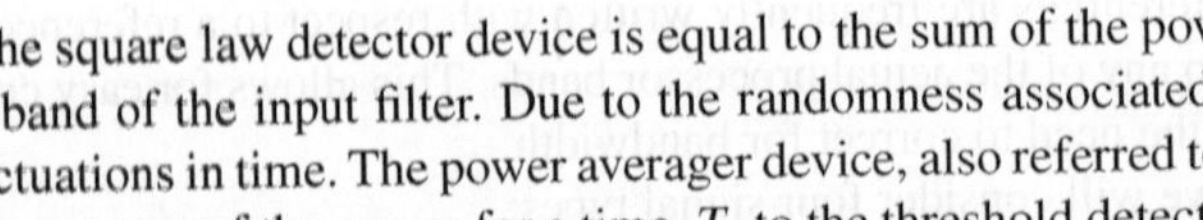

Figure 13.1 Basic broadband processor

13.1.1 PBB Step 1: Theoretical Broadband Nrd

Assuming a flat or white Gaussian signal and noise over a band, W, for a time, T:

$$\text{Nrd} = -5\log(WT) + D \tag{13.2}$$

where D is the detection index defined in Chapter 12, expressed in dB. The detection index is a function of the probability of false alarm ($P(\text{fa})$) and the number of independent samples ($n = 2WT$).

As seen in Chapter 12, the probability of false alarm is given by

$$P(\text{fa}) = \frac{1}{\sqrt{2\pi}}\int_{y_2}^{\infty} \mathrm{e}^{-y^2/2}\,\mathrm{d}y \tag{13.3}$$

where

$$y_2 = \sqrt{2\alpha} - \sqrt{2n-1} \tag{13.4}$$

The required signal-to-noise ratio or Nrd is given by

$$\frac{S}{N} = \frac{2\alpha^2}{2n-1} - 1 = \frac{y_2^2 + 2\sqrt{2n-1}\,y_2}{2n-1} \tag{13.5}$$

For large values of n ($\ll {y_2}^2$), this becomes

$$\frac{S}{N} = \frac{2y_2}{\sqrt{2n}} = \frac{y_2}{\sqrt{WT}} \tag{13.6}$$

$$Nrd = 10\log\left[\frac{y_2}{\sqrt{WT}}\right] = 10\log(y_2) - 5\log(WT) \tag{13.7}$$

The value of y_2 can be determined given the probability of the false alarm rate. This is done by using Microsoft Excel© or by using reference tables such as those of Abramowitz and Stegun [2]. Figure 13.2 shows the detection index, D, for a Gaussian signal and noise versus probability of false alarm assuming that the bandwidth time, WT, is large. The receiver operating curves for these signals are given in Figure 13.3.

13.1.2 PBB Step 2: Correction for Noise Spectrum

The assumption in Step 1 is that the noise spectrum is flat over the bandwidth of interest and zero elsewhere. The bandwidth is simply the difference between the upper and lower limits of frequencies. Since the noise spectrum is virtually never flat over the bandwidth of interest, it is necessary to determine the actual or effective noise bandwidth, and compute the resulting

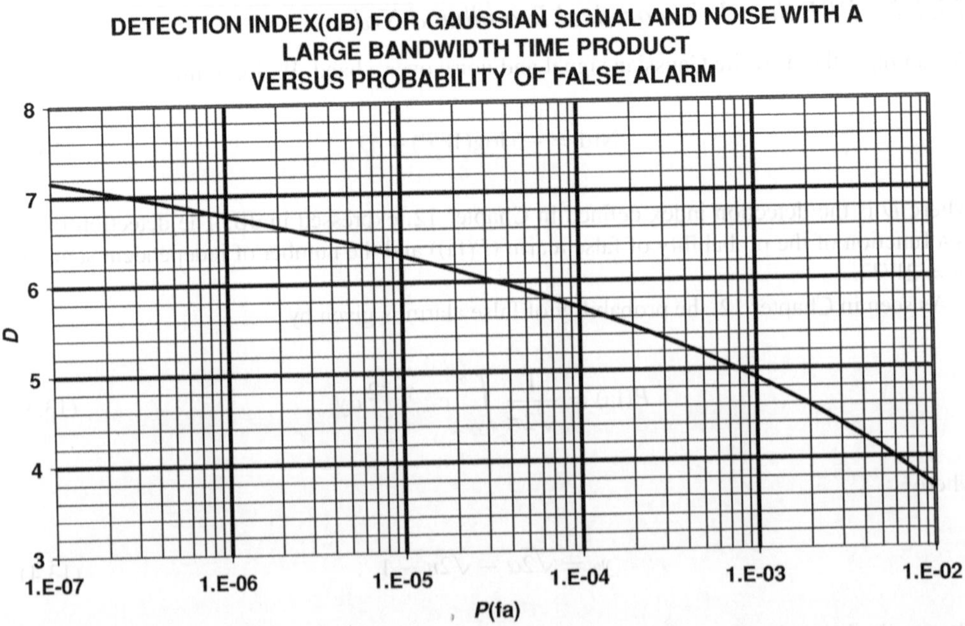

Figure 13.2 Detection index for a Gaussian signal and noise with a large bandwidth versus probability of false alarm

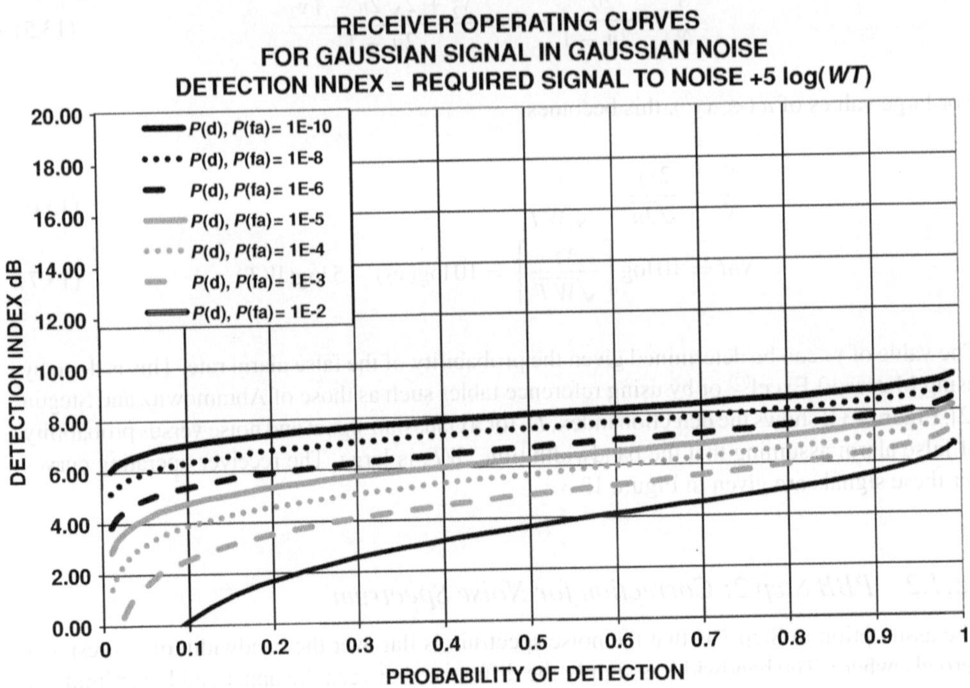

Figure 13.3 Receiver operating curves for Gaussian signal and noise with a large bandwidth

difference in Nrd. The effective bandwidth, W_{eff}, is given by

$$W_{\text{eff}} = \frac{\left[\int_0^{\infty} H(f)N(f)\,\mathrm{d}f\right]^2}{\int_0^{\infty} H(f)^2 N(f)^2\,\mathrm{d}f} \tag{13.8}$$

The adjustment to Nrd is

$$\Delta\text{Nrd} = 5\log\left(\frac{W}{W_{\text{eff}}}\right) + D(W_{\text{eff}}) - D(W) \tag{13.9}$$

As an example, consider a 1 to 2 kHz frequency band. Recalling from Chapter 7, Ambient Noise, that sea state noise is not flat, but slopes downward about 6 dB per octave or, in power terms, one over frequency squared. In Chapter 3, Transducers, Directionality, and Arrays, we saw that area arrays have a directivity index that increases by 6 dB per octave. Combining these two would indicate that the noise coming out of the beamformer would slope downward at 12 dB per octave or, in power terms, one over frequency to the fourth. Assuming that a flat bandpass filter was used, $H(f) = 1$ from 1–2 kHz and $H(f) = 0$ elsewhere. The effective noise bandwidth is given by

$$W_{\text{eff}} = \frac{\left[\int_{1000}^{2000} f^4\,\mathrm{d}f\right]^2}{\int_{1000}^{2000} f^8\,\mathrm{d}f} \tag{13.10}$$

This is about 600 Hz instead of the 1000 Hz for a flat signal. The correction to the recognition differential, assuming $T = 1$ s, is

$$\begin{aligned} \Delta\text{Nrd} &= 5\log\left(\frac{W}{W_{\text{eff}}}\right) + D\,(W_{\text{eff}}) - D(W) \\ &= 5\log\left(\frac{1000}{600}\right) + 6.374 - 6.335 = 1.148\ \text{dB} \end{aligned} \tag{13.11}$$

It is important to remember from Chapter 12 that false alarms are caused by random fluctuations in the noise field, not by other signals not of interest. For example, when using sonar to search for submarines, the false alarm rate is established by setting a threshold. Unfortunately, a whale might cause a signal that looks enough like a submarine to cause a threshold crossing. Whales are not random noise; they are false contacts. This is an important distinction. D is determined by the probability of false alarm, but the number of whales per square mile determines the probability of getting a whale false contact. The probability of false alarm is calculated for a single look direction and for a time interval equal to the length of the signal beginning at the precise moment the signal arrives. Normally, a search sonar operator will be looking at many different bearings and time intervals, so the false alarm rate of the system must be calculated as $1 - (1 - P(\text{fa}))^m$, where m is the number of choices needed to be made in a period of time.

13.1.3 PBB Step 3: Correction for Processor Implementation

As noted above, the optimum detection system is a frequency filter, followed by a square law detector, a power averager, and a threshold detector. The effects of the filter will be considered in Step 2 and Step 4. In Step 3, we need to adjust for other factors that cause the process to deviate from the optimum. This includes detectors, other than square law, the use of displays or an averager and thresholder, the requirement to estimate background noise, the possibility of OR-ingOR-ing, and many other possible implementation choices.

13.1.3.1 Correction for Detector Type

The equation for D assumes a square law detector, but many systems adopt more easily implemented detectors, so corrections must be applied.

Linear detectors produce an output that is proportional to the mean absolute value of its input. For large time–bandwidth products, the difference has been determined to be approximately 0.19 dB degraded [3].

Correlation detectors multiply inputs from two sources. These detectors are available in two types: auto- and cross-correlators. The autocorrelator takes inputs from the same source and time delay before multiplication. This type of processing is frequently used to detect multipath arrivals. The cross-correlator takes two different sources of inputs, frequently two halves of an acoustic aperture or two separate arrays. When the noise from the two sources is independent and the signal is identical, the correlator outperforms the square law detector by 1.5 dB. However, it should be noted that the signal to noise is 3 dB worse than if the two halves of the array had been added together. Thus, despite the Nrd being better, the performance is 1.5 dB worse. Many choices of time delay are displayed for autocorrelation, thus determination of the Nrd requires careful analysis of the number of display bins generated and their effect on the probability of false alarm (P(fa)).

13.1.3.2 Displays and Post-Detection Integration

Step 1 assumes a perfect running average over the last T seconds of the detector output. This is the optimum when detection is required for a signal lasting T seconds. For many reasons, this is rarely implemented for passive sonar. The analysis for the theoretical system proceeds as though there is only one detection channel. Realistic systems frequently have many beams and perhaps several focusing ranges per beam. The target's detectable signal duration is not known and it may change beams and level as it moves. Therefore, it is desirable to have displays that take these factors into account.

In World War II and earlier, the data were supplied to an operator aurally who swept the beam around to perform a search. While the human ear is a remarkable signal processor, it

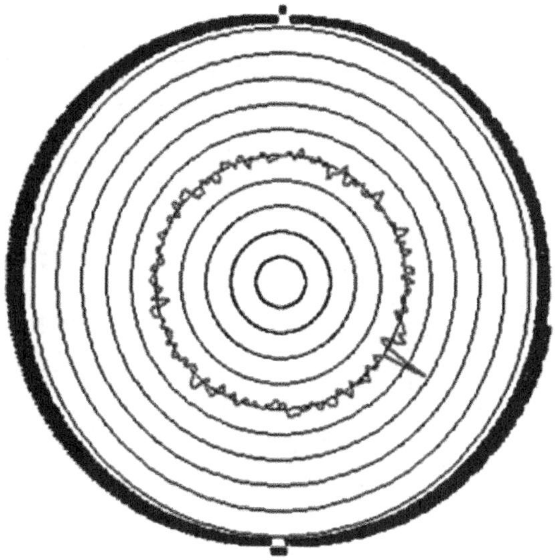

Figure 13.4 A-scan display, output versus look direction, with a target at about 122°

suffers from several defects:

1. It can only listen in one beam at a time.
2. Its memory or integration time is limited to about 1 second.
3. The bandwidth is wider than optimal for some signals.
4. There is a great deal of variability among operators.

Today, visual displays and computer-aided processes are used as the primary means of detection, with aural used as a confirmation and classification tool. The first visual displays for sonar were similar to radar displays. For example, an A-scan (see Figure 13.4) displayed a circle if no energy was coming in and deflections represented the level of energy or correlation versus direction. The radial component was the signal or noise deflection and the angular dimension was the look direction. Typically, these systems swept in bearings at a selectable rate. What was displayed on the screen usually represented a time constant or integration time matched to the scan rate. To avoid spreading the signal across several beams, the integration time would be set to the time a target would spend in a beam during one sweep. Because it was desirable to obtain an all-bearing look frequently, the sweep rates were measured in seconds to a minute, making the time on target, or signal time, the sweep time multiplied by the beam width divided by 360°. For a one minute sweep with a 6° beam, the signal time would be 6 s:

$$T_{\text{signal}} = T_{\text{sweep}} \times \text{Beamwidth}/360 = 60\,\text{s} \times 6^\circ/360\,\text{s} = 6\,\text{s} \qquad (13.12)$$

In addition, the corresponding integration time would be 6 s. The effect of a failure to match T_{signal} to the integration time will be discussed below.

Modern displays strive to allow for longer visual integration times. Typical broadband displays are referred to as bearing time displays or bearing time recorders (BTR). These are

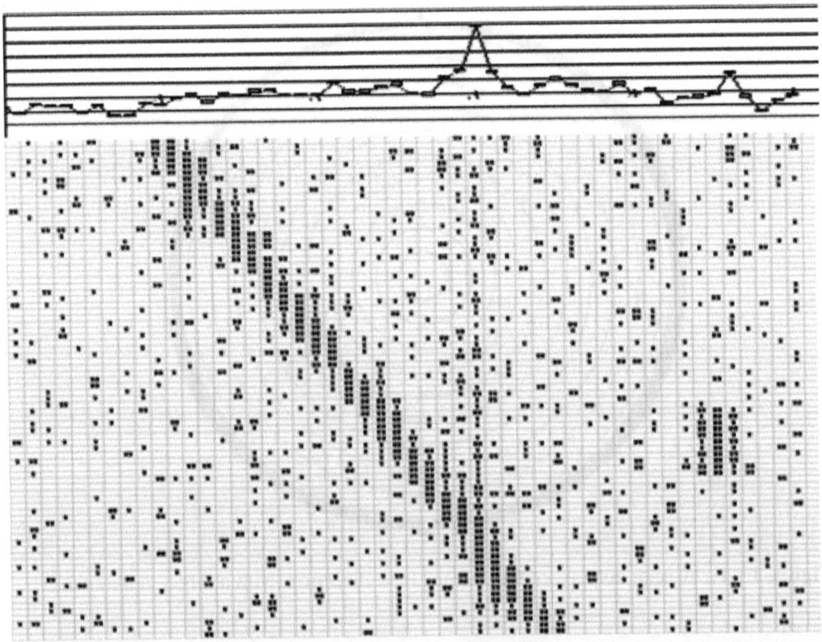

Figure 13.5 Example of a broadband display

typically rectangular displays, also called grams, DIMUS (a misnomer since DIMUS stands for DIgital MUltibeam Steering), or waterfall displays, because data flows down the screen like water with the bottom data eliminated as a new set is inserted at the top (see Figure 13.5). The bottom portion of this figure is a B-scan, BTR, waterfall, or DIMUS display that shows time vertically, bearing horizontally, with the intensity or a gray scale coloring representing the energy or correlation level. This technique allows targets to be seen over a period of time and allows for the visual integration of multiple lines to achieve a large processing time, T. On the top of Figure 13.5 is an A-scan or Automatic Line Integration Threshold (ALIT) display, which is the sum of a vertical row minus the screen mean divided by the screen standard deviation. This type of display allows operators visually to match the processing time to the signal duration, within the limits of the averaging time used for each pixel (T_p), to the total time displayed (nT_p), where n is the number of lines of data displayed. Older scanned systems marked a line at a given time so that it was many of the A-scan lines presented all together. Modern sonars, with pre-formed beams, allow continuous monitoring of all directions simultaneously. The advantage of pre-formed beams is that the signal time processed is equal to the total time, not just that fraction of time the scanning beam is on the target. For the example above, this means 60 s is available instead of 6 s, or a gain of 5 log(60/6) or 5 dB.

Occasionally, a sum of amplitude display (ALIT display) is added to the bearing time display, as is shown at the top of Figure 13.5. This is simply the sum of the energy displayed in each of the lines directly below it, similar to summing the vertical columns of an Excel© spreadsheet. If the target bearing is constant, then we the integration time is equal to the sum of each row.

A perfect "boxcar" integrator that exactly matches the signal duration is about 2.5 dB better than a good BTR display [4]. This is primarily because the BTR allows as many false alarm choices as possible. Some gram displays have a limited dynamic range, few gray scale levels or colors, a one bit gram, i.e., it is on or off, and has 1 dB additional degradation. As can be seen in Figure 13.5, the steady signal in bearing and level just to the right of center is very obvious on the ALIT display yet faintly visible on the gram display. However, the other two signals clearly visible on the gram are not obvious on the ALIT display. In one case, a strong short signal below center right was presented. The ALIT summed many cells of noise and only a few of signal. The other strong signal that changed bearing from left to right was smeared over many bearing cells for the ALIT display. Considerations such as this can very quickly give the advantage to a BTR display.

It is easier to implement an exponential integrator than a boxcar integrator. A boxcar integrator has to remember all the values that went into the average and, at each time step, remove the oldest data and insert a new value. Exponential integrators are like a Markov process; at each time step the old value and the new value are weighted and added together to result in a running average:

Boxcar

$$SUM_{\text{boxcar}}(I) = \frac{[mSUM_{\text{boxcar}}(I-1) + LEVEL(I) - LEVEL(I-m-1)]}{m} \tag{13.13}$$

where m is the number of samples in the running average and I represents the latest value:

Exponential

$$SUM_{\text{exp}}(I) = aSUM_{\text{exp}}(I-1) + bLEVEL(I) \tag{13.14}$$

where $a + b = 1$, $a = \exp(-\Delta t/T_{\text{int}})$, T_{int} is the desired integration period, and Δt is the time interval between level measurements. The exponential integrator does not suffer a penalty if the signal has been present for an infinity length of time (an uninteresting case). If the signal is present and steady for one integration period, then at the end of that time, the total signal energy accumulated by the exponential integrator will be $1 - \exp(-1)$, or about 63 % of what the boxcar integrator would have. Given this, the signal-to-noise ratio for the exponential integrator will be down by about 2 dB.

13.1.3.3 Display OR-ing

OR-ing is simply taking several inputs and displaying the largest of them. This reduces the number of pixels needed to display the data. Most broadband systems do not employ OR-ing because the number of channels needing to be displayed easily fits on a modern display. However, if a system had many vertical beams, as well as horizontal beams, then OR-ing might be employed to fit all the data on a single screen. This would result in a loss. If the mean noise level is assumed to be the same for each of the channels to be OR-ed (Section 13.1.3.3), a strong assumption probably implying that a sector space average (SSA) has been applied

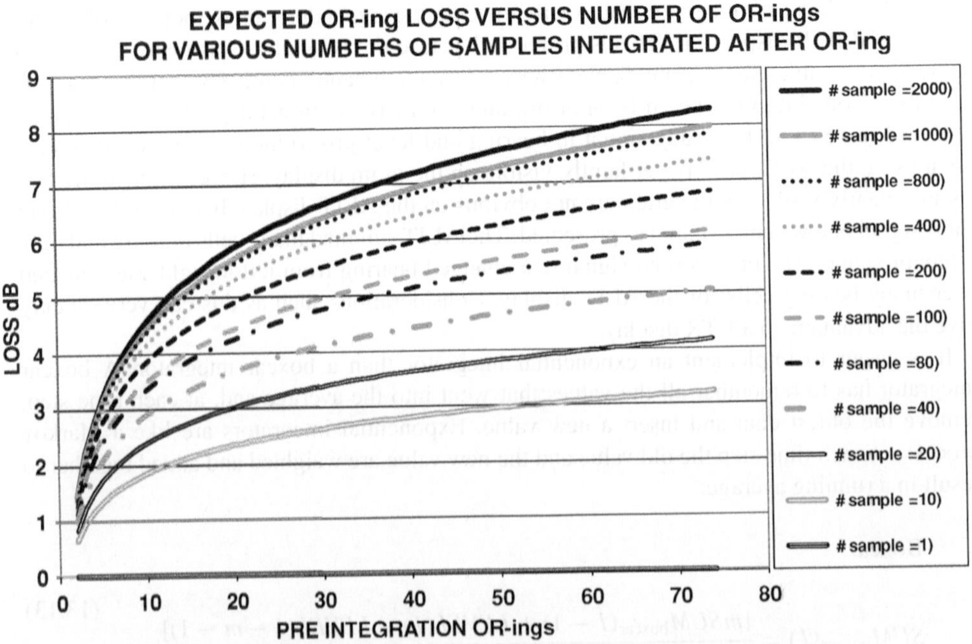

Figure 13.6 OR-ing loss versus number of OR-ings for various sample sizes

across the channels, then the OR-ing loss can be estimated as [5]

$$\Delta\text{Nrd} = 2.164\ln(\text{Number of Channels OR-ed})\frac{x}{\left(1+x^4\right)^{0.25}}$$
$$x = 0.189\frac{\ln(\text{Number of Samples Integrated})}{[1.443\ln(\text{Number of Channels OR-ed})]^{0.125}} \tag{13.15}$$

This equation is valid for pre-integration OR-ing. Post-integration OR-ing is less desirable and is not usually employed. For a discussion of this, see Nuttal [6]. Figure 13.6 shows OR-ing loss versus the number of OR-ings for various sample sizes.

13.1.3.4 Background Noise Estimation

In Chapter 12, we assumed that the noise level was known a priori. In reality, the noise for most real systems varies and must be estimated during the processing. This results in a penalty to the theoretical Nrd. Typically, the noise received by sonar varies with time and with bearing. It is not uncommon for ambient noise to vary by 10 dB in azimuth when shipping level-limited or to vary by a similar amount in the depression/elevation (D/E) angle. To process the data, it is necessary to take out this structure and to estimate the underlying mean noise level. A sector space averager (SSA) is used to accomplish this, by using adjacent beams to estimate the noise. Simply stated, because a statistical estimate of noise has to be used, it is necessary to set the threshold higher to allow for the imprecision with which the noise is known. Since the bandwidth time product is assumed to be large, the precision associated with the noise estimate improves as the square root of the number of samples available to make the estimate. The correction is given by

$$\Delta \text{Nrd} = 5 \log \left(1 + \frac{1}{M} \right) \tag{13.16}$$

where M is the number of independent estimates available.

In practice, there are several methods of accomplishing a sector space averager (SSA). A common implementation is called a split window sliding normalizer. In this process, the noise for a cell is estimated from the average of a number of cells on either side of it, perhaps with additional guard cells to help prevent the signal from affecting the noise estimate.

For example, a 4–3–4 window would estimate the mean of the center cell by excluding that cell and its adjacent cells, averaging the eight cells, four on either side. At the edges of the display, we have to either wrap around if the other edge is the adjacent beam or substitute a one-sided estimate. As we will see later, a similar process is used for narrowband displays, as only adjacent frequency bins are averaged instead of bearings; this is called a noise spectrum equalizer (NSE). This process assumes that the noise field is well behaved, which is frequently not the case for broadband systems, except in the laboratory.

Figure 13.7 shows a display of data before and after the split window sliding normalizer is applied. The improvement seen on the right in Figure 13.7 is caused by the use of a local average for the noise estimate for each cell or a single noise estimate for the whole screen on the left. In this example, there are two broad noisy directions interfering with performance. The same three targets shown on Figure 13.5 are present here and can be seen on the right after the normalizer is applied.

In addition to the simple split window normalizer, many other SSA schemes have been developed, e.g., a two-pass split window with peak shearing. Here the first step is the same as described above, but then each cell is compared to its mean estimate. If the value is above

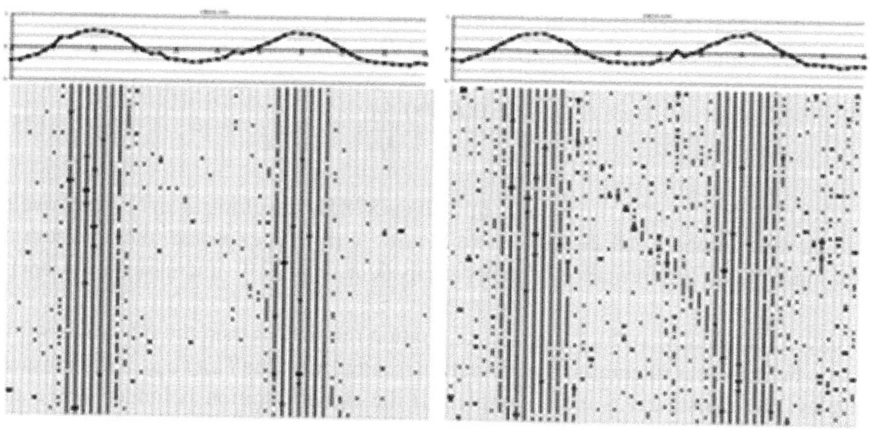

Figure 13.7 Broadband display before (left) and after application of a split window normalizer

the mean by a threshold amount, it is replaced by the mean and the sheared value is stored off for later reinsertion. The split window averager is then applied to the new data, followed by reinserting the sheared values back in their original positions. This process reduces the white banding that will occur around strong signals with a single-pass normalizer.

Some systems do not have multiple beams to play with, such as many sonobuoys. For sonobuoys, we can either use an estimate obtained from prior time, when hopefully the target was not present, or average groups of sonobuoys that are reasonably close together and would be expected to have similar background levels.

13.1.4 PBB Step 4: Correction for Nonideal Signal Characteristics

If we view the recognition differential as the required signal to noise in the band for a 50% probability of detection in one integration period with the specified false alarm rate, then the spectral shape of the signal does not affect the Nrd. However, if we intend to do a single frequency calculation by using the signal and noise at one frequency to represent the band of signal to noise, then the signal spectral shape will affect the Nrd.

Continue with the example from PBB Step 2 of an area array with input noise sloping at 6 dB per octave and with a flat bandpass filter from 1 to 2 kHz. If we assume that the signal has the same shape in the water as the noise, then the signal to noise in the band is given by

$$\frac{\int_0^\infty S(f)H(f)\,\mathrm{d}f}{\int_0^\infty N(f)H(f)\,\mathrm{d}f} \tag{13.17}$$

with $H(f) = 1$ if $1000 < f < 2000$ and 0 otherwise. The signal to noise in power (not dB) is

$$\frac{12\,000\,000}{7}\frac{S_0}{N_0} \tag{13.18}$$

where S_0 is the signal power level and N_0 is the noise power level, both at 1000 Hz.

If, instead, an Ekhart filter was used $[H(F) = S(f)/N^2(F)]$, then the signal to noise would be

$$\frac{186\,000\,000}{7}\frac{S_0}{N_0} \tag{13.19}$$

This is 1.9 dB higher. The effective noise bandwidth is 878 Hz, resulting in an Nrd improvement over the flat filter of about 0.8 dB, for a net gain of 2.7 dB. As expected, the optimum filter provides much better performance, roughly equivalent to a 37 % increase in the aperture size.

The only other adjustment for a signal is a duration that is less than the integration time of the system. For a boxcar integrator, the correction is simply $10 \log(T_{\text{int}}/T_{\text{signal}})$. For a waterfall type display, the adjustment is more complex. If the signal duration is longer than a pixel, but less than the total (T_{int}), the adjustment is approximately $5 \log(T_{\text{int}}/T_{\text{signal}})$. If the signal duration is a single pixel or less, then there may be infinite loss for a single bit display. If many levels of intensity can be displayed, then a detection could be made based on a single very intense pixel. The loss in this case is approximately $5 \log(T_{int}/T_{pixel}) + 10 \log (T_{pixel}/T_{signal})$.

13.1.5 PBB Step 5: Adjustment for Additional At-Sea Losses

The preceding discussion has tried to take into account the spectral and temporal characteristic of both the signal and noise. Underlying assumptions are that they can be described as stationary Gaussian processes. In the operational environment at sea, the noise field may arise largely from distant shipping sources that move with time through the beams of a sonar or through acoustic caustic that cause brief bursts of signal or transients caused by human activity, such as hatch slamming. An analysis of at-sea recorded noise shows a myriad of such effects; i.e., the ambient noise has too much noise in its tails for the Gaussian assumption to be true and it contains many impulsive events. A distant ship is a correlated noise source and does not behave like a distributed source either in the beamformer or in the signal processor. Typical at-sea loss varies widely depending on the frequency region processed and the dominant sources of noise. For systems that are dominated by flow noise or by sea state noise, the corrections required are a few dB. For low-frequency broadband systems, in highly cluttered environments, 10–20 dB loss has been experienced. At-sea measurements are the only reliable method of estimating these losses.

13.2 Continuous Narrowband Signals (PNB)

Narrowband signals are those that have a relatively narrow frequency spread and therefore can be approximated by a sine wave. Generally, we are interested in the detection of these signals in the presence of wide band noise. While we showed in Chapter 12, Statistical Detection Theory, that the optimum detector for this case is a replica correlator, this is only true if we know exactly for what frequency we are searching. Most passive narrowband (PNB) search systems try to search a wide frequency region looking for narrowband signals. This approach results in a system very much like the broadband processor, except that many are run in parallel, with each searching a portion of the total band, as in Figure 13.8.

The detection system is a frequency filter, followed by a square law detector, a power averager, and a threshold detector. Today, the filters are typically accomplished using FFTs (fast Fourier transforms). The square law detector is a device whose output is equal to the sum of the powers of the signal plus noise in the band of the input filter. Because of the randomness associated with the inputs, there are large fluctuations in time; thus the power averager, also referred to as an integrator, provides a running average of the power for a time period, T, to the threshold detector. The threshold detector simply tests the output of the averager with a fixed threshold, declaring a signal when the threshold is exceeded. This is a statistical test with the value chosen to reflect the desired false alarm rate when no signal is present. As we saw in the Chapter 12, this threshold is a function of the expected output when only noise is present and the desired false alarm rate.

With multiple beams or multiple sonobouys, this must be done for each beam/buoy. The result, which we discuss in more detail later, has two problems: a processing problem and a display problem. The processing problem is because this process needs to be accomplished for (1) each beam (proportional to frequency for a line array or frequency squared for an area

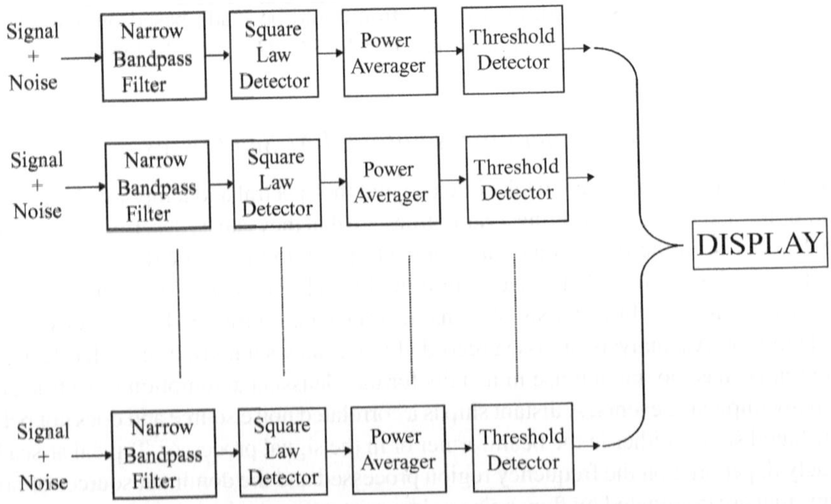

Figure 13.8 Basic narrowband processor

array), (2) each frequency bin, and (3) each focusing range (for large array proportional to frequency again). Second, for a display problem, a multidimensional display surface is needed (intensity versus bearing, frequency, time, and possibly range) as compared to bearing, time, and possibly range for the broadband case. Because of the large number of detection bins or channels involved, the false alarm rate must be chosen to reflect the entire processing set not just a single channel.

As an example, suppose that we wished to process a line array with a design frequency of 1000 Hz (the upper limit on frequency processed) which was 50 wavelengths long (at 1000 Hz) and we wanted to process data from 10 Hz to 1 kHz for all beams at a resolution of one percent of frequency. Assuming that we are only interested in long-range contacts (one focusing range), the processing shown in Figure 13.8 may require 46 300 chains (100 beams and 463 frequency bins). This means that the probability of false alarm must be set not for a single channel but for 46 300 or where a probability for a single channel of 1×10^{-4} per integration period might seem reasonable. The corresponding probability of false alarm for this set of channels must be set to 2×10^{-9} [$= 1 - (1 - 0.0001)^{1/46300}$]. Clever design can reduce the number of channels somewhat, but the number of calculations required is still extremely large. If we sample the data at 5 kHz, the computer must be able to handle about $5000 \times 46\,300 = 230$ million FFTs per second plus all the beamforming, power averaging, threshold detecting, and display normalizations.

By convention, the recognition differential for narrowband processing is defined as the signal power required to achieve a probability of detection of one half (50 %) in a specified time, T, at the selected probability of false alarm (P(fa)) divided by the noise power in a 1 Hertz band in the vicinity of the signal. This definition can be confusing because the processing band is usually not 1 Hertz. The signal power is defined as the total power concentrated near the frequency of interest. The signal is generally referenced to a different bandwidth than the noise. Thus, Nrd of 0 dB means that the signal is detectable when its power is just equal to the noise in a 1 Hertz band. When solving the sonar equation, this dimensionality issue is addressed by using the noise in a 1 Hertz band. We will see a similar convention arise with active sonar.

We will use a five-step process to compute the recognition differential. Step 1 is the computation of the idealized case where the signal is a pure sine wave in white Gaussian noise. The signal starts at time zero and detection is considered at a time T. The filter is perfectly rectangular with a flat response over a band of W Hz and zero everywhere else. The power averager produces a running average of the previous time period, T, of the input, called a boxcar averager. The detection threshold is set to the exact value required to achieve the desired probability of false alarm, which as noted previously, assumes exact knowledge of the mean noise. The result of Step 1 leads to a value that is frequently referred to as a theoretical or textbook Nrd, in that it applies to a very idealized situation.

The remaining steps are used to adjust the theoretical Nrd to compensate for nonideal systems. Step 2 corrects for nonideal noise characteristics, Step 3 deals with the actual processor to be used, Step 4 deals with the signal not being a perfect sine wave, and Step 5 adjusts for additional at-sea losses.

13.2.1 PNB Step 1: Theoretical Narrowband Nrd

Assuming a sine wave signal and flat or white Gaussian noise over a band W for a time, T, then

$$\text{Nrd} = -5\log(WT) + 10\log(W) + D = 5\log(W) - 5\log(T) + D \tag{13.20}$$

Comparing this result with the broadband case in Section 13.1.1, the difference is the second term ($10\log(W)$), which is caused by the narrowband Nrd being referenced to a 1 Hertz band as opposed to the actual processed band. D is the detection index (d) defined in Chapter 12, expressed in decibels, and is a function of the probability of false alarm ($P(\text{fa})$) and the number of independent samples (WT).

Robertson [3] carried out numerical calculations for the performance of linear detectors as a function of the number of independent samples, relating the probability of detection, probability of false alarm, and the signal-to-noise ratio. An excellent review of this is contained in Whalen and McDonough [7]. An approximation to Robertson's curves has been given by Albersheim [8]:

$$\text{Nrd}(P(\text{d})) = -5\log(m) + \left[6.2 + \frac{4.54}{\sqrt{m+0.44}}\right]\log(A + 0.12AB + 1.7B) \tag{13.21}$$

where

$A = \ln(0.62/P(\text{fa}))$
$B = \ln(P(\text{d})) - \ln(1 - P(\text{d}))$
m = number of independent samples averaged

Over the regions $10^{-3} < P(\text{fa}) < 10^{-7}$, $1 < m < 8096$, and $0.1 < P(\text{d}) < 0.9$, this expression is accurate to 0.2 dB.

The second term of Albersheim's equation is D. Note that for $P(\mathrm{d}) = 0.5$, $B = 0$, and $m \gg 1$, D becomes

$$D = 6.2\log(A) = 2.693\ln(A) = 2.693\ln\left[\ln\left(\frac{0.62}{P(\mathrm{fa})}\right)\text{ (linear detector)}\right] \tag{13.22}$$

Robertson [9] also showed that for $m \gg 1$ a square law detector is about 0.19 dB better than a linear detector. Therefore,

$$D = 2.693\ln\left[\ln\left(\frac{0.62}{p(\mathrm{fa})}\right)\right] - 0.19 \tag{13.23}$$

However, this form assumes the *WT* is large, as it was for broadband, which is normally not the case. Given this, it is necessary to make an adjustment for smaller *WT* products. Edleblute [10] developed an equation to make this adjustment:

$$D = 2.693\ln\left[\ln\left(\frac{0.62}{P\,(\mathrm{fa})}\right)\right] - 0.19 + 10\log\left(1 + \frac{A}{\sqrt{WT}}\right)$$
$$A = 0.22\log\left(\frac{100}{P(\mathrm{fa})}\right) \tag{13.24}$$

Figure 13.9 shows a graph of an approximation for the detection index, D, as a function of both the probability of false alarm (P(fa)) and the product of bandwidth and time (*WT*), referred to as the bandwidth time product.

As noted previously, Albersheim's equation can be used to compute values for probabilities other than 50 %. Figure 13.10 shows the ROC curves for a sine wave signal (CW) in Gaussian noise. The graph shows the required signal to noise versus probability of detection for various probabilities of false alarm. E/N_0 is the signal energy (power × time) divided by the noise in a 1 Hertz band.

13.2.2 PNB Step 2: Correction for Noise Spectrum

The bandwidth being processed is assumed to be small; therefore there is no need to adjust for nonflatness in the band. As noted above and in Chapter 12, we cheated in calculating the theoretical Nrd, in the sense that we assumed that the noise level was known a priori. In reality, the noise for most real systems varies and must be estimated during the processing. This results in a penalty to the theoretical Nrd. Given that we are searching for narrowband signals over a wide range of frequencies, the underlying noise will almost certainly vary from frequency bin to frequency bin (FFT). To process the data, it is necessary to take out this structure and estimate the underlying mean noise level. A noise spectrum equalizer (NSE) is used to accomplish this by using nearby frequency bins to estimate the noise. Simply stated, because a statistical estimate of noise has to be used, it is necessary to set the threshold higher to allow for the imprecision with which the noise is known.

A common implementation for an NSE is the split window sliding normalizer. In this process, the noise for a cell is estimated from the average of a number of cells on either side

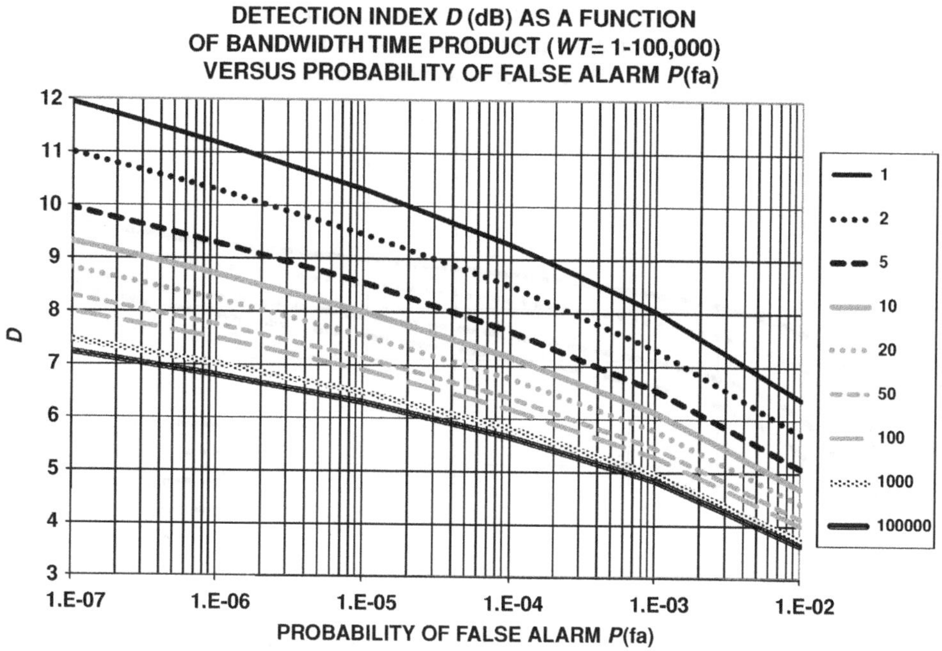

Figure 13.9 Detection index, D, as a function of the bandwidth time product versus probability of false alarm

of it, perhaps with additional guard cells to help prevent the signal from affecting the noise estimate (for this case, in frequency or bearing in the PBB case). For example, a 4–3–4 window would estimate the mean of the center cell by excluding that cell and its adjacent cells (for a total of three cells) and averaging the other eight cells, four on either side. At the edges of the display or processed band, we have to substitute a one-sided estimate. This process assumes that the noise field is well behaved, at least within the width of the normalizer. Figure 13.11 shows a display of data before and after the split window sliding normalizer is applied. The left side of the figure is created from the raw data. The right side is after the use of a 4–3–4 split window normalizer to estimate local noise. In this figure, time runs vertically (most current at the top) and frequency runs horizontally (low to high frequency equates to left to right).

The improvement seen on the right side of Figure 13.11 is caused by the use of a local average for the noise estimate for each cell or a single noise estimate for the whole screen on the left. In this example, the background noise is steadily decreasing with frequency from left to right, just as we would expect from an ambient limited situation. Two narrowband signals are present and only clearly visible on the right. At the top of each gram is a corresponding ALIT plot. In this case, where the signal is on continuously and does not change frequency bin, the ALIT provides a much stronger argument for the signals being present than does the gram. Interestingly enough, Figure 13.12 is a simple linear three-dimensional surface plot of the data from the left-hand side of Figure 13.11. Notice that the weaker signal on the right is clearly visible without the use of a normalizer. This may suggest that multicolor surface plots (this one was done with Microsoft Excel©) are superior to grams. However, to achieve this level of granularity required plotting several different combinations of perspectives and orientations

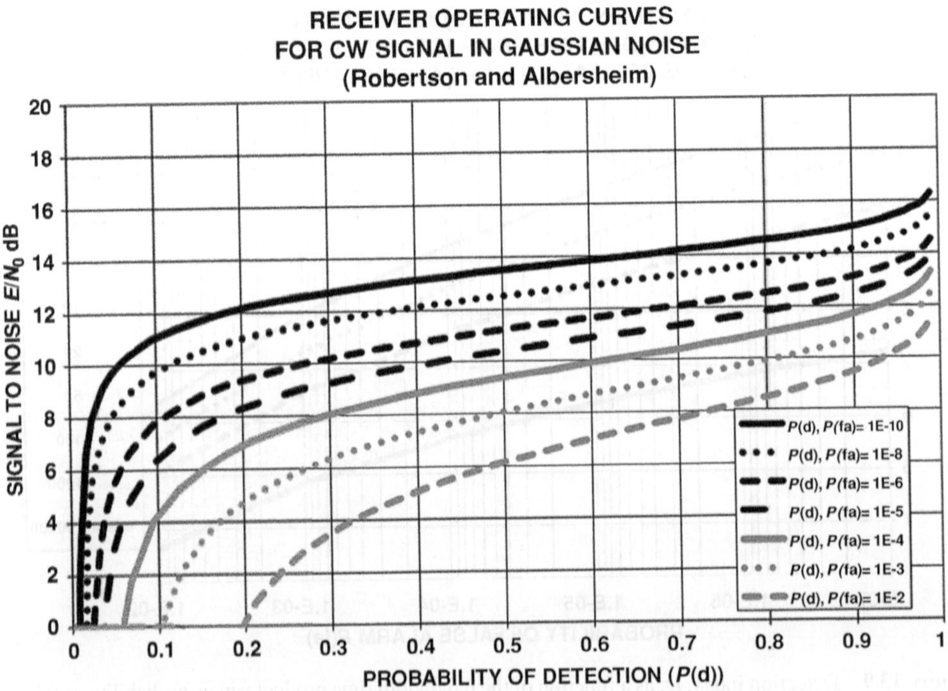

Figure 13.10 Receiver operating curves for sine wave signal in Gaussian noise

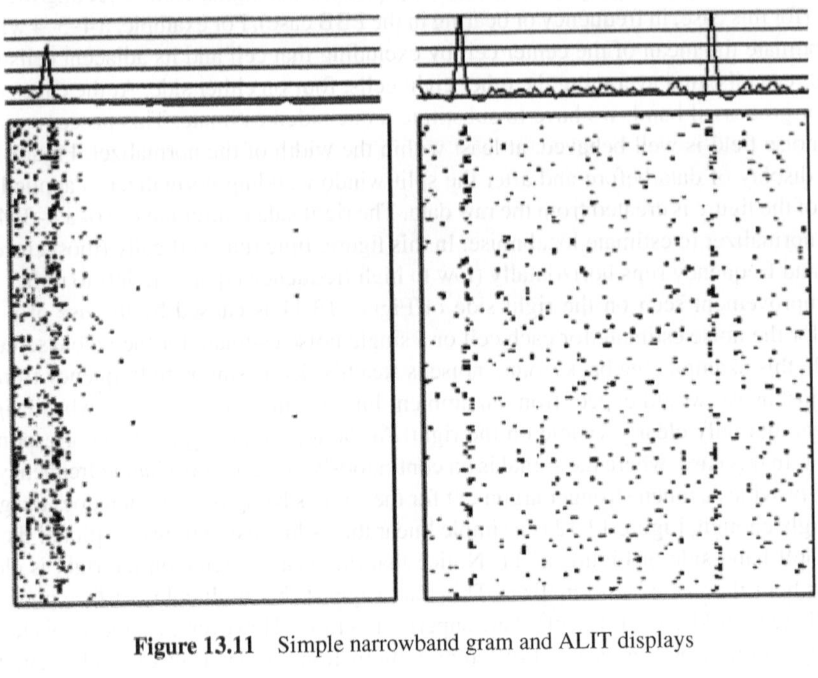

Figure 13.11 Simple narrowband gram and ALIT displays

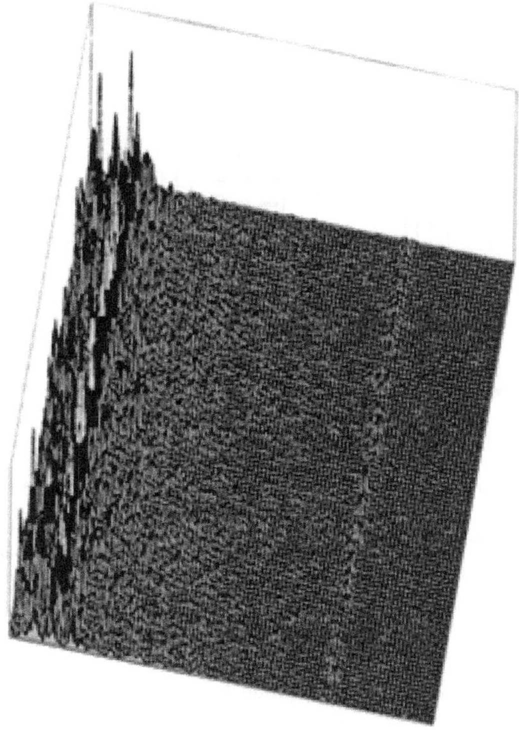

Figure 13.12 A three-dimensional linear plot of the same data without normalization

of the plot so that both signals could be seen. While a properly constructed three-dimensional plot is frequently superior, high signals or noise can hide lower signals. Given this, the data need to be reviewed from multiple directions to find everything. This extra effort makes the use of three-dimensional plots in real time difficult at best.

The design of NSE algorithms involves tradeoffs. Using a small number of bins limits the statistical precision of the estimate, while the use of many bins may violate the assumption that the noise is well behaved.

Pryor [5] developed a methodology for estimating the NSE loss. The results are shown in Figures 13.13 through 13.15. The equations used are provided later in Section 13.2.4.3.

13.2.3 PNB Step 3: Correction for Processor Implementation

This step adjusts for (1) nonideal input bandpass filters, (2) detectors other than square law, (3) OR-ing loss, (4) nonideal integration over time, and (5) some miscellaneous losses.

13.2.3.1 Nonideal Bandpass Filters

The calculation of the theoretical Nrd in Step 1 assumed a set of perfect rectangular frequency filters of width, W, covering the entire frequency range of interest with a unit response inside and zero response outside. Real filters generally differ from perfect filters in the following

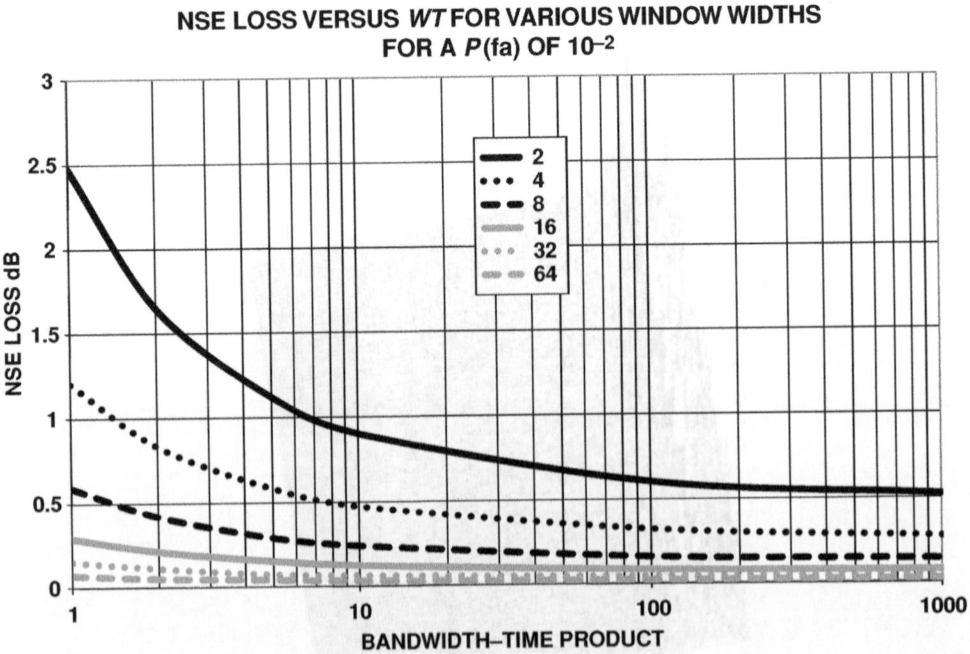

Figure 13.13 NSE loss versus bandwidth–time product for a P(fa) of 0.01, for various window widths

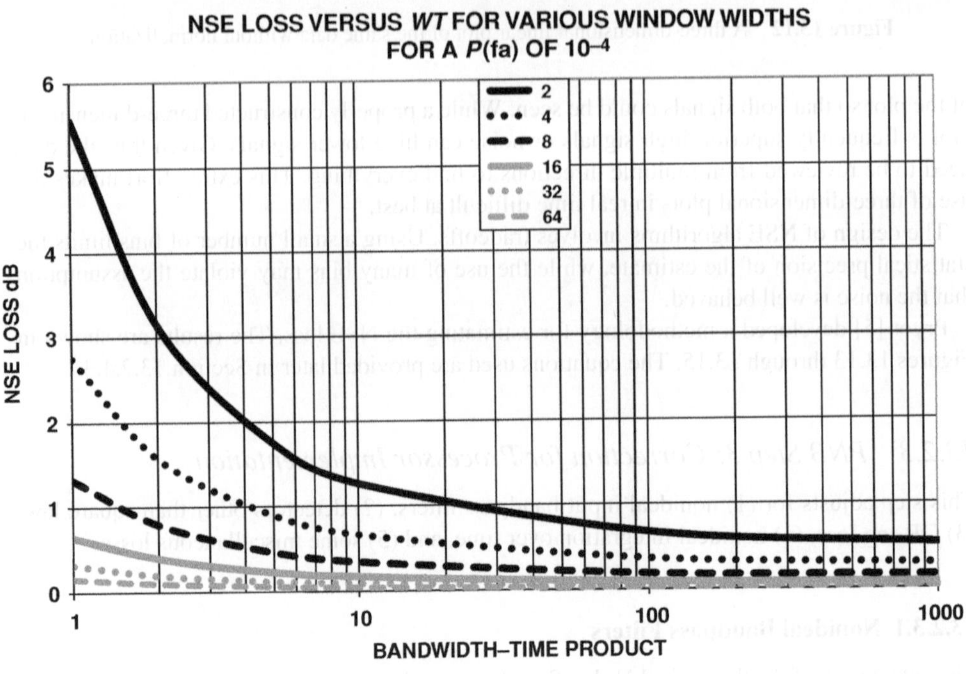

Figure 13.14 NSE loss versus bandwidth–time product for a P(fa) of 10^{-4}, for various window widths

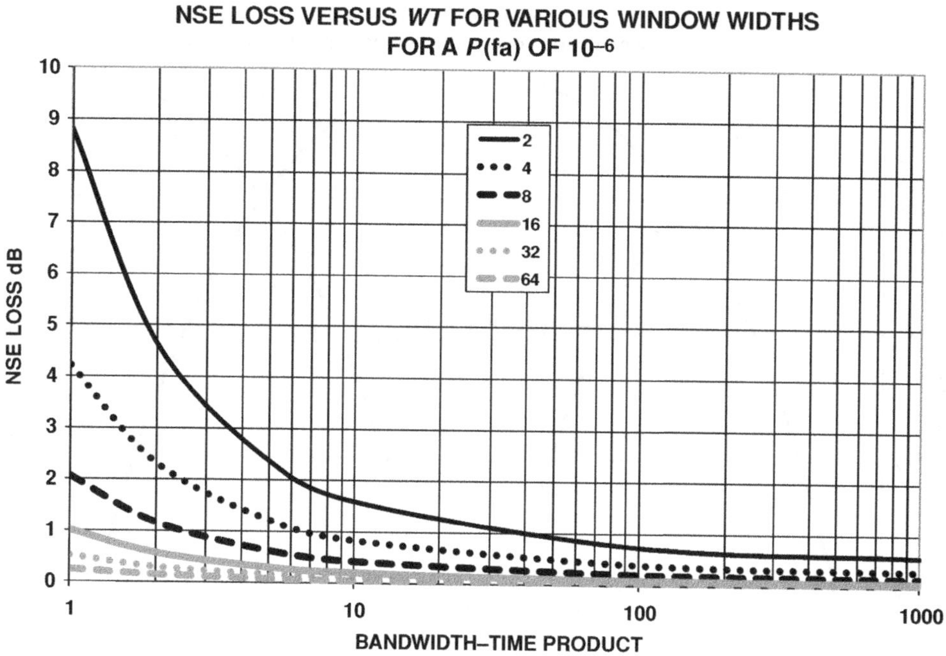

Figure 13.15 NSE loss versus bandwidth–time product for a P(fa) of 10^{-6}, for various window widths

ways: (1) the tops of real filters are not flat, so unless the signal is exactly centered it will suffer a "scalloping loss," (2) the effective noise bandwidth of the real filter is not exactly W, and (3) the weighting and overlap ratios of the FFT (length of the data block to the interval between the output of the FFT) are different from those of perfect filters. Table 13.1 gives the corrections associated with a variety of filters. Figure 13.16 shows an example of typical scalloping loss.

Table 13.1 Corrections to Nrd due to real filters

	Idea (boxcar)	Unweighted FFT	Half cosine weighting FFT	Hanning weighting FFT	Hamming weighting FFT
ΔNrd scalloping loss	0	1.1	0.7	0.5	0.6
Noise bandwidth	W	W	1.23W	1.50W	1.36W
ΔNrd (noise W)	0	0	0.91 dB	1.76 dB	1.34 dB
Smoothing bandwidth	W	1.5W	1.70W	2.08W	1.90W
ΔNrd smoothing	0	−0.88 dB	−1.16 dB	−1.59 dB	−1.39 dB
ΔNrd (1:2 overlap)	1.5 dB	2.38 dB	2.66 dB	3.09 dB	2.89 dB
ΔNrd (1:1 overlap)	0 dB	0.88 dB	1.16 dB	1.59 dB	1.39 dB
ΔNrd (2:1 overlap)	0 dB	0.25 dB	0.05 dB	0.20 dB	0.11 dB
ΔNrd (4:1 overlap)	0 dB	0.07 dB	0.0 dB	0.0 dB	0.0 dB
ΔNrd (8:1 overlap)	0 dB	0.0 dB	0.0 dB	0.0 dB	0.0 dB

Where ΔNrd (noise W) = $10\log(W_N/W)$ and ΔNrd (smoothing) = $-5\log(W_s/W)$

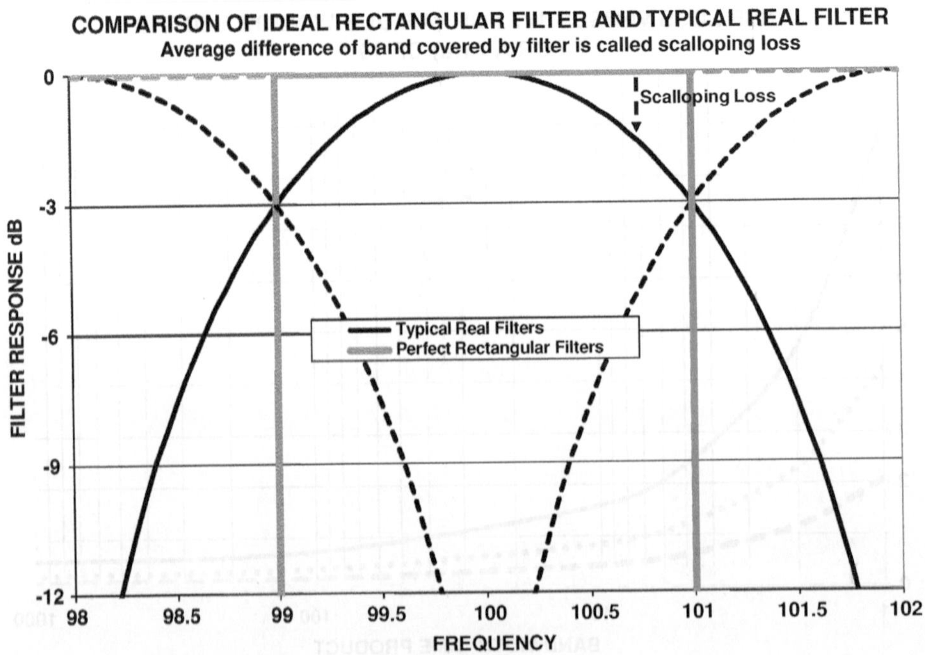

Figure 13.16 Example of scalloping loss by a real world filter

13.2.3.2 Correction for Nonsquare Law Detector

The use of the square law detector in the theoretical calculation was not based on this being the most commonly implemented detector, but rather three considerations: (1) it is the detector used in most textbooks, (2) it is fairly easy to describe and analyze mathematically, and (3) it is nearly an optimum detector.

The most commonly used detector is the linear or envelope detector, the output of which is the average absolute value of the voltage (or power). In older analog systems, this was very easy to implement by using a rectifier and a lowpass filter. For modern digital systems, this detector has the advantage of requiring a smaller dynamic range, and thus fewer bits to represent the output. For small bandwidth times, this detector is slightly better than the square law detector, whose performance, as mentioned above, has been extensively studied by Robertson [3]. The difference between the linear detector and the square law is in the range of about ±0.2 dB. An expression developed by Pryor [5] for the correction is

$$\Delta\mathrm{Nrd} = \frac{\dfrac{0.2N^3}{64} - 0.12N}{\left(1 + \dfrac{N^3}{64}\right)} \qquad (13.25)$$

$$N = \log_2(WT) = 1.443\ln(WT)$$

Figure 13.17 shows a plot of this function versus the bandwidth–time product (WT).

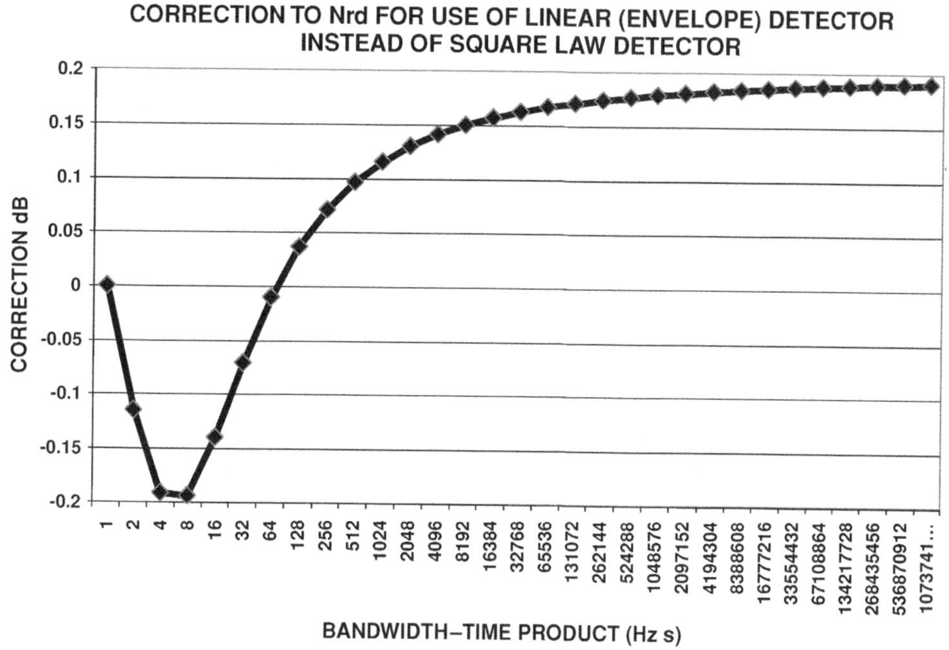

Figure 13.17 Correction to Nrd due to the use of a linear detector instead of a square law detector versus the bandwidth–time product (*WT*)

13.2.3.3 Correction for OR-ing

As mentioned above, a multidimensional display surface is needed to completely display output, intensity versus bearing, frequency, time, and possibly range. One method for fitting this on a CRT screen is to OR the data (see Section 13.1.3.3). This is simply comparing two or more output values and building a display using only the strongest. For example, consider a system with 100 look directions. If each look direction was displayed on a LOFAR((LOw Frequency Analysis and Ranging) display (frequency versus time plot like Figure 13.11), then an operator would have to look through 100 such displays in order to do an all-around search. Not surprisingly, this could take a considerable amount of time. If we assumed it took an operator 30 seconds to evaluate each display, then a full search would take 3000 seconds or 50 minutes, a very long time if we are in a threatened situation. An alternative would be to create, in an extreme case, a single display where all 100 beams have been OR-ed, resulting in the operator only having to study a single display. It would be desirable also to provide the operator with a tool that would allow him or her to find out which beam is being displayed in each bin. This ability would facilitate shifting to a non-OR-ed display once a target was detected.

It is important to consider at what point in the processing chain to do the OR-ing. For example, if the outputs required to display each of the 100 beams were formed and OR-ed, the loss in performance would be modest if there was only a single target somewhere and random noise everywhere else. However, in a very busy environment, massive losses could result, with hundreds of nontarget signals cluttering the screen.

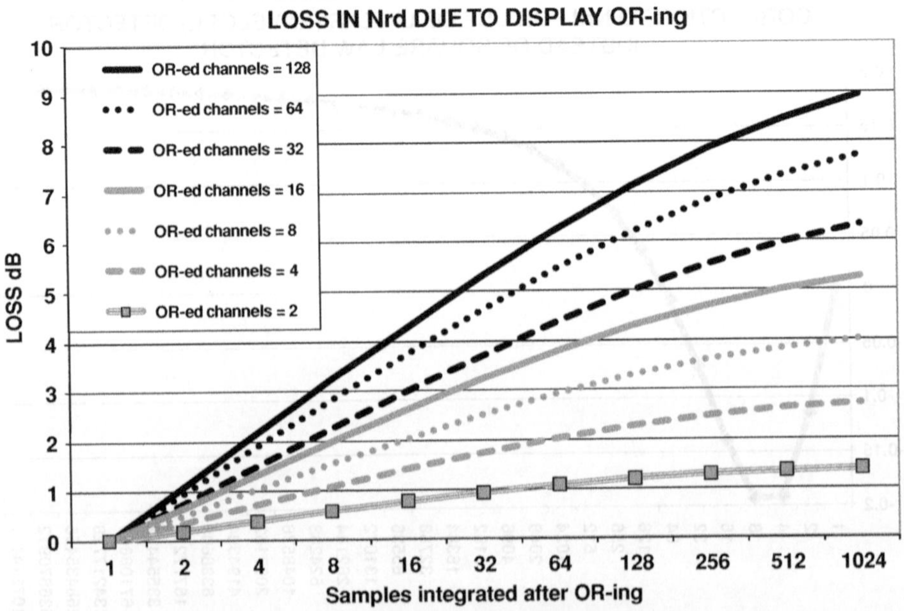

Figure 13.18 Loss in Nrd due to display integration of OR-ed data

There is an OR-ing loss for any system that does time integration after OR-ing, as is the case for the gram (Figure 13.11) display/operator. OR-ing losses have been extensively studied by Nuttal [6, 11] and Struzunski [12]. Unfortunately, all three references are classified or have limited distribution. However, an equation developed by Pryor [5] as a fit to the data is

$$\Delta \text{Nrd} = 1.5 \log 2(\text{Number of OR-ed channels}) \frac{x}{\sqrt[4]{1+x^4}}$$
$$x = 0.131 \frac{\log_2(\text{Number of samples combined after OR-ing})}{\log_2(\text{Number of OR-ed channels})^{1/8}} \tag{13.26}$$

Figure 13.18 shows this expression for various choices of OR-ing and integration.

OR-ing earlier in the processing chain can lead to massive losses. For example, if the outputs of each beam have not been normalized to have the same mean, then the highest mean is likely to dominate.

13.2.3.4 Nonideal Integration over Time

Another method used to reduce the data that need to be displayed is to integrate or power sum the data over a time interval for each frequency bin. As noted above, a perfect boxcar integrator that exactly matches the signal duration is about 2.5 dB better than a good gram display. This may be accomplished with an ALIT, as shown above in Figure 13.11, or with a FRequency AZimuthal (FRAZ) display. A FRAZ display shows frequency in one dimension and direction in the other for a fixed or selectable averaging/integration time, for example. If the boxcar integrator does not match the signal duration, then an Nrd loss is incurred.

As an example, assume that the signal duration is actually half of the integration time; then the signal to noise out will be 3 dB less than when it was on (noise for the full period and signal for half the period). To achieve the same probability of detection, the signal would have to increase by 3 dB. Therefore, we can model this as a 3 dB loss in Nrd even though it is actually a 3 dB loss in signal, not Nrd. If the signal duration was actually twice the duration of the integrator, then the maximum gain will not be achieved, but performance will be better than if the signal only lasted one integration time. This is because there will be two chances for the detection to be made.

13.2.4 PNB Step 4: Correction for Nonideal Signal Characteristics (Signal Is Not a Perfect Sine Wave)

There are four nonideal signal characteristics to be considered: (1) the signal has a defined spectral density, (2) the signal's center frequency wanders at a rate that is rapid when compared to one over the processor bandwidth ($1/W$), (3) the center frequency wanders at a rate that is slower than $1/W$, but less than the display time, T, and (4) the center frequency wanders at a rate longer than T. Since a rapidly fluctuating signal frequency is the definition of a broadening of the spectral line, case (2) is guaranteed to have a bandwidth wider than W. This is the only difference between cases (1) and (2).

Three adjustments must be made to correct for these nonideal signal characteristics. First, the calculation of D assumes the statistics of a sine wave. As the signal broadens, it is more appropriate to use a narrowband random process (a Rayleigh distribution), as was done with noise. Second, as the signal bandwidth increases beyond a single bin width, its power is divided among multiple analysis bins, resulting in a splitting loss. Finally, if the bandwidth becomes large enough, it will contribute power to the bins used by the NSE to estimate noise and suppress the signal on the display (NSE loss).

13.2.4.1 Rayleigh Signal Correction

When the signal bandwidth is equal to or greater than the analysis bin width, it is appropriate to use Rayleigh statistics for the signal and the noise. The mean value in both cases is the same, but the median on which a probability of detection of 50 % is based is different. An approximation has been developed by Prior [5] for this case:

$$\Delta \mathrm{Nrd}_{\mathrm{Rayleigh}} = \frac{1.56}{WT} - \frac{\left[DP(\mathrm{fa}), 1) - 10\log\left[\ln(P(\mathrm{fa})) - 1\right]\right.}{\sqrt{WT}} \tag{13.27}$$

For the case when the signal bandwidth is less than the analysis bandwidth, the correction is usually quite small, except in the case of small values of WT. In this region, Pryor recommends interpolating between the full adjustment and no adjustment.

13.2.4.2 Signal Splitting Loss

Signal splitting loss may seem like an easy process of simply computing how much power remains in the bin, but this would be an overestimate. Just as we assume that an operator can visually integrate along the time axis, it is also assumed that he or she can mentally combine adjacent frequency bins. This is not to say that there is no loss, but it results in 5 log of the ratio of the signal bandwidth divided by the analysis bandwidth, instead of 10 log of the ratio.

13.2.4.3 NSE Loss

The NSE loss due to the signal spreading across multiple analysis bins is caused by the signal power being used in the estimate of background noise, thereby making the actual signal appear closer to the noise. The change in recognition differential is given by

$$\Delta \mathrm{Nrd}_{\mathrm{NSE}} = -10 \log \left[1 - \beta \frac{\left(W_s + \frac{S}{N} \right)}{2KW} \right] \tag{13.28}$$

where

W = analysis bandwidth
W_s = signal bandwidth
β = fraction of signal power in the noise estimation windows
S/N = signal to noise at the detector input
K = number of bins used in each of the two-sided noise estimation windows

13.2.5 PNB Step 5: Adjustment for Additional At-Sea Losses

The preceding discussion has tried to take into account the spectral and temporal characteristic of both the signal and the noise. Underlying assumptions are that the signal and noise can be described as stationary Gaussian processes. In the operational environment at sea, the noise field may arise largely from distant shipping sources that move with time through the beams of a sonar, through acoustic caustics that cause brief bursts of signal, or transients caused by human activity, such as hatch slamming. An analysis of at-sea recorded noise shows a myriad of such effects; i.e., the ambient noise has too much noise in its tails for the Gaussian assumption to be true and it contains many impulsive events. A distant ship is a correlated noise source and does not behave like a distributed source, either in the beamformer or in the signal processor. Typical at-sea loss varies widely depending on the frequency region processed and the dominant sources of noise. For systems that are dominated by flow noise or sea state noise, the corrections required are a few decibels for low-frequency broadband systems. In highly cluttered environments, 10–20 dB loss has been experienced. At sea, measurements are the only reliable method of estimating these losses.

13.2.6 Nrd Calculation Example

Consider a processor with a linear detector, a bandwidth of 0.1 Hz, a 200 second integration time, a probability of false alarm of 10^{-5}, Hamming weighting, and a 2:1 overlap. The theoretical recognition differential is

$$\mathrm{Nrd}\,(P(\mathrm{d})) = 10\log(W) - 5\log(m) + D(0.5, 10^{-5}, 0.1 \times 200)$$

$$\mathrm{Nrd}\,(P(\mathrm{d})) = 10\log(W) - 5\log(m) + \left(6.2 + \frac{4.54}{\sqrt{m + .044}}\right)\log\left[\ln\left(\frac{0.62}{P(\mathrm{fa})}\right)\right] \quad (13.29)$$

$$\mathrm{Nrd}\,(P(\mathrm{d})) = 10\log(0.1) - 5\log(0.1 \times 200) + 7.51$$

$$\mathrm{Nrd}\,(P(\mathrm{d})) = -8.99\,\mathrm{dB\ re:1\,Hz}$$

The corrections are:

Scalloping loss (average) = 0.6 dB

Noise bandwidth ΔNrd = 1.34 dB

Smoothing ΔNrd = −1.39 dB

2:1 Overlap correction = 0.11 dB

$D(0.5,10-5, 20) = -7.56$ dB

$D(0.5,10-5,1.9 \times 20) = 7.24$ dB

Total correction = 0.34 dB

The corrected Nrd is −8.99 + 0.34 = −8.66 dB.

13.3 Active Sonar

For active sonar, there is an additional complication that may arise in that some authors and programs use the energy form of the sonar equation where the source level is input in units of energy or power. For consistency of units, we would need an energy detection threshold (EDT) that is the normal *DT* plus 10 log of the effective pulse length, since energy is power × time. Similarly, an energy recognition differential is defined as the normal Nrd plus 10 log of the effective pulse length.

Four types of signals are of primary interest: (1) constant frequency, continuous wave (CW), (2) frequency modulated or coded pulse (FM or CP), (3) impulsive (short duration) sources, such as sparkers (electrical discharges in water or explosives that by their nature are broadband), and (4) pseudo random pulses (PRN). Each of these pulse types has advantages and disadvantages. CW allows Doppler discrimination that not only allows estimation of target speed in the line of sight but, for high Doppler targets, can eliminate interference from reverberation. Unfortunately, CW has, for the same pulse length as other types, poor range accuracy.

FM or CP has excellent range resolution and generally works better than CW against targets with low Doppler, but cannot estimate target speed in the line of sight from Doppler. The principal advantage of impulsive sources is low-cost transmitters. These sources have good range resolution, but lack both Doppler discrimination and good processing gain. PRN pulses attempt to capture all the good qualities of CW and CP by sending a known broadband signal. Good detection, range accuracy, Doppler resolution, and high processing gain are all possible. Given this, why are not all sonars using PRN? First, it is technically difficult to get transducers that work well in this mode and, second, the processing is difficult because a replica correlator working over many ranges and Doppler bins is needed.

Active sonar processing falls into two broad categories: replica correlation or coherent processing and energy detection or incoherent processing. However, hybrid systems do exist that incorporate both on the same signal by using replicas for the chirps making up a signal and then combining the outputs incoherently.

Again, the five-step process is used to compute the recognition differential. Step 1 is the computation of the idealized case. Almost all of the adjustments required in the subsequent steps have been discussed above, so they will not be repeated in this section. Instead, we will consider Step 1 for each type of signal and note any special considerations needed for each in adjusting the recognition differentials.

13.3.1 CW Active Pulse Active Step 1: Theoretical Nrd

Continuous wave (CW) pulses come in two forms: unshaped and shaped. Unshaped means that the amplitude is flat over the duration of the pulse; this is also referred to as a rectangular pulse. If the pulse is rectangular, of duration τ, then the frequency domain power density can be obtained by using a Fourier transform:

$$F(f) = \frac{[\sin(\pi f \tau)]^2}{\pi^2 f^2 \tau^2} \tag{13.30}$$

The bandwidth, W, of the pulse is

$$W = \frac{\left[\int_0^\infty \left(\frac{\sin(\pi f \tau)}{\pi f \tau}\right)^2 \mathrm{d}f\right]^2}{\int_0^\infty \left[\frac{\sin(\pi f \tau)}{\pi f \tau}\right]^4 \mathrm{d}f} \tag{13.31}$$

$$W = \frac{3}{4\tau}$$

The bandwidth of a rectangular pulse multiplied by its bandwidth is a constant (0.75). Therefore, long CW pulses have narrow bandwidths, which means they are better able to resolve Doppler shifts in a return. It is interesting to note that the standard deviation and variance in the frequency domain is infinite.

Table 13.2 Fourier transform of modulated signals

Modulation	Fourier transform
$f(t)e^{i\omega_0 t}$	$[F(\omega - \omega_0)]$
$f(t)\sin(\omega_0 t)$	$i/2[F(\omega + \omega_0) + F(\omega - \omega_0)]$
$f(t)\cos(\omega_0 t)$	$1/2[F(\omega_0 + \omega_0) + F(\omega - \omega_0)]$

Consider a sine wave signal ($\sin(\omega t)$) that is modulated by $\cos(\omega_0 t)$:

$$\sin(\omega t)\cos(\omega_0 t) = \frac{1}{2}\{\sin[(\omega + \omega_0)t] + \sin[(\omega - \omega_0)t]\} \tag{13.32}$$

The result is the sum and difference frequency. This can be generalized as if the Fourier transform of a signal is shifted by an amount, ω_0. The corresponding effect in the time domain is described by multiplying the original signal by $\exp(i\omega t)$, as shown in Table 13.2.

Pure incoherent processing (filtering and energy detection) is used primarily for CW. CW processing looks much like the narrowband processing described in Section 13.2. Frequency bin widths are set to match the expected signal bandwidth (a combination of transmitted signal bandwidth, environmental spreading, and contributions from source and receiver motion as seen through the beam patterns). Additional bins on either side allow for Doppler shift caused by the target. Own ship Doppler nullification (ODN) is generally used, where the display for each beam is frequency corrected for the contribution of the source and receiver in the center of the beam. This results in zero target Doppler in the center corresponding to a target with no speed in the line of sight. The second dimension, time, is translated to range with the bin widths being set to the expected signal duration or pulse resolution. Instead of the frequency–time display for each bearing that was shown for PNB, there is a Doppler–range display (Figure 13.19) for each beam. Given that modern active sonars can search multiple bearings simultaneously, frequently the display shown is bearing–range with the Doppler values being OR-ed and/or with the zero Doppler bin notched out to remove reverberation.

There is an additional consideration for CW, as can be seen in Figure 13.19. While the ambient and flow noise is usually relatively flat across the frequency band of interest, the reverberation generally is not. In particular, as discussed in Chapter 3, Transducers, Directionality, and Arrays, the returned reverberation has a shape in frequency space that is the convolution of the original pulse shape, environmental frequency spreading, and frequency spreading due to transmitter and receiver motion as seen through the beam patterns. As detection frequency bins move away in frequency for zero Doppler targets, they have lower and lower reverberation, as shown in Figure 13.20. This effect could, and probably should, be handled in the reverberation term of the sonar equation. In practice, this effect is handled as an adjustment to the reverberation signal or recognition differential by applying a Doppler gain curve, which is zero at zero target Doppler and improves performance as Doppler increases.

If the processor applied to Figure 13.20 had 2 Hz wide boxcar filters, then the Doppler gain curve would look like Figure 13.21. Here, a 10 dB gain against reverberation is obtained against a target with 4 knots of speed in the line of sight. Amplitude shading is frequently used on CW signals to improve Doppler gain by reducing the side lobes in the frequency domain.

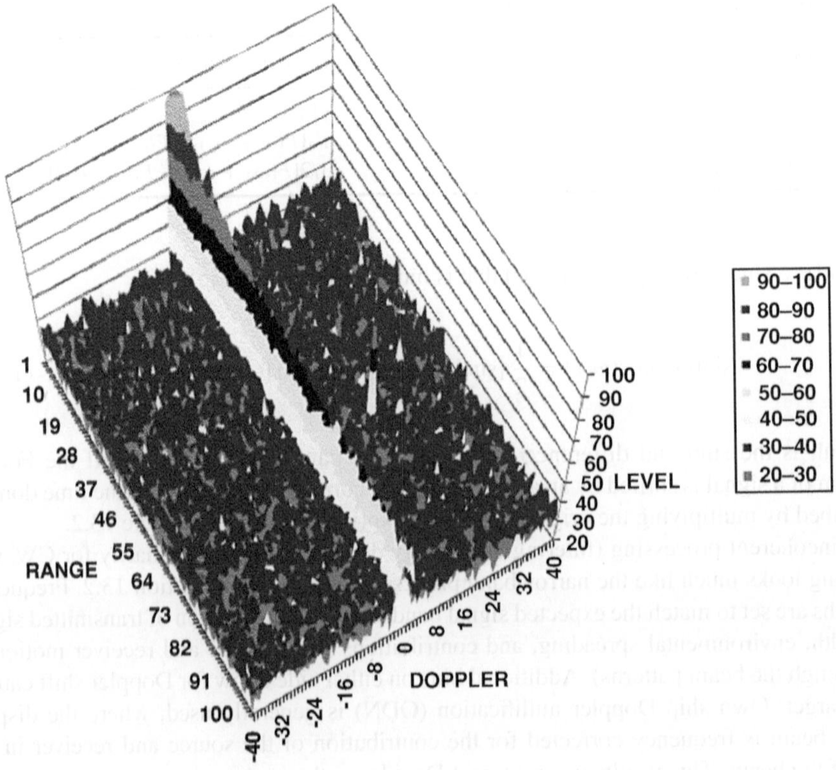

Figure 13.19 Example of a range–Doppler map. Note the reverberation in bins near zero Doppler decaying with increasing range and a target return around the range of 70 and Doppler = +16 knots

However, for the example above, shading on the receiver beams would be needed because they are dominant at high Doppler.

By convention among sonar designers, the recognition differential for active processing is defined as the signal power required to achieve a probability of detection of one half (50 %) for a specified pulse length, T, at the selected probability of false alarm (P(fa)) divided by the noise power in a reference band in the vicinity of the signal. This definition can be confusing as the processing band is frequently not the reference band. The signal power is defined as the total power concentrated near the frequency of interest; therefore, the signal is generally referenced to a different bandwidth than the noise. An Nrd of 0 dB means the signal is detectable when its power in the processing band is just equal to the noise in the reference band. When solving the sonar equation, this issue is addressed by using the noise in the reference band.

13.3.1.1 CW Active Step 1: Theoretical Narrowband Nrd

Step 1 is the computation of the idealized case where the signal is a pure sine wave in white Gaussian noise. The signal starts at time zero and detection is considered at a time, T, equal to the pulse length. The filter is perfectly rectangular with a flat response over a band of W Hertz

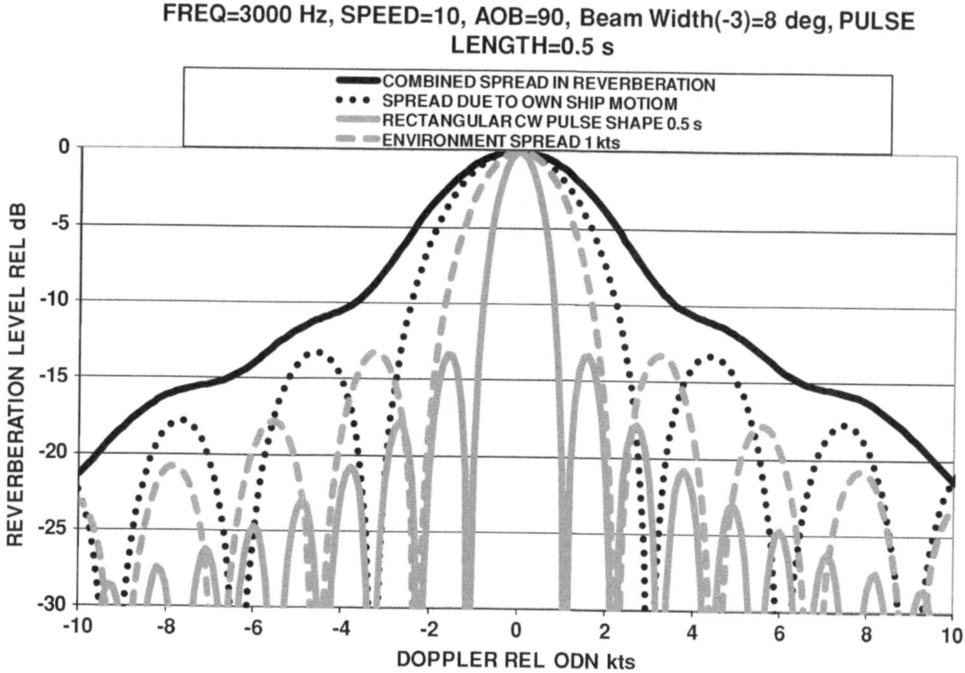

Figure 13.20 Relative strength of reverberation versus Doppler shift, showing the contributions of components

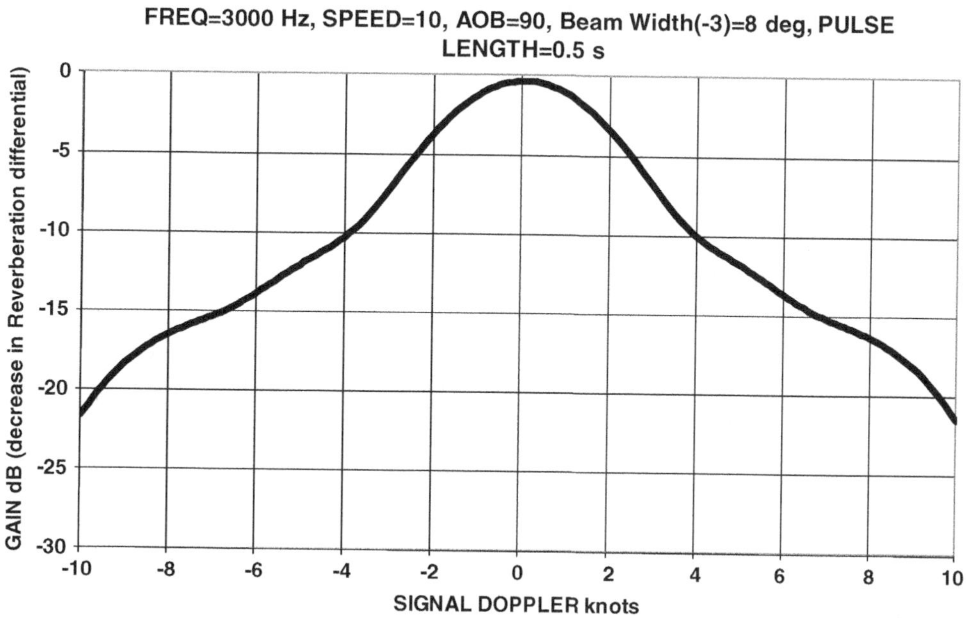

Figure 13.21 Example of Doppler gain curve against reverberation for a 2 Hz boxcar filter bank

and zero everywhere else. The power averager produces a running average of the previous time period, T, of the input, called a boxcar averager. The detection threshold is set to the exact value required to achieve the desired false alarm probability, which as noted previously assumes exact knowledge of the mean noise. The result of Step 1 leads to a value that is frequently referred to as a theoretical or textbook Nrd, in that it applies to a very idealized situation.

The calculation of a false alarm rate frequently starts with a specification, like one false alarm per four hour period for naval systems. This specification can result in a very small P(fa) since P(fa) is per decision. Given a 1 second pulse, 50 beams, and 60 Doppler bins, in four hours there are 4 h × 3600 s/h × 50 beams × 60 Doppler bins × 1 decision/s/bin or 43 000 000 decisions and a requirement for a P(fa) of 2.3×10^{-8}. This is identical to Section 13.2.1 above. Assuming a sine wave signal and flat white Gaussian noise over a band, W, for a time, T, then

$$\mathrm{Nrd} = -5\log(WT) + 10\log(W) + D = -5\log(W) - 5\log(T) + D \qquad (13.33)$$

Comparing this result with the broadband case in Section 13.1.1, the difference is the second term (10 log(W)), which, as mentioned above, is caused by the definition of narrowband Nrd as being referenced to a 1 Hertz band as opposed to the actual processed band. D is the detection index, d, defined in Chapter 12, Statistical Detection Theory, expressed in decibels. It is a function of the probability of false alarm (P(fa)) and the number of independent samples (WT).

Robertson [3] carried out numerical calculations for the performance of linear detectors as a function of the number of independent samples. Here, he related the probability of detection, probability of false alarm, and the signal to noise. An excellent review of this is contained in Whalen and McDonough [7]. An approximation to Robertson's curves has been given by Albersheim [8]:

$$\mathrm{Nrd}\,(P(\mathrm{d})) = -5\log(m) + \left[6.2 + \frac{4.54}{\sqrt{m+0.44}}\right]\log(A + 0.12AB + 1.7B) \qquad (13.34)$$

where

$A = \ln(0.62/P(\mathrm{fa}))$
$B = \ln(P(\mathrm{d})) - \ln(1 - P(\mathrm{d}))$
m = number of independent samples averaged

For unshaded CW pulses (rectangular amplitude shape), the time bandwidth product for the match filter is 1, since the pulse length is one over the bandwidth (Fourier). Over the region from $10^{-3} < P(\mathrm{fa}) < 10^{-7}$, $1 < m < 8096$, and $0.1 < P(\mathrm{d}) < 0.9$, this expression is accurate to 0.2 dB.

The second term of Albersheim's equation is D. Note that for $P(d) = 0.5$, $B = 0$, and $m \gg 1$, D becomes

$$\begin{aligned} D &= 6.2\log(A) = 2.693\ln(A) \\ &= 2.693\ln\left[\ln\left(\frac{0.62}{P\,(\mathrm{fa})}\right)\right] \text{(linear detector)} \end{aligned} \qquad (13.35)$$

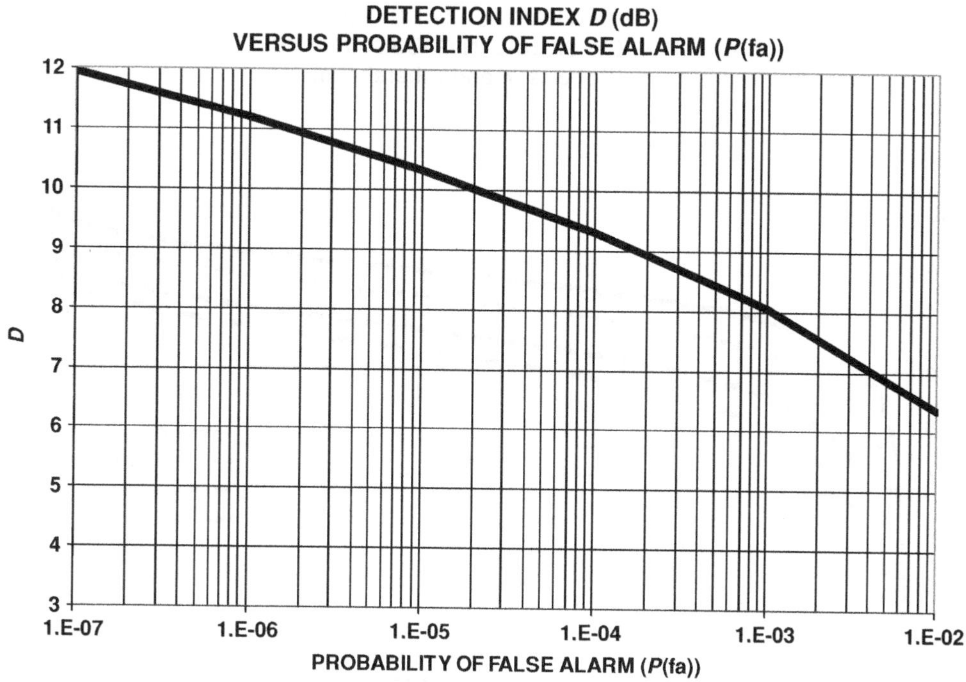

Figure 13.22 Detection index versus probability of false alarm for a CW signal in Gaussian noise

Robertson [9] also showed that for $m \gg 1$, a square law detector is about 0.19 dB better than a linear detector. Therefore

$$D = 2.693 \ln \left[\ln \left(\frac{0.62}{P\,(\mathrm{fa})} \right) \right] - 0.19 \tag{13.36}$$

This form assumes that WT is large, as it was for broadband, which is not the case here. Therefore, it is necessary to make an adjustment for smaller WT products. Edleblute [10] developed an equation to make this adjustment:

$$\begin{aligned} D &= 2.693 \ln \left[\ln \left(\frac{0.62}{P\,(\mathrm{fa})} \right) \right] - 0.19 + 10 \log \left(1 + \frac{A}{\sqrt{WT}} \right) \\ A &= 0.22 \log \left(\frac{100}{P\,(\mathrm{fa})} \right) \end{aligned} \tag{13.37}$$

Figure 13.22 shows a graph of this approximation for D as a function of probability of false alarm (P(fa)). This graph shows the required signal to noise versus probability of detection for various probabilities of false alarm. E/N_0 is the signal energy (power × time) divided by the noise in a 1 Hertz band.

As previously noted, Albersheim's equation can be used to compute values for probabilities other than 50 %. Figure 13.23 shows ROCs for detecting a single sine wave pulse in white

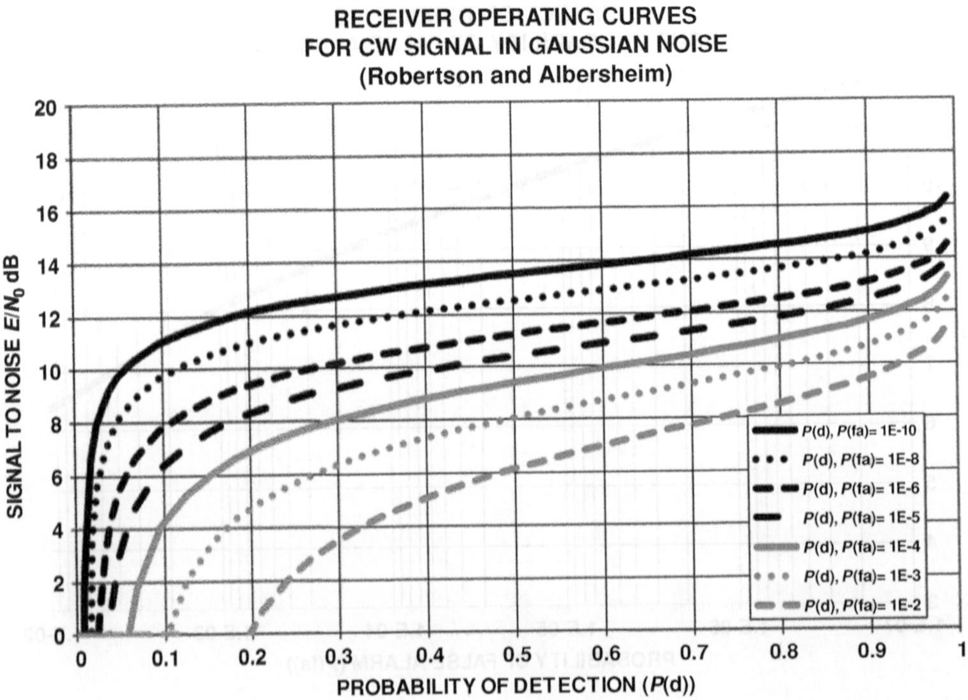

Figure 13.23 Receiver operating curves for detecting a single sine wave pulse in white Gaussian noise

Gaussian noise. The required signal to noise versus probability of detection for various probabilities of false alarm is shown. E/N_0 is the signal energy (power × time) divided by the noise in a 1 Hertz band.

An alternative assumption is to assume that the signal is slowly fading (Rayleigh process). The result from Whalen and McDonough [7] is

$$P(\text{d}) = P(\text{fa})^{1/(1+E/N_0)} \tag{13.38}$$

The receiver operating curves are shown in Figure 13.24.

13.3.1.2 CP/FM/PRN Active Pulse Step 1: Theoretical Nrd

These pulses are "broadband" in the sense that the pulse length multiplied by the band width is greater, usually much greater, than the 0.75 for CW pulses. Coded pulses (CP) or frequency modulated (FM) pulses come in a wide variety of forms. There are two general categories: frequency stepped pulses and frequency swept pulses. The first pulse type changes frequency discontinuously and the latter continuously. Figure 13.25 shows examples of various CP pulses.

CP, FM, and PRN pulses uses replica correlator processing, corresponding to the signal known case in Chapter 12. In general, the assumption is that the signal will have the same shape as the transmitted pulse. As will be shown in Chapter 18, Tracking, Target Motion

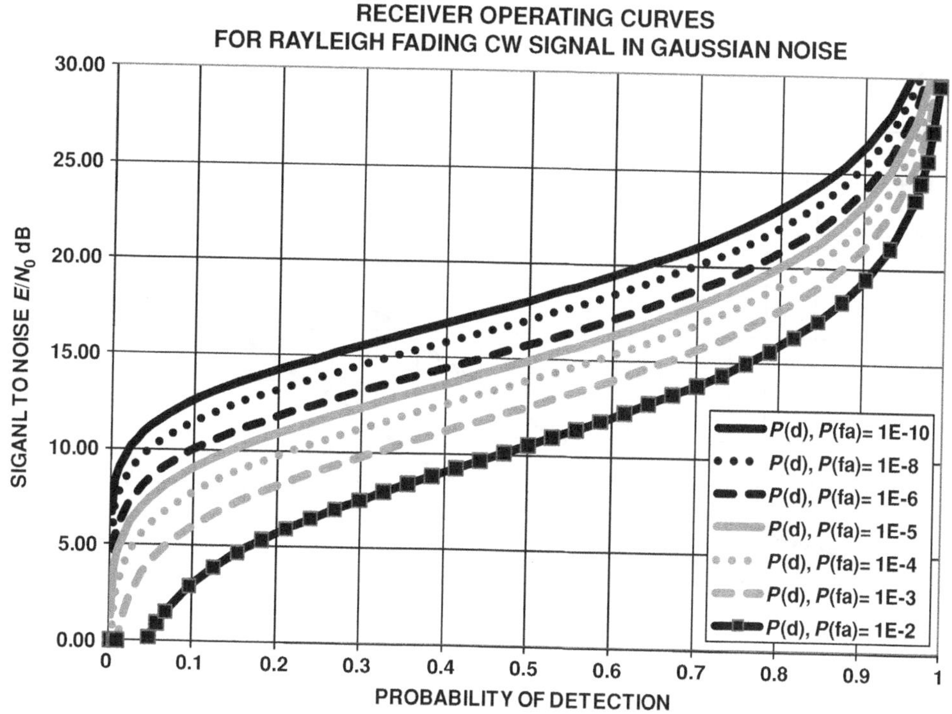

Figure 13.24 Detection index for a Rayleigh fading signal

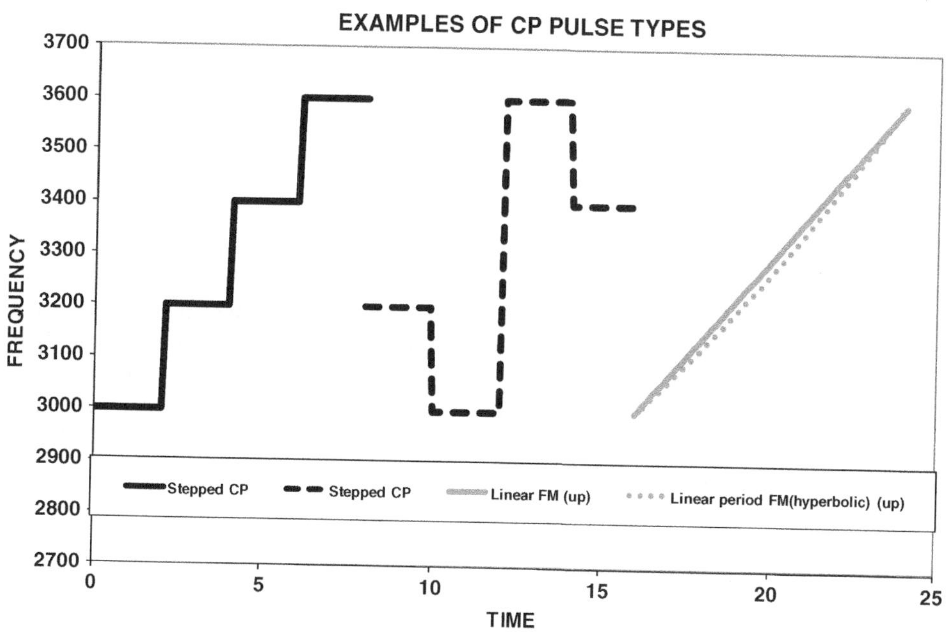

Figure 13.25 Examples of various CP pulse types

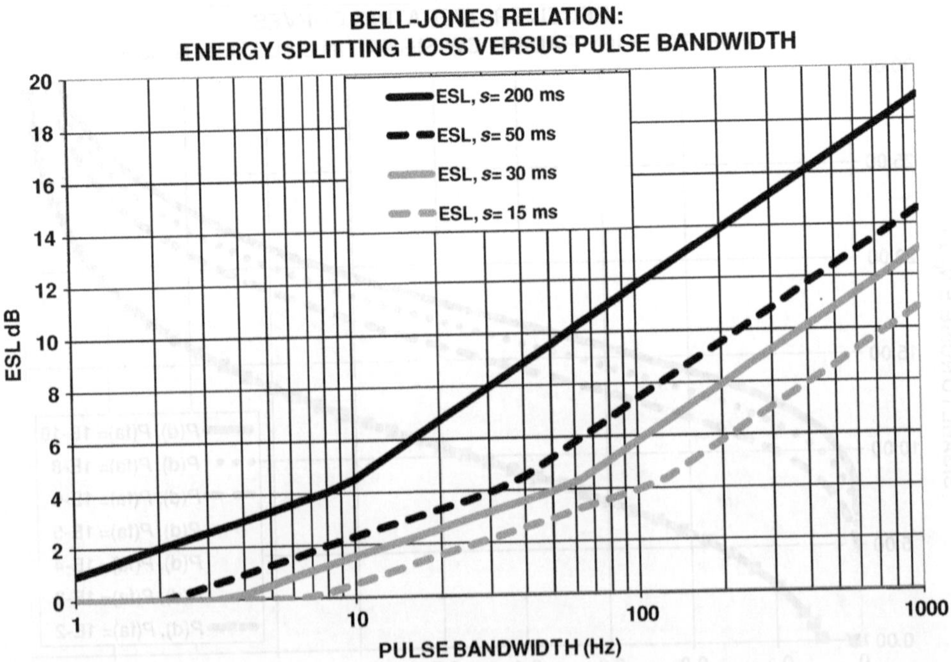

Figure 13.26 Energy splitting loss from the Bell–Jones relation for various time spreads versus pulse bandwidth

Analysis, and Localization, the range resolution of these pulses is inversely proportional to the bandwidth. At high SNR,

$$\sigma(R) = \frac{0.35}{W} \tag{13.39}$$

where

R = range resolution (nmi)
W = bandwidth (Hz)

A pulse 200 Hz wide would have a range resolution of 0.001 75 nmile, or about 10.5 ft. This has two effects. First, since the reverberation is proportional to the range resolution, reverberation is reduced substantially relative to a CW pulse of the same length. Second, for typical bandwidths (hundreds of Hertz), the range (time) resolution is so small that multiple receive paths and features of the targets are resolved individually rather than being combined. This latter effect results in energy splitting loss (ESL) caused by echo time spreading, which is also present for very short CW or impulsive sources. A standard loss for ESL is the Bell–Jones relation [13] (Figure 13.26):

$$\begin{aligned} &\mathrm{ESL} = 0 && \text{for } s/r < 0.1 \\ &\mathrm{ESL} = 3.4 + 3.4\log\left(\frac{s}{r}\right) && \text{for } 0.1 < s/r < 1.82 \\ &\mathrm{ESL} = 2.36 + 7.2\log\left(\frac{s}{r}\right) && \text{for } s/r > 1.82 \end{aligned} \tag{13.40}$$

Table 13.3 Environmental time spreading values for use with the Bell–Jones relation

Environment	s(ms)
Surface duct/CZ	15
Bottom bounce MGS 1–4	15
Bottom bounce MGS 5	30
Bottom bounce MGS 6	50
Bottom bounce MGS 7–8	200
Shallow water	50

where

s = standard deviation of the signal spreading
r = time resolution ($1/W$) of the signal

Table 13.3 shows reasonable values for environmental spreading. Keep in mind that there may be some double counting if a program is used to account for multipath time spreading, such as the Comprehensive Acoustic System Simulation (CASS) program. In addition to environmental spreading, there is target spreading, which is aspect dependent (see Figure 13.27). The total signal spreading is usually taken as the rms (square root of the sum of the squares) combination of the two sources.

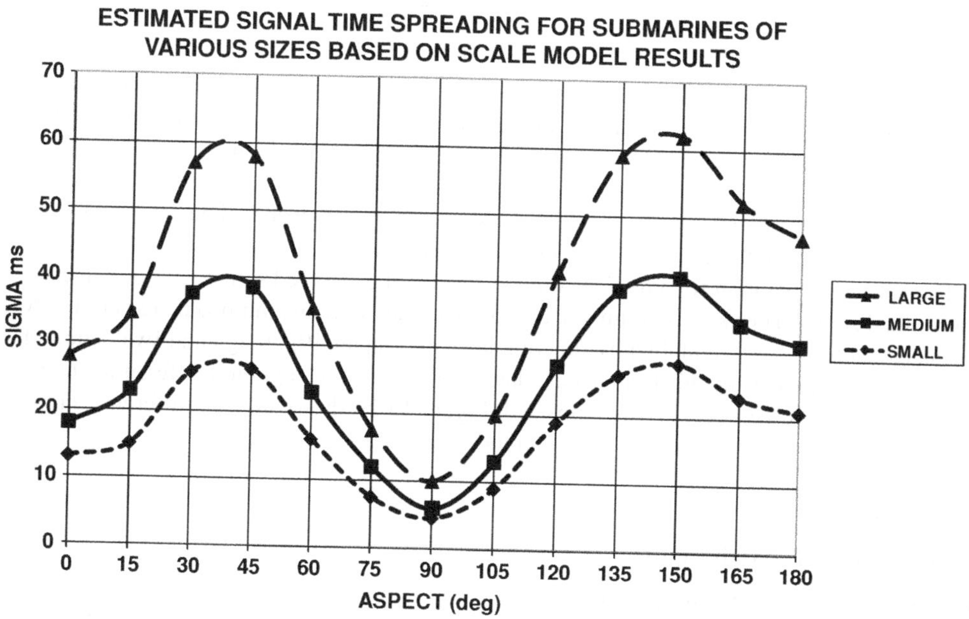

Figure 13.27 Target time spreading values for use with the Bell–Jones relation

13.3.1.3 FM/CP/PRN Active Step 1: Theoretical Nrd

Depending on the assumptions made, there are different methods for calculating the theoretical Nrd. Starting with the result from Chapter 12 for a known signal:

$$P\,(\mathrm{d}) = \frac{1}{\sqrt{2\pi}} \int_{a\,-\,d/\sqrt{d}}^{\infty} \mathrm{e}^{-z^2/2}\,\mathrm{d}z$$

$$P\,(\mathrm{fa}) = \frac{1}{\sqrt{2\pi}} \int_{a/\sqrt{d}}^{\infty} \mathrm{e}^{-z^2/2}\,\mathrm{d}z \tag{13.41}$$

The choice of $\alpha/\sqrt{d}$ determines the probability of a false alarm. Similarly, if $\alpha = d$, then the probability of detection is 50 %. This signal to noise is the recognition differential (Nrd or RD) or the detection threshold (DT), depending on the authors (see the beginning of this chapter for a discussion of a stricter definition of these terms). Since α is a constant chosen to set the false alarm rate, the recognition differential can be written in dB as

$$\text{Nrd or RD} = \text{DT} = \text{SNR}((P\,(\mathrm{d}) = 0.5)) = 10\log\left[\frac{d}{2WT}\right] = 10\log\left[\frac{\alpha}{2WT}\right]$$

$$\text{Nrd or RD} = \text{DT} = \text{SNR}((P)\,(d) = 0.5)) \tag{13.42}$$

$$= 4.584\ln\left[\ln\left(\frac{0.25}{P(\mathrm{fa})}\right)\right] - 1.02 - 10\log(WT)\quad \text{dB}$$

This is the signal to noise input power to the processor required to achieve a 50 % probability of detection for a known signal of bandwidth, W, and duration, T, with the specified probability of false alarm (corresponding to the α chosen). Figure 13.28 shows the value of D ($10\log(\alpha/2)$) for this case. This graph shows the required signal to noise versus probability of detection for various probabilities of false alarm. E/N_0 is the signal energy (power × time) divided by the noise in a 1 Hertz band. This is equivalent to the match filter result.

The normal assumption for a replica correlator is that the received signal will look like the transmitted signal. This is very optimistic because Doppler shift and multipath are very likely to change its form (see the ambiguity function later in Chapter 19, Design and Evaluation of Sonars). Rather than using this form to calculate the theoretical recognition differential, many practitioners use Robertson curves [7] instead. Figure 13.29 shows that the difference in the detection index is small between the two methods.

13.3.1.4 Impulsive Active Pulse

Impulsive Active Step 1: Theoretical Nrd

For impulsive sources, which are intrinsically broadband, the processing looks much like the passive broadband described in Section 13.1 above (intensity for bearing versus time).

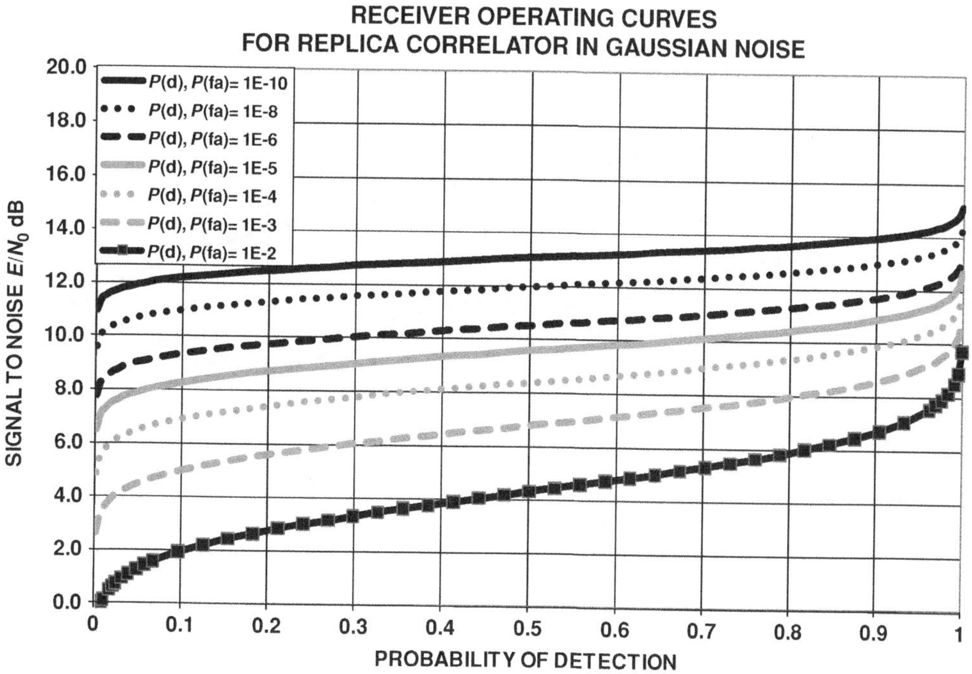

Figure 13.28 Receiver operating curves for detecting a known pulse in white Gaussian noise

13.3.2 Active Step 2: Correction for Noise Spectrum

Since the bandwidth being processed is, in most cases, small there is no need to adjust for nonflatness noise other than the reverberation in the band. As noted above and in Chapter 12, when calculating the theoretical Nrd, we assumed that the noise level was known a priori. In reality, the noise for real systems varies and must be estimated during the processing. This results in a penalty to the theoretical Nrd. As can be seen in Figure 13.19, the reverberation associated with a CW pulse is anything but flat; therefore a normalizer, the equivalent of a PNB NSE, is applied to smooth the data and estimate the underlying noise (including reverberation) in each detector bin.

Pryor [5] developed a methodology for estimating the NSE loss based on the F test:

$$\text{NSE loss} = 10\log\left[\frac{\left(\dfrac{F\,(\text{pfa})}{F\,(\text{pd})}-1\right)}{\left(\dfrac{F_0\,(\text{pfa})}{F_0\,(\text{pd})}-1\right)}\right] \tag{13.43}$$

where F(pfa) and F(pd) are values for the number of windowing bins, W, on each side used in the estimation and F_0(pfa) and F_0(pd) are for an infinite set of windowing bins, where

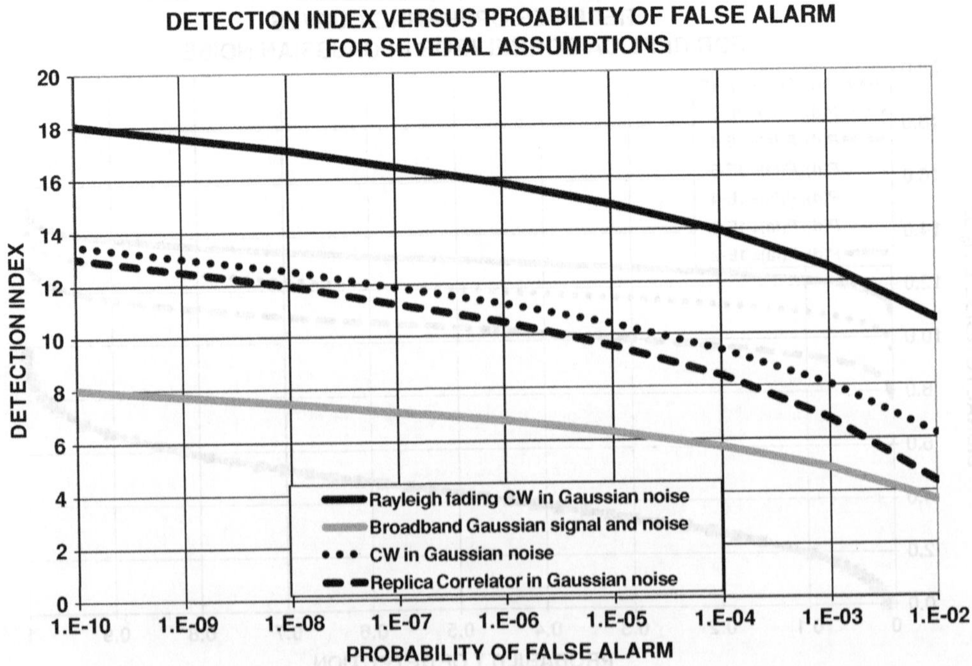

Figure 13.29 Detection index for various assumptions versus probability of false alarm

$$
\begin{aligned}
F(\mathrm{p}) &= \exp(2w) \\
w &= Y_{\mathrm{p}}\sqrt{\frac{h+\lambda}{h}} - \left(\frac{1}{m-1} - \frac{1}{n-1}\right)\left(\lambda + \frac{5}{6} - \frac{2}{3}h\right) \\
h &= \frac{2}{\dfrac{1}{m-1} + \dfrac{1}{n-1}} \\
\lambda &= \frac{Y_{\mathrm{p}}^2 - 3}{6} \\
Y_{\mathrm{p}} &= QI(\mathrm{p})
\end{aligned}
\tag{13.44}
$$

where

$m = 2 \times \mathrm{BT}$
$n = 4W \times \mathrm{BT}$
$W =$ number of window bins on each side
$\mathrm{BT} =$ bandwidth time product in each bin

$$
Y_{\mathrm{p}} = t - \frac{C_0 + C_1 t + C_2 t^2}{D_0 + D_1 t + D_2 t^2 + D_3 t^3} \tag{13.45}
$$

where

$C_0 = 2.515\,517$
$C_1 = 0.802\,853$
$C_2 = 0.010\,332\,8$
$D_1 = 1.432\,788$
$D_2 = 0.189\,269$
$D_3 = 0.001\,308$

This correction is shown in Figure 13.15.

The signal has a shape similar to the reverberation. Therefore, processing bins may exclude some signal and a similar amount of reverberation. Ambient and self noise are generally much flatter and do not suffer loss. This results in a slightly different Nrd for reverberation-limited or noise-limited operation (see Table 13.3).

13.3.3 Active Step 3: Correction for Processor Implementation

As discussed in Sections 13.1 and 13.2, compromises are generally made in the design of any active sonar system. These compromises bring about losses in performance. The mechanisms for active sonars are very similar to those already discussed. Table 13.4 shows the correction as a function of the type of processing employed.

13.3.3.1 Bandpass Filter Loss

This correction applies to nonmatched filter implementations. The corrections are the same as the filter correction in Section 13.2.3.1.

13.3.3.2 Detector Type

Matched filter receivers convolve the received signal with a replica of the expected signal. There are two implementations commonly used: (1) a true replica correlator and (2) an envelope correlator. The true replica correlator samples at a sufficient rate that all phases are sampled at slightly different time delays. The envelope correlator processing is done by convolving the received signal with two replicas, with carriers 90° apart and taking the quadrature sum of these as the matched filter output. This essentially matches to the modulation of the carrier and allows sampling at a rate comparable with the modulation, instead of the carrier frequency. The tradeoff is a few decibels in Nrd for a much lower computation rate. *D* for the two correlators

Table 13.4 Corrections needed for differnet processor types

	Matched filter processor	Energy detection processor
Bandpass filter		x
Overlap loss	x	x
Detector type		x
Display/integrator		x
Display OR-ing		x
Noise estimation	x	x
Other losses	x	x

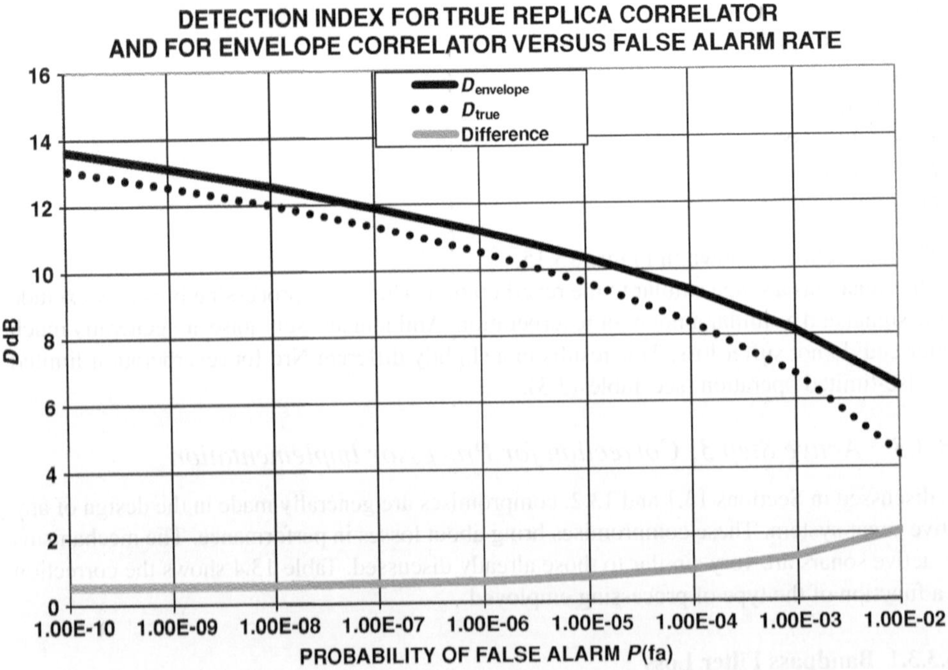

Figure 13.30 Detection index for true replica correlator and envelope correlator

is given by

$$D_{true} = 4.584 \ln\left[\ln\left(\frac{0.25}{P(\mathrm{fa})}\right)\right] - 1.02$$

$$D_{envelope} = 2.292 \ln\left[\ln\left(\frac{0.25}{P(\mathrm{fa})}\right)\right] + 0.989 + 10\log\left[1 + 0.22\log\left(\frac{100}{P(\mathrm{fa})}\right)\right] \tag{13.46}$$

The difference would be the correction needed. Figure 13.30 shows these as a function of probability of false alarm.

Replica correlators are typically used for coded pulses. For CW pulses, the processing is much like narrowband; therefore, corrections for linear or square law detection may be necessary (see Section 13.2.3.2).

13.3.3.3 OR-ing Correction

A multidimensional display surface is needed to completely display to output, intensity versus bearing, Doppler, sequential (multiple) ping, and range. One method for fitting this on a CRT screen is to OR the data (see Section 13.1.3.3). This is simply comparing two or more output values and building a display using only the strongest. For example, you might need to form the outputs required to display each of the 100 beams, 90 Doppler bins, and 200 range bins. If the Doppler bins were simply OR-ed it could fit on a typical CRT display. If the Nrd has been calculated using the full number of processed bins, there is essentially no loss in performance.

It is important to consider at what point in the processing chain to do the OR-ing. There is an OR-ing loss for any system that does time integration after OR-ing, as is the case for the gram display/operator. OR-ing losses have been extensively studied by Nuttal [6, 11] and Struzunski [12]. Unfortunately, all three references are classified or have limited distribution. However, an equation developed by Pryor [5] as a fit to the data is

$$\Delta \mathrm{Nrd} = 1.5 \log 2(\text{Number of OR-ed channels}) \frac{x}{\sqrt[4]{1+x^4}}$$
$$x = 0.131 \frac{\log_2(\text{Number of samples combined after OR-ing})}{\log_2(\text{Number of OR-ed channels})^{1/8}} \tag{13.47}$$

Figure 13.18 shows this expression for various choices of OR-ing and integration.

13.3.3.4 Nonideal Integration over Time Correction

In theory, replica correlators do not suffer from this because the processing time equals the signal time. However, in practice, the number of range bins may be difficult to display. For example, a 200 Hz wide FM pulse operated on a 60 kyd scale has about 17 000 independent range bins to display. Short CW signals suffer from the same issue. All CWs suffer from the possibility of the signal not being in the center in a processing bin. Splitting loss (ESL), discussed in Section 13.3.2, is another form of loss due to a time mismatch.

13.3.4 Active Step 4: Correction for Nonideal Signal Characteristics

The assumption for the match processor is that the signal is known exactly. If this is not the case, e.g., if fading is occurring during the arrival of the signal, performance is reduced. The effective length of a return is a function of target length and aspect, so the exact nature of the signal is not known a priori.

13.3.5 Active Step 5: Adjustment for Additional At-Sea Losses

The proceeding discussion has tried to take into account the spectral and temporal characteristic of both the signal and noise. Underlying assumptions are that both the signal and the noise can be described as stationary Gaussian processes. In the operational environment at sea, the noise field may arise largely from distant shipping sources that move with time through the beams of a sonar, through acoustic caustics that cause brief bursts of signal, or transients caused by human activity, such as hatch slamming. An analysis of at-sea recorded noise shows a myriad of such effects; i.e., the ambient noise has too much noise in its tails for the Gaussian assumption to be true and it contains many impulsive events. A distant ship is a correlated noise source and does not behave like a distributed source, either in the beamformer or in the signal processor. Typical at-sea loss varies widely depending on the frequency region processed and the dominant sources of noise. For systems that are dominated by flow noise or sea state noise, the corrections required are a few decibels for low-frequency broadband systems. In highly cluttered environments, 10–20 dB loss has been experienced. At sea, measurements are the only reliable method of estimating these losses. Under some circumstances, reverberation can have signal-like characteristics that could confuse a replica correlator.

13.3.6 Nrd Calculation Examples

Table 13.5 shows sample calculations of active Nrd with system losses included. Note that for CW pulses, two sets are needed, one for the noise-limited case and a second set versus Doppler for the reverberation-limited case. For CP pulses, it is the signal/replica mismatch that causes a change with Doppler; therefore, it applies to both the noise and reverberation cases.

13.4 Aural Detection

The above discussion does not address a human operator listening for either passive signals or active returns. For older sonars, this was frequently the only detection device. In modern times, the dependence on the human ear for initial detection has diminished because the human ear cannot listen to every sonar beam simultaneously. Instead, it has a limited time memory (integration time) and has a frequency response that may not match the signals of interest. That said, it is still a primary, if not the primary, method of investigating signals within its frequency range that have been detected by other displays. In some systems, the frequency is shifted for aural presentation to match the human ear better. Therefore, it is appropriate to discuss the characteristics of the human ear as a detector.

The human ears, thanks to evolution, are a remarkable detection and signal analysis system. The ear can be modeled as a set of narrow filters covering the audio region (see Figure 13.8); the parameters to be determined are the bandwidth of the filters and the integration time. The bandwidth of these filters is referred to in the literature as the critical bandwidth and is a function of the center frequency. Figure 13.31 shows a plot of the approximate critical bandwidth of the average human ears versus frequency. The curve is given by

$$W_{\text{crit}} = 10^{(af^2+bf+c)/10} \tag{13.48}$$

where

$a = 6.75$
$b = 35.1$
$c = 62$

The literature in this field is much too large to innumerate; the curve fit represents a fit to the following *Journal of the Acoustical Society of American* (JASA) articles: Schafer *et al.* [14]; Swets *et al.* [15]; Green *et al.* [16]; Hawkins and Stevens [17]; and the NDRC Technical Report 9 [18]. Measurements of the detection threshold for tones in broadband noise level off for long tones (Figure 13.32) (see Green *et al.* [16], Garner *et al.* [19], Hamilton [20], and

Table 13.5 Sample calculations of Nrd for various active signals

Type	CW	CW	CW	FM	FM	FM	Explosive
Pulse length (s)	0.01	0.5	2	0.01	0.5	2	0.0005
Processor bandwidth (Hz)	100	2	0.5	300	300	300	300
Reference bandwidth (Hz)	500	500	500	500	500	500	500
System false alarm rate (number/h)	0.25	0.25	0.25	0.25	0.25	0.25	0.25
Number of beams	50	50	50	50	50	50	50
Number of Doppler bins	3	80	320	1	1	1	1
Number of decisions/h	5.4E+07	2.9 E+07	2.9 E+07	5.4 E+07	5.4 E+07	5.4 E+07	3.6 E+08
P(fa)	5.33E-09	9.99E-09	9.99E-09	5.33E-09	5.33E-09	5.33E-09	7.99E-10
Albershein index	12.7	12.5	12.5	12.7	12.7	12.7	13.1
Nrd	12.7	12.5	12.5	7.9	−9.1	−15.1	13.1
System losses							
ESL	5	0	0	5	10	10	13
Transmitter scalloping	0	0	0	0	0	0	0
Receiver scalloping	0.17	0.17	0.17	0.17	0.17	0.17	0.17
Normalizer	1	2.7	4	0.8	0.8	0.8	1
FFT overlap	0.28	0.28	0.28				
Frequency scalloping	0.3	0.3	0.3				
Doppler mismatch							
0 kts				0	0	0	
1 kts				0.1	0.1	0.1	
2 kts				0.2	0.2	0.2	
4 kts				0.3	0.3	0.3	
8 kts				0.6	0.6	0.6	
16 kts				1.2	1.2	1.2	
Total system loss							
0 kts	6.5	3.2	4.5	6.0	11.0	11.0	14.2
1 kts	6.5	3.2	4.5	6.1	11.1	11.1	14.2
2 kts	6.5	3.2	4.5	6.2	11.2	11.2	14.2
4 kts	6.5	3.2	4.5	6.3	11.3	11.3	14.2
8 kts	6.5	3.2	4.5	6.6	11.6	11.6	14.2
16 kts	6.5	3.2	4.5	7.2	12.2	12.2	14.2
Nrd + loss (in processor band)							
0 kts	19.1	15.7	17.0	13.9	1.9	−4.1	27.3
1 kts	19.1	15.7	17.0	14.0	2.0	−4.0	27.3
2 kts	19.1	15.7	17.0	14.1	2.1	−3.9	27.3
4 kts	19.1	15.7	17.0	14.2	2.2	−3.8	27.3
8 kts	19.1	15.7	17.0	14.5	2.5	−3.5	27.3
16 kts	19.1	15.7	17.0	15.1	3.1	−2.9	27.3

(Continued)

Table 13.5 (*Continued*)

Type	CW	CW	CW	FM	FM	FM	Explosive
Nrd + loss (in reference band)	Noise only	Noise only	Noise only	Noise and reverb.	Noise and reverb.	Noise and reverb.	Noise and reverb.
0 kts	12.1	−8.3	−13.0	11.6	−0.3	−6.4	25.0
1 kts	12.1	−8.3	−13.0	11.7	−0.2	−6.3	25.0
2 kts	12.1	−8.3	−13.0	11.8	−0.1	−6.2	25.0
4 kts	12.1	−8.3	−13.0	11.9	0.0	−6.1	25.0
8 kts	12.1	−8.3	−13.0	12.2	0.3	−5.8	25.0
16 kts	12.1	−8.3	−13.0	12.8	0.9	−5.2	25.0
CW Doppler gain versus reverberation	Reverb. only	Reverb. only	Reverb. only				
0 kts	−0.1	−0.1	−0.1				
1 kts	−0.1	−15.8	−31.0				
2 kts	−0.1	−17.9	−31.9				
4 kts	−0.2	−31.0	−36.0				
8 kts	−0.6	−31.9	−48.9				
16 kts	−2.3	−36.0	−49.9				
CW Nrd + loss versus reverberation	Reverb. only	Reverb. only	Reverb. only				
0 kts	19.0	15.6	16.9				
1 kts	19.0	−0.1	−14.0				
2 kts	19.0	−2.2	−14.9				
4 kts	18.9	−15.3	−19.0				
8 kts	18.5	−16.2	−31.9				
16 kts	16.8	−20.3	−32.9				

the NDRC Report [18]. This suggests that the effective integration time is a little less than a second. The equation plotted in Figure 13.32 is

$$\text{Nrd} = \text{DT} = 5\log\left(d\frac{W_{\text{crit}}}{t}\right) + \left|5\log\left(\frac{T}{t}\right)\right| \tag{13.49}$$

where

$dW_{\text{crit}} = 600$

$T = 0.6$ s

The human operator has the capability to process time-varying signal behavior that is not matched in sonar signal processors. If a song is played into a broadband sonar processor, a line will appear on the display. If it is played into a narrowband sonar processor, a smear across many frequencies will appear on the display. If a human listens, they will unmistakenly recognize the national anthem, thus conveying enormously more information than the sonar displays. The human brain and ear have evolved to perform complex processing on sound. One has only to recall having a conversation in a night club, with

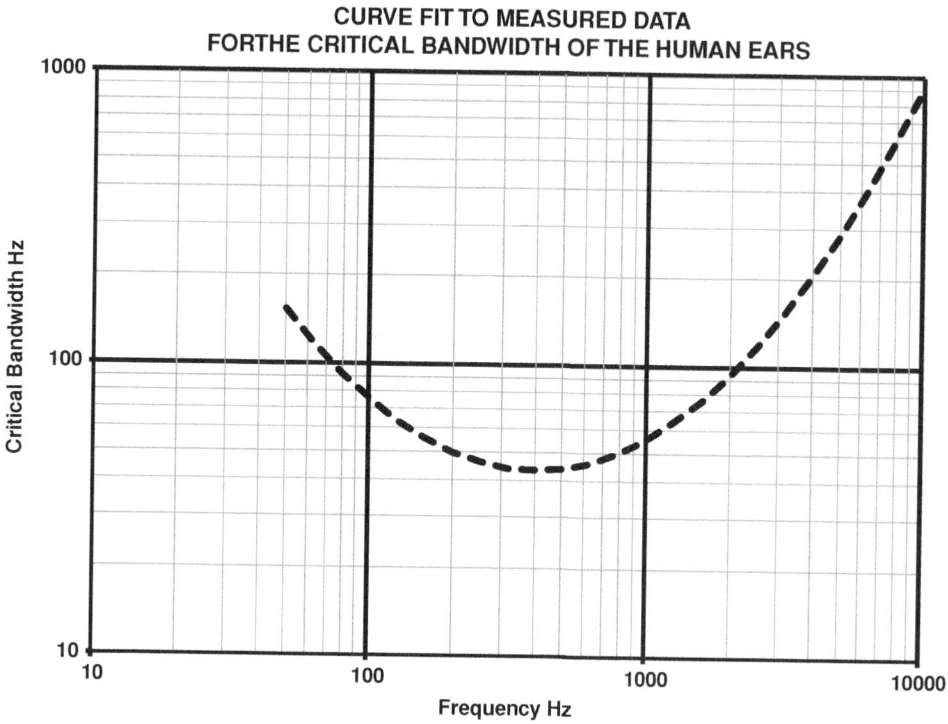

Figure 13.31 Human aural critical bandwidth versus frequency

multiple simultaneous ongoing conversations and loud background music, to realize that you can understand even when the signal to interference is very unfavorable. The human ear can also discern beat frequencies between closely spaced tones or amplitude modulation on a signal, again identifying a quality that might be important, but not obvious for a sonar display.

13.5 Display Nomenclature

A-display or scan or scope or indicator

A display in which targets appear as vertical deflections from a line representing a time base.

B-display or scan or scope or indicator

A rectangular display in which targets appear as blips with bearing indicated by the horizontal coordinate and distance by the vertical coordinate.

C-display or scan or scope or indicator

A rectangular display in which targets appear as blips with bearing indicated by the horizontal coordinate and angles of elevation by the vertical coordinate. Also:

D-display or scan or scope or indicator

A C-display or scan or scope or indicator in which the blips extend vertically to give a rough estimate of distance.

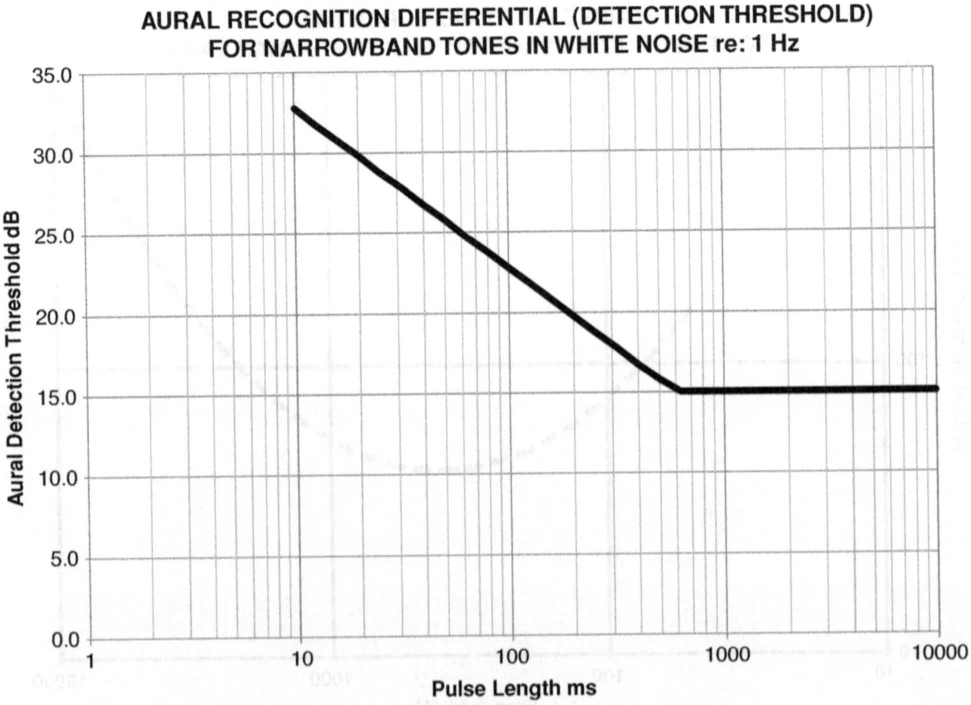

Figure 13.32 Human aural recognition differential versus pulse length for narrowband tones in white noise; this is a fit to measured data

E-display or scan or scope or indicator
: A rectangular display in which targets appear as blips with distance indicated by the horizontal coordinate and elevation by the vertical coordinate.

F-display or scan or scope or indicator
: A rectangular display in which a target appears as a centralized blip when the radar antenna is aimed at it. Horizontal and vertical aiming errors are respectively indicated by the horizontal and vertical displacement of the blip.

G-display or scan or scope or indicator
: A rectangular display in which a target appears as a laterally centralized blip when the radar antenna is aimed at it in azimuth and wings appear to grow on the blip as the distance to the target is diminished. Horizontal and vertical aiming errors are respectively indicated by horizontal and vertical displacement of the blip.

H-display or scan or scope or indicator
: A B-display or scan or scope or indicator modified to include an indication of the angle of elevation. The target appears as two closely spaced blips which approximate a short bright line, the slope of which is in proportion to the sine of the angle of elevation.

I-display or scan or scope or indicator
: A display in which a target appears as a complete circle when the radar antenna is correctly pointed at it and in which the radius of the circle is proportional to the

target distance. When not correctly pointing at the target, the circle reduces to a segment of a circle, the segment length being inversely proportional to the magnitude of the pointing error and its angular position being reciprocal to the direction of the pointing error.

J-display or scan or scope or indicator

A modified A-display or scan or scope or indicator in which the time base is a circle. The target signal appears as a radial deflection from the time base.

K-display or scan or scope or indicator

A modified A-display or scan or scope or indicator in which a target appears as a pair of vertical deflections or blips instead of a single deflection. When the radar antenna is correctly pointed at the target in azimuth, the blips are of equal height. When not correctly pointed, the difference in blip height is an indication of direction and magnitude of the azimuth pointing error.

L-display or scan or scope or indicator

A display in which a target appears as two horizontal blips, one extending to the right and one to the left, from a central vertical time base. When the radar antenna is aligned in azimuth at the target both blips are of equal amplitude. When not correctly pointed the relative blip amplitude indicates the pointing error. The position of the signal along the baseline indicates the target distance. The display may be rotated 90° when used for elevation instead of azimuth aiming.

M-display or scan or scope or indicator

A display in which the target distance is determined by moving an adjustable blip along the baseline until it coincides with the horizontal position of the target signal deflections. The control that moves the blip is calibrated in distance.

N-display or scan or scope or indicator

A display similar to the K-display or scan or scope or indicator in which the target appears as a pair of vertical deflections or blips from the horizontal time base. Direction is indicated by the relative amplitude of the vertical deflections; target distance is determined by moving an adjustable signal along the baseline until it coincides with the horizontal position of the vertical deflections. The horizontal control is calibrated in distance.

Plan position indicator (PPI)

1. A cathode-ray indicator in which a signal appears on a radial line. Distance is indicated radially and bearing as an angle.
2. In the radar technique, a cathode-ray indicator on which blips produced by signals from reflecting objects and transponders are shown in plan position, thus forming a map-like display.

   A north-upward plan position indicator has north at the top of the indicator regardless of the heading; a heading-upward plan position indicator has the heading of the craft maintained at the top of the indicator. On a delayed plan position indicator the start of the sweep is delayed so that the center represents a selected range. This allows distant targets to be displayed on a short range scale, thus providing larger scale presentation. An open-center plan position indicator has no signal displayed within a set distance from the center. An off-center plan position indicator is one modified so that the center about which the trace rotates can be moved from the center of the screen to provide a larger scale for distant targets. A master plan position indicator controls remote indicators or repeaters.

References

[1] Urick, R. J., *Principles of Underwater Sound for Engineers*, New York: McGraw-Hill, 1967.

[2] Abromowitz, M., and Stegun, I. A. (Eds), *Handbook of Mathematical Functions, with Formulas, Graphs, and Mathematical Tables*, New York: Dover Publications, 1 June 1965.

[3] Robertson, G. H., "Operating Characteristics for a Linear Detector of CW signals in Narrow-band Gaussian Noise," *Bell Systems Technical Journal*, **46** (4), 1967, 755–774.

[4] Pryor, C. N., and Kendall, P. A., " Effects of Hard Limiting on Recognition Differential of Gram Type Displays," Naval Ordnance Laboratory, Report NOLTR 73-189, 2 October 1973.

[5] Pryor, C. N. *A Common Methodology for the Calculation of Recognition Differential*, Vol. 1, Space and Naval Warfare Systems Command, Arlington, VA, PMW 182, 15 September 1997.

[6] Nuttal, A., "Detection Performance Characteristics for a System with Quantizer, OR-ing and Accumulator," Naval Underwater Systems Center, Technical Report 6815, October 1982.

[7] Whalen, A. D., and McDonough, R. N., *Detection of Signals in Noise*, 2nd edition, New York: Academic Press, 1995.

[8] Albersheim, W. J., "A Closed-Form Approximation to Robertson's Detection Characteristics," *Proceedings of the Institute of Electrical and Electronic Engineers*, **69** (7), July 1981, 839.

[9] Robertson, G. H., "Comparison of Square Law and Linear Detectors for *M* Independent Pulses," *Bell Systems Technical Journal*, **47**, March 1968.

[10] Edleblute, D., " Passive Sonar Equations for Broadband and Narrow-Band Detection," Naval Ocean Systems Center, TD 805, June 1985.

[11] Nuttal, A. " Detection Performance Characteristics for a System with Quantizers, OR-ing, and Accumualtor," *Journal of Underwater Acoustics*, **73** (5), May 1983, and Naval Underwater Systems Center, Technical Report 6815, 1 October 1982.

[12] Struzunski, W. A. " A Method for Treating Statistical Dependent Inputs to an OR-ing Device," *Journal of Underwater Acoustics*, **74** (1), July 1983.

[13] Dawe, R. L., "Detection Threshold Modeling Explained," DSTO Aeronautical and Maritime Research Laboratory, Melbourne, Victoria, Australia, Available from http://stinet.dtic.mil/index.html, AD Number: ADA335337, 1997.

[14] Schafer, T. H., Gales, R. S., Shumaker, C. A., and Thompson, P. O., " The Frequency Selectivity of the Ear as Determined by Masking Experiments," *Journal of the Acoustical Society of America*, **22**, 1950, 490.

[15] Swets, J. A., Green, D. M., and Tanner Jr., W. P., " On the Width of Critical Bands", *Journal of the Acoustical Society of America*, **34**, 1962, 108.

[16] Green, D. M., Birdsal, T. G., and Tanner Jr., W. P., " Signal Detection as a Function of Signal Intensity and Duration," *Journal of the Acoustical Society of America*, **29**, 1957, 523.

[17] Hawkins, Jr., J. E., and Stevens, S. S., " The Masking of Pure Tones and Speech by White Noise," *Journal of the Acoustical Society of America*, **22**, 1950, 6.

[18] National Defense Research Committee (NDRC), " The Masking of Pure Tones by White Noise," NDRC Technical Report 9, 1946.

[19] Garner, W. R., "Auditory Thresholds of Short Tones as a Function of Repetition Rate," *Journal of the Acoustical Society of America*, **19**, 1947, 600.

[20] Hamilton, P.M., "Noise Masked Thresholds as a Function of Tonal Duration and Band Width," *Journal of the Acoustical Society of America*, **29**, 1957, 506.

[21] Nutta,l A., " Operating Characteristics for Indicator Oring of Incoherently Combined Matched-Filter Outputs," New London, CT: Naval Underwater Systems Center, 21 September 1987.

14

False Alarms, False Contacts, and False Targets

Confusion and disagreement on this subject stems from differences in the definitions in use by different groups involved with sonars. Detection to a detection theorist means that a certain statistical test has been passed. To an officer on a submarine, detection frequently means that a contact has been classified as a particular type of object. Accordingly, a false alarm to a display designer may be a mark on a screen not caused by a target's signal. A false alarm to a detection theorist means passing a certain statistical test when the signal of interest is not present, to a sonar operator it may mean any contact that is not a target of interest, and to the skipper of a ship it may mean something he ordered attacked that turned out to not be a target of interest.

For this discussion, we will define a false alarm as a mark, or an unusually high level, on a sonar display. Sources of false alarms may be random noise, reverberation, biologics, or nonthreat platforms. Sensor engineers usually include only random noise in the probability of False Alarm (P_{fa}) calculation. The frequency of occurrence for false alarms can be enormous to extremely low depending on the sources included and the environmental sources. The level of random noise false alarms is one of the determining factors in the recognition differential, N_{RD}, of a given system.

14.1 Sea Story

When I first entered this field in 1967, I was involved in a program whose purpose was to communicate acoustically with submarines. One of the things I was asked to do, primarily as a learning experience, was to review the false alarm rate calculation that had been set by specification to 1 per 100 years per 100 receivers. The calculations were straightforward. The only difficulty was finding tables that went out to low enough probabilities ($\sim 7\sigma$ was required). These calculations were done shortly after the first operational test was performed on a deployed submarine. When the submarine returned, the first question asked was how many messages did the submarine receive? The answer was four messages were received. This was a cause for celebration because four messages had been sent over the two-month

Underwater Acoustics: Analysis, Design and Performance of Sonar Richard P. Hodges

test period. Rejoicing did not last long as everything turned to doom and gloom the next day. Only one message receipt time matched a message transmission time. The received message that did match up in time was not the message sent. One incorrect message was received and three others were failures to receive. As a consequence, we were faced with at least three false alarms on one receiver in two months, or roughly 200 000 times what the system was specified and designed for. The question to be answered was "what went wrong?"

14.2 Failure to Detect

The analysis of the failure to receive messages was fairly easy. First of all, reconstruction of the circumstances when these messages had been sent indicated that the signal to noise at the receiver should have been more than adequate. Second, the signals that had been received had been recorded so we had something to work with. Analysis showed the failure to detect was caused by two factors: (1) we were using a set of replica correlators to detect the messages and (2) the signals where suffering heavy multipath arrivals whose characteristics were different than earlier technical tests. Therefore, the signals were probably present, but the multipath arrivals made the detector ignore them in three cases and to interpret one incorrectly in the other case. Two things were learned from this test: (1) replica correlators, which were chosen because they theoretically gave better detection performance for a given false alarm rate, also reject real signals if the transfer function of the ocean is incorrect (i.e., have the wrong replica) and (2) a better set of codes was needed to prevent misinterpretation.

14.3 Detection Theory

In order to understand the problem with false alarms, it is necessary to walk through a discussion of statistical detection theory. The objective of this chapter is not to provide a rigorous discussion of detection theory, but instead to highlight the important aspects of detection theory so the subsequent discussion can be appreciated. In addition to Chapter 12, Statistical Detection Theory, two excellent sources for a more rigorous discussion are *Detection, Estimation, and Modulation Theory, Part 1*, by Van Trees [1], and *Detection of Signals in Noise*, by Whalen and McDonough [2]. A less rigorous, yet very well presented, discussion can be found in "Lecture Notes on Underwater Acoustics", by Bartberger [3].

14.3.1 Hypothesis Testing

There are two hypotheses concerning a received sequence, $x(n)$. The first, the null or H_0 hypothesis, assumes that the received sequence consists only of noise, that is $x(n) = g(n)$. The second hypothesis, H_1, assumes that a (the) signal is present or $x(n) = s(n) + g(n)$. Let D_0 be the choice of H_0 and D_1 be the choice of H_1. Four outcomes are possible based on the sample $x(n)$:

1. Correct decision that the signal is present. The probability of this outcome, called the probability of detection (P_D), is written $P(D_1/H_1)$.
2. Incorrect decision that the signal is present. The probability of this outcome, called the probability of false alarm (P_{fa}), is written $P(D_1/H_0)$.

3. Incorrect decision that the signal is not present. The probability of this outcome, called the probability of false rest (P_{fr}), is written as $P(D_0/H_1)$.
4. Correct decision that the received sequence consists only of noise. The probability of this outcome, called the probability of a correct rest (P_{cr}), is written as $P(D_0/H_0)$.

It can be seen that outcomes 1 and 3, as well as outcomes 2 and 4, are mutually exclusive, exhaustive events. If the signal is present, either outcome 1 or 3 must happen, $P_D + P_{fr} = 1$. If the signal is not present outcome 2 or 4 must happen, $P_{fa} + P_{cr} = 1$. Thus, there are really only two probabilities to calculate, one from each group, typically P_D and P_{fa}. In some situations, such as with a communication system, the a priori probability that a signal is present may be known. In general, for sonars, the a priori probabilities are not known.

In order to make a decision, a receiver must measure some property that is suitably related to the probability of detection. Peterson and Birdsall [4] showed that the property that should be measured by an optimum receiver is the likelihood ratio. This is the ratio of the conditional probability density of the observed input to the receiver, under the hypothesis that the signal is present, to the conditional probability density of the input under the hypothesis that only noise is present. This is called a Neyman–Pearson (NP) criterion and can be applied because only knowledge of the signal and the noise probability density functions (PDFs) is required.

14.3.2 Probability Density Function

An input to a receiver might consist of a continuously varying signal voltage to which has been added, usually not intentionally, a continuously varying random noise voltage. Assuming that the input lasts a time, T, and is band limited in frequency to a bandwidth, W, any discrete sample of the input voltage taken at an instant is a random variable because noise is a random process. If x_i is the ith sample, the probability density $p(x_i)$ is a function such that $p(x_i)\,dx_i$ represents the probability that the voltage has a value between x_i and $x_i + dx_i$. Sampling theory shows that if the input is sampled at regular intervals of one half watt-seconds (the Nyquist rate [5]), it can be completely reconstructed from the $2WT$ samples. The actual continuous input, $x(t)$, and the sequence of measurements, $x_1, x_2, \ldots, x_n$, are interchangeable (i.e. they contain the same information). Sampling at this rate, the samples are statistically independent. Therefore, the joint probability density function of all the $n(= 2WT)$ samples is the product of the individual density functions.

The NP detector, which maximizes the P_D for a specified P_{fa}, uses a likelihood ratio test of the following form:

$$L_t = \frac{p_1(y)}{p_0(y)} \geq \beta \tag{14.1}$$

where

L_t = test statistic
y = single observation of $x(n)$
$p_1(y)$ = probability density of y given H_1 is true
$p_0(y)$ = probability density of y given that H_0 is true

The decision rule is:

$$\text{Choose } H_1 : \text{when } L_t \geq \beta$$
$$\text{Choose } H_0 : \text{otherwise}$$

14.3.3 Detection of Constant Level

The signal is a constant level, c, and the noise is a sample drawn from a zero mean, with unit variance Gaussian, g:

$$H_1 : y = c + g$$
$$H_0 : y = g$$

An optimum detection rule, in the NP sense, is desired for a P_{fa} of 0.001. Then

$$L_{\text{t}} = \frac{p_1(y)}{p_0(y)} = \frac{\text{Gaussian(mean} = c\text{, var} = 1)}{\text{Gaussian(mean} = 0\text{, var} = 1)} \tag{14.2}$$

$$L_{\text{t}} = \frac{\frac{1}{\sqrt{2\pi}} \text{e}^{-(y-c)^2/2}}{\frac{1}{\sqrt{2\pi}} \text{e}^{-(y)^2/2}} = \text{e}^{(2cy - c^2/2)} \tag{14.3}$$

The desired P_{fa} (0.001) is used to determine the threshold value, y_{t}:

$$P_{\text{fa}} = \frac{1}{\sqrt{2\pi}} \int_{y_{\text{t}}}^{\infty} \text{e}^{-y^2/2} \, \text{d}y \tag{14.4}$$

The desired value of y_{t} is calculated as 3.09 [6], i.e. the number of sigma on a Gaussian probability distribution to get 0.999. The probability of detection (P_{D}) is

$$P_{\text{D}} = \frac{1}{\sqrt{2\pi}} \int_{y_{\text{t}}}^{\infty} \text{e}^{-(y-c)^2/2} \, \text{d}y \tag{14.5}$$

Table 14.1 shows the results of Equation 14.5 for several values of c.

Table 14.1 Probability of detection for selected values of c

c	$P_{\text{D}}(c)$
1	0.018
2	0.138
3	0.464
4	0.819
5	0.972

14.4 False Alarm Probability Calculation

In the example above, the specification of P_{fa} requires the determination of the proper lower bound for the appropriate integral. Tables are readily available that cover the cases discussed [6]. For more complex systems, a numerical integration might be required. Therefore, setting the probability of false alarm appears to be a straightforward procedure, albeit potentially messy.

The false alarm rate recognizes that each detector has an assumed time and number of channels associated with it, so the number of false alarms per hour, day, or 100 years can be calculated. Normally, the probability of false alarm is set to low values because P_{fa} is for a single instance of an opportunity. For example, Whalen and McDonough [2] identifies typical values as between 10^{-2} and 10^{-10}. For example, an active search system might have an opportunity to make a detection or a false alarm every 500 ms on any of 30 bearings. In a day, that is

$$\frac{24\ \text{h}}{\text{day}} \times \frac{60\ \text{min}}{1\ \text{h}} \times \frac{60\ \text{s}}{1\ \text{min}} \times \frac{2\ \text{time periods}}{\text{s}} \times \frac{30\ \text{bearings}}{\text{time period}}$$
$$= \text{over 5 million opportunities per day!}$$

If P_{fa} is 10^{-4}, or 500 false alarms per day, that would be one false alarm every three minutes, definitely high enough for the Commanding Officer to request the system is shut down.

Revisiting the Sea Story, several things could go wrong that would produce the same results. Consider a simple detector such as the square law detector described in Chapter 13, Methodology for Calculation of the Recognition Differential. This detector measures the total energy arriving during a fixed period of time. If the energy exceeds the established threshold, a detection is declared. When building comparable devices, it is necessary to know at how many watt-seconds to set the detection threshold. To compute this, a noise level must be known or assumed, like sea state 3. If the threshold is set for SS3, then each time a storm forms, the detector will go off continuously. Complicating estimates of an appropriate threshold is that the ambient noise is not a stationary Gaussian process, uncorrelated from sample to sample as assumed in the example. Figure 14.1 shows a set of ambient noise measurements taken over a period of many days. Note the variation in noise power over a range of 20 dB, which is a factor of 100 in actual power terms. As a consequence for practical systems, a method must be found to adjust the threshold to follow the mean noise level. If the threshold estimate is missed by even a fraction of a decibel, the false alarm rate for many systems will be significantly changed. (See Whalen [7], pp. 256, Figure 8.12. Notice that a change of 4 dB is sufficient enough to change the P_{fa} by a factor of 1 000 000.)

A similar variability is true for shipping-related noise. Consider a calculation of noise from surface traffic to a sensor at a depth, d. The level received will be given by

$$N = \int_{d}^{R_{\max}} \frac{2\pi\rho S R}{R^2}\, dR \tag{14.6}$$

$$N = 2\pi\rho S \ln\left(\frac{R_{\max}}{d}\right) \tag{14.7}$$

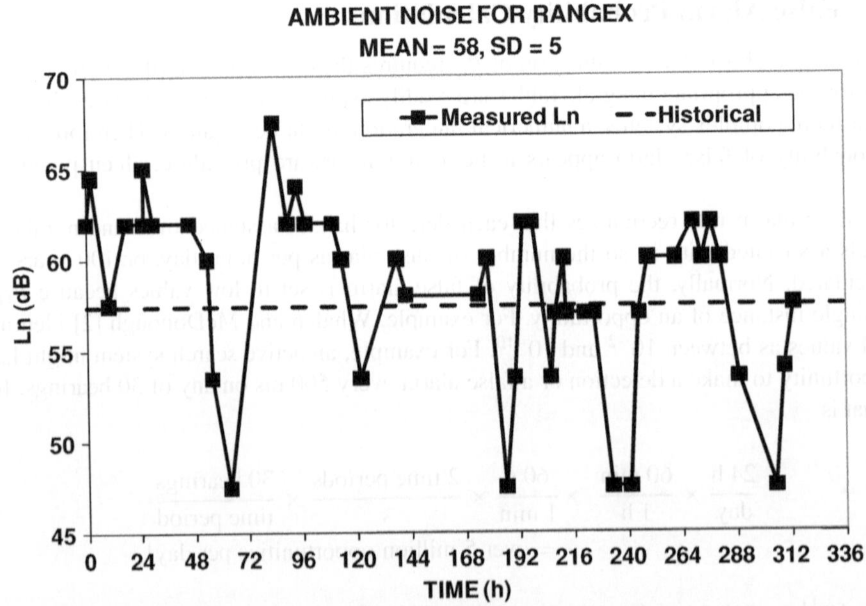

Figure 14.1 Variability of ambient noise with time

where

ρ = traffic density (typically 0.001 ships per unit square nautical mile)
S = signal radiated in the band of interest
$1/R^2$ = propagation loss

Assuming $R_{max} = 1500$ nmile, $d = 100$ ft, $\rho = 0.001$ ships per nmile^2, and $S = 100$:

$$N = 7.2 \text{ or } 8.6 \text{ dB}$$

A quick investigation will show that 50 % of the noise comes from inside 4.93 nmile. For the ship density, ρ, chosen and assuming a uniform distribution, there is a 93 % chance that no ships are in the region, a 7 % chance of one ship, and 0.3 % chance of two or more ships. The contribution we added in the integration for this region corresponds to one ship at 5.3 nmile. Therefore, if no ships are inside 4.93 nmile, the level will be too low and if one ship or more is inside 4.93 nmile the level will be too high. In reality, it is unlikely to observe a value very close to the average calculated; instead widely varying values caused by the discrete nature of the sources will be observed. A Monte Carlo simulation using these parameters and a sample size of 1000 gave a minimum value of 3.3 dB, a maximum value of 25.8 dB, a median of 5.2 dB, and a mean of 5.8 dB; 85 % of the values were within 2 dB of the median, but the other 15 % were distributed over a 20 dB region. Notice that the mean from the simulation is quite different from the calculated value of 8.6 dB. This difference is because the noise level is dominated by what is occurring at short range and a sample size of 1000 is not large enough to get good convergence on the mean.

14.5 False/Nonthreat Contacts

The detection theory analysis above regards signal as anything that is not the white band limited noise analyzed. The oceans are full of discrete sources or reflectors that are not the targets or signals of interest and may not be well described by a Gaussian probability density curve. These include merchant traffic, fish, marine mammals, shrimp, fishing boats, oil rigs, own ship, and the ocean surface.

> The term "ambient noise" refers to the noise that remains after all easily identifiable sound sources are eliminated. For instance, the presence of many ships randomly distributed over the ocean surface results in a component of ambient noise ascribed to "distant shipping" or "ship traffic". However, the noise produced by a single nearby ship is easily identified and localized, and is therefore treated as an acoustic signal rather than as a part of the ambient noise (Burdic [8]).

False or nonthreat contacts are physical objects that are incorrectly classified as the target of interest. The ability of a false or nonthreat contact to cause detections is easily calculated if the detector characteristics and the object's signal characteristics are known. To compute the probability of false or nonthreat contacts, it is necessary to know the frequency at which they occur. This frequency is highly dependent on location, season, sensor characteristics, and operation. Due to the discrete nature of the sources, it would be a mistake to degrade all sensor beams to try to reduce the probability of detecting them (this holds true for all but the most unique signals). Instead, the proper technique is to check each of the detections and eliminate them as the target of interest by classifying them as nontargets. Active systems encounter an additional problem in that the bottom, surface, merchant ships, and schools of fish are all sources of false or nonthreat contacts because they reflect the transmitted signal back to the receiver.

14.6 False Targets

A false target is a detection that has been incorrectly classified as a contact of interest resulting in actions that may include the use of deadly force. Sources of false targets include, but are not limited to, nonthreat traffic, biologics, and reverberation. The frequency of their occurrence ranges from a fraction of detections to one per day. When modeling false targets, targets are generated that are misclassified at a prescribed rate to allow for a platform to react appropriately.

For sonars, which by design have a low probability of a false alarm, a false target is almost always a false or nonthreat contact. Classification is a more complex process than detection and is not easily treated by this same kind of approach.

Revisiting the Sea Story, the recordings taken during the test showed that the three "false alarms" were false or nonthreat contacts that were mistaken for the actual target message. Reviews of both the sound and the log book suggest that two of the contacts were most probably own ship, given that the submarine was shallow doing noise evolutions. The third contact was never identified with any certainty as to whether it was a biologic or man-made sound. Based on this, it was concluded that the false alarm rate of the system was actually very low, but what was really desired was a very low probability of a false target and that is not what was specified, designed, or delivered. This example illustrates the communication gap between engineers, system managers, and operators.

14.7 Summary and Conclusions

By the definitions defined here, a low false alarm rate makes a good display for operators to use because it does not impose a high burden of spurious contacts. Consider the system described in Section 14.4, which with a P_{fa} of 10^{-4} would give a false alarm every three minutes. An Officer in Charge may tire of a false alarm every three minutes. Operators may adapt by simply reporting only contacts that were present for two successive pings. This change in behavior would drop the reporting of false alarms to around one or two a year, but would do almost nothing in eliminating false or nonthreat contacts.

False contacts are not false, but are real contacts that a sonar operator must classify into nonthreat contact categories such as biologics, merchants, fishing boats, reverberation, the bottom, the surface, etc. The principal effects of false or nonthreat contacts on sonar performance are masking contacts of interest at the same location on the display and tying up operators by creating a queue so that they are delayed in, or may never get around to, looking at or listening to contacts that are of actual interest.

Finally, false targets are, by the definition given here, false or nonthreat contacts that have been incorrectly classified as contacts of interest. These may result in anything from the break in the normal search plan, to launching of a helicopter, or to attack by use of a weapon, etc.

References

[1] Van Trees, H. L., *Detection, Estimation, and Modulation Theory, Part 1*, New York: John Wiley & Sons, Inc., 1968.

[2] Whalen, A. D., and McDonough, R. N., *Detection of Signals in Noise*, 2nd edition, New York: Academic Press, 1995.

[3] Bartberger, C. L., "Lecture Notes on Underwater Acoustics," U.S. Naval Air Development Center, Johnsville, PA, NADC Report NADC-WR-6509, 17 May 1965 (AD 468 869).

[4] Peterson, W. W., and Birdsall, T. G., "The Theory of Signal Detectability," Electronic Defense Group, Engineering Research Institute, University of Michigan, Technical Report 13, June 1953.

[5] Bendat, J. S., *Principles and Applications of Random Noise Theory*, New York: John Wiley & Sons, Inc., 1958.

[6] Zwillinger, D., *CRC Standard Mathematical Tables and Formulae*, 31st edition, London: Chapman & Hall, 27 November 2002.

[7] Whalen, A. D., *Detection of Signals in Noise*, New York: Academic Press, 1971.

[8] Burdic, W. S., *Underwater Acoustic System Analysis*, Englewood Cliffs, NJ: Prentice-Hall Inc., 1984.

15

Variability and Uncertainty

When the passive or active sonar equations are solved, as in Figure 15.1, values for each term are needed. In general, these inputs are estimates of the median or mean values expected. Unfortunately, at any moment at sea, the target will not be the expected class average value, the ambient noise will not be the median for that month in that area, the sound speed profile will not match exactly what we retrieved from a database, the propagation loss will come from a model and the operator performance will not be exactly the median. As a result, each term in the sonar equation is off by an amount that we do not know at any given moment at sea. In addition, there is "temporal variability" that alters each term with time. Given this, the result from a sonar equation calculation is something of a "global mean or median," in other words, an expected value.

The actual observed signal excess will be different by an unknown quantity from the expected value:

$$XS_{\text{obs}} = XS_{\text{exp}} + e$$
$$SE_{\text{obs}} = SE_{\text{exp}} + e \tag{15.1}$$

where e includes the temporal variability and all differences between databases, models, and the real world.

Research into the development of a model for e includes three major approaches: (1) comparison of expected detection ranges with at-sea tests and real world encounters; (2) time step by time step comparison of the sonar equation and measured signal to noise from at-sea tests; and (3) studies of the variability of each sonar equation term.

For the first approach, comparison with at-sea exercise results, the model for e currently in use by the Naval Undersea Warfare Center (NUWC) is the sum of three components: (1) a long-term (long time constant between independent samples) correlated random variable, (2) a long-term uncorrelated random variable, and (3) a short-term random variable. The parameters for the distributions are shown in Table 15.1. The correlated variable takes on the same value for all signals being considered by a platform. The uncorrelated variable takes

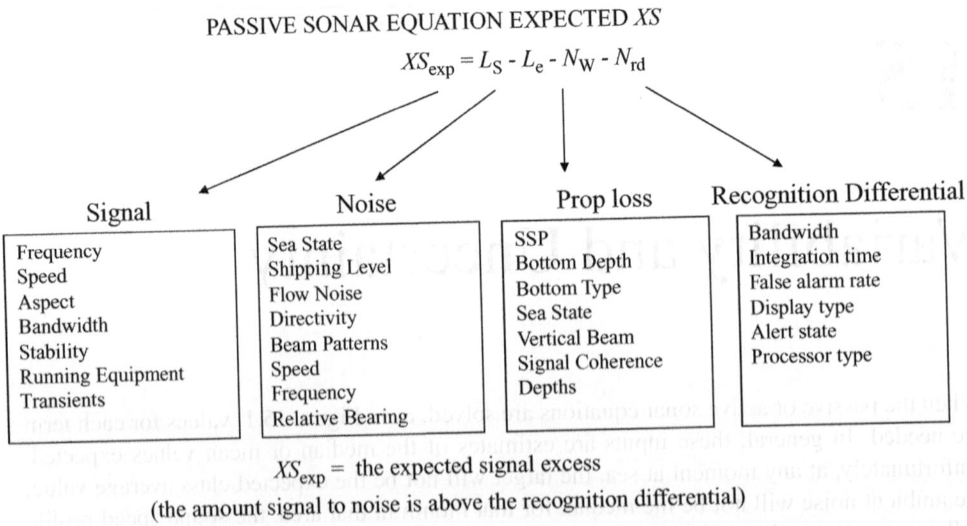

Figure 15.1 The expected signal excess equation components

on different, statistically independent values for each signal. The short-term random variable, which is also uncorrelated, is independent with each time step or ping for an active system for a cumulative probability of detection (CPD) calculation (see Chapter 17, Cumulative Probability of Detection).

The Gauss–Markoff process has the following form:

$$R_i = \alpha R_{i-1} + \beta \mathrm{GRV}_i \tag{15.2}$$

where

R_i = ith random variable
$\alpha = \exp(-t/T)$
t = time step
T = correlation time associated with the process
$\beta = \sqrt{(1-\alpha^2)}$
GRV_i = ith independent Gaussian random variable with zero mean

Table 15.1 Stochastic processes parameters for sonar (passive/active)

Variable	Type	Median	Standard deviation	Correlation time
Rlc	Gauss–Markoff	0.0	4.5 / 9.0	12 h
Rlu	Gauss–Markoff	0.0	4.7 / 6.0	4 h
R_short	log Rayleigh	0.0	2.7/5.4	0
	Total	0.0	7.0/12.1 dB	5.6/6.6 h

The resulting series of random numbers is Gaussian, or normally distributed, with a mean of zero, a standard deviation equal to that of the GRV used, and a correlation time of T.

In classic detection theory, R_short has the shape of the receiver operator curve (ROC) if the time step is equal to the processor integration time (the receiver operator curve or the probability of detection versus signal to excess is discussed in Chapter 13). The distribution curves for R_short depend on the details of the signal processing. In practice, its variance is small compared to the long-term random variables. As long as the variance is approximately correct, the shape of R_short does not have a large effect on the probability of detection. Worse, for passive sonar, the processor integration time is generally long compared to the speed in which the tactical situation is changing. For example, range, and consequently transmission loss, may change significantly during this time. For that reason, NUWC generally uses a generic distribution, which for most purposes can be made into an adequate ROC. The generating function for this distribution is

$$z = -5p\log\left[\frac{-\log(\mathrm{URV})}{\log(2)}\right] \tag{15.3}$$

where

z = a random sample from distribution
p = an input parameter (1 for passive and 2 for active)
URV = uniform random variable, i.e., a random number drawn from a uniform distribution between 0 and 1

An additional advantage of using this distribution function is that it allows the calculation of the cumulative probability of detection to be independent of the time step chosen, provided the actual changes in the sonar equation are small over the time step. The distribution for Rs can be shown to be a log Rayleigh as follows. The Rayleigh density function is

$$P(x) = \frac{x}{v}\mathrm{e}^{-x^2/(2v)} \tag{15.4}$$

The generating function for this is

$$x = \pm\sqrt{-2v\ln(\mathrm{URV})} \tag{15.5}$$

Comparing $-10\log(x)$ with z:

$$z = -5p\log\left[\frac{\log(\mathrm{URV})}{\log(2)}\right] = -5p\log\left[\frac{\ln(\mathrm{URV})}{\ln(10)\log(2)}\right] \tag{15.6}$$

$$-10\log(x) = -5\log[-2v\ln(\mathrm{URV})] \tag{15.7}$$

If $p = 1$ and $2v = 1/[\ln(10) \times \log(2)]$ then the expressions are identical.
For $p = 2$:

$$z = -10\log\left[\frac{-\log(\text{URV})}{\log(2)}\right] \tag{15.8}$$

Let $-10\log(y) = z$, which gives

$$y = \frac{-10\log(1-P)}{\log(2)} \tag{15.9}$$
$$-\log(2)y = \log(1-P)$$

or

$$1 - P = 10^{-\log(2)y}$$
$$P = 1 - \mathrm{e}^{-\ln(2)y} \tag{15.10}$$

and

$$\rho(p) = \ln(2)\mathrm{e}^{-\ln(2)y} \tag{15.11}$$

The underlying distribution is exponential (chi-Squared with 2 degrees of freedom) or what might be expected for a narrowband process (a signal with a Gaussian spectral shape passing through a filter that also has a Gaussian spectral shape).

15.1 Random Variability of a Sonar

The random variability associated with a sonar represents what is left over when the best acoustic models are used to predict signal to noise and that value is compared to measured data. If all components of the sonar equation are taken into account, and the values observed at sea are subtracted from the calculated values, the result can be viewed as a random variable with a measurable mean, standard deviation, correlation time, etc. Given this, the actual observed detections and signal-to-noise ratios (1) have a median value given by the sonar equation and (2) the distribution about which cannot be statistically distinguished from the random processes described with the data currently available (60+ exercises/technical evaluations/real encounters). An accurate prediction of the outcome of an exercise or series of encounters can and has been made by using the sonar equation with these random processes. These predictions include, but are not limited to, the correct distribution of detection ranges, detecting sensors, and detection signals.

15.2 Sources of Variability

Is the error an error in the model or is it due to variations in the real world? The answer is both. The underlying models, such as propagation loss models, may include errors because

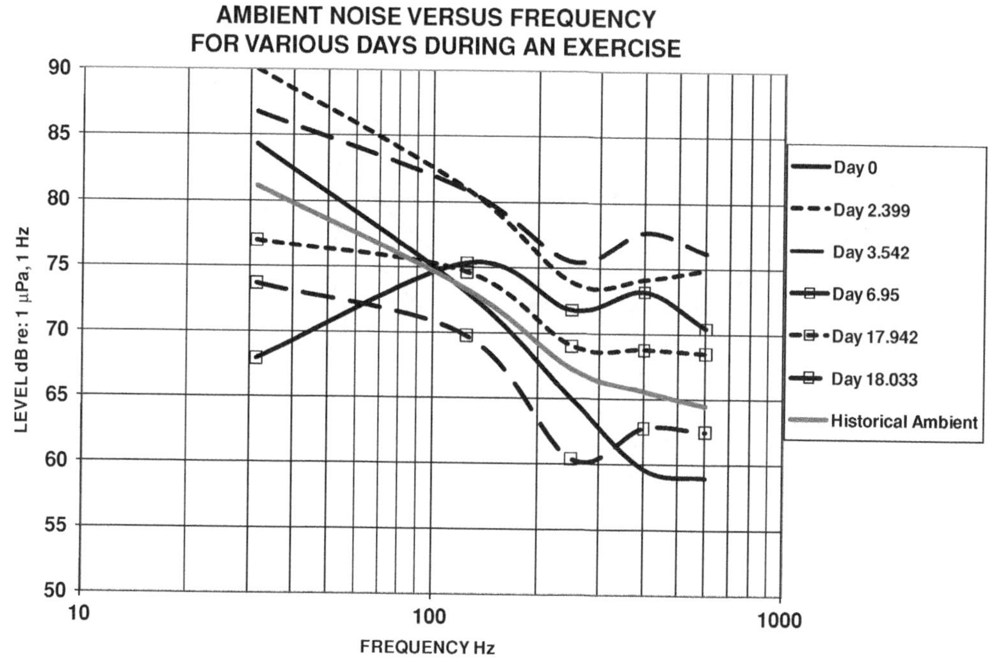

Figure 15.2 Measured ambient noise in the North Atlantic at various times during an exercise

these models (1) use approximations in their calculations, (2) have inaccurate submodels (e.g., volume attenuation, surface loss, or bottom loss), (3) have inputs that may be in error (e.g., SSPs, bottom characteristics, or wave height), and (4) assume conditions are static (i.e., waves, SSP, etc., generally do not vary with time, although most modern models allow SSP to vary with location) and target and own ship are stationary.

At-sea measurements suggest that the dominant sources of propagation loss error are from assumptions (3) and (4). In reality, the terms in the sonar equation vary from moment to moment, i.e., have temporal variations. For example, Figure 15.2 shows the ambient noise measured during an exercise. The predictions for this operating area and season were sea state (SS) 3 and shipping level (SL) V, which on average was correct, but during the exercise the actual ambient noise varied by at least ± 10 dB.

In *Principles of Underwater Sound for Engineers*, Urick [1] discusses the "afternoon effect," which was an observation that sonar performed better in the morning than in the afternoon. The explanation for this was that the sound speed profile (SSP), among other things, was changing as the sun warmed the ocean surface. Any discussion of detection theory should explicitly include noise, which will vary from moment to moment. As an example, for the calculation of passive narrowband (PNB) recognition differentials, it is common to assume that signal to noise of interest in the short term is a chi-squared distribution with $2N$ degrees of freedom (see Pryor [2]). As such, random processes are real, observable, and, unfortunately, very complex.

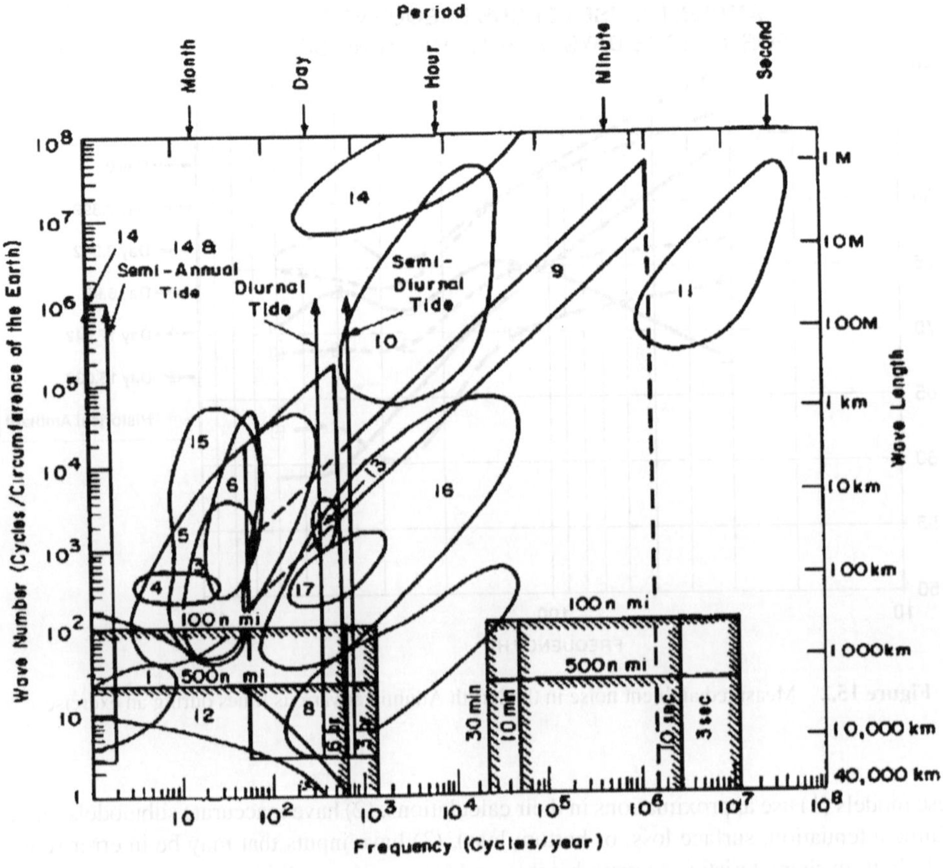

Figure 15.3 Major ocean phenomena

For a given scenario, the random variables represent the variability that one might expect to encounter if a series of engagements were run at that location, in the season indicated, with the platforms specified. While the nominal sea state might be 3 for this area, one morning it might be SS 2 and the next morning SS 5.

Normally, tides are ignored in deep water, but are of major importance in shallow water. Measurements of propagation loss using fixed hydrophones in Long Island Sound showed a 10 dB variation that was highly correlated with the tidal cycle.

Figures 15.3 through 15.6 show a rough approximation of the size and time constants associated with elements of variability in the oceans [3].

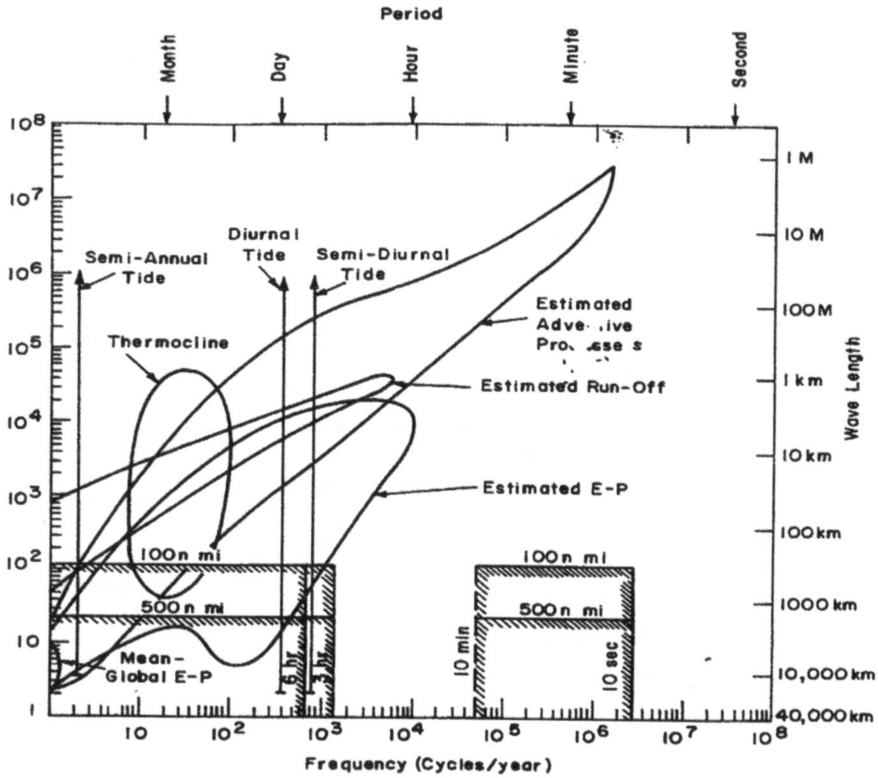

Figure 15.4 Phenomena and processes affecting the variability of salinity

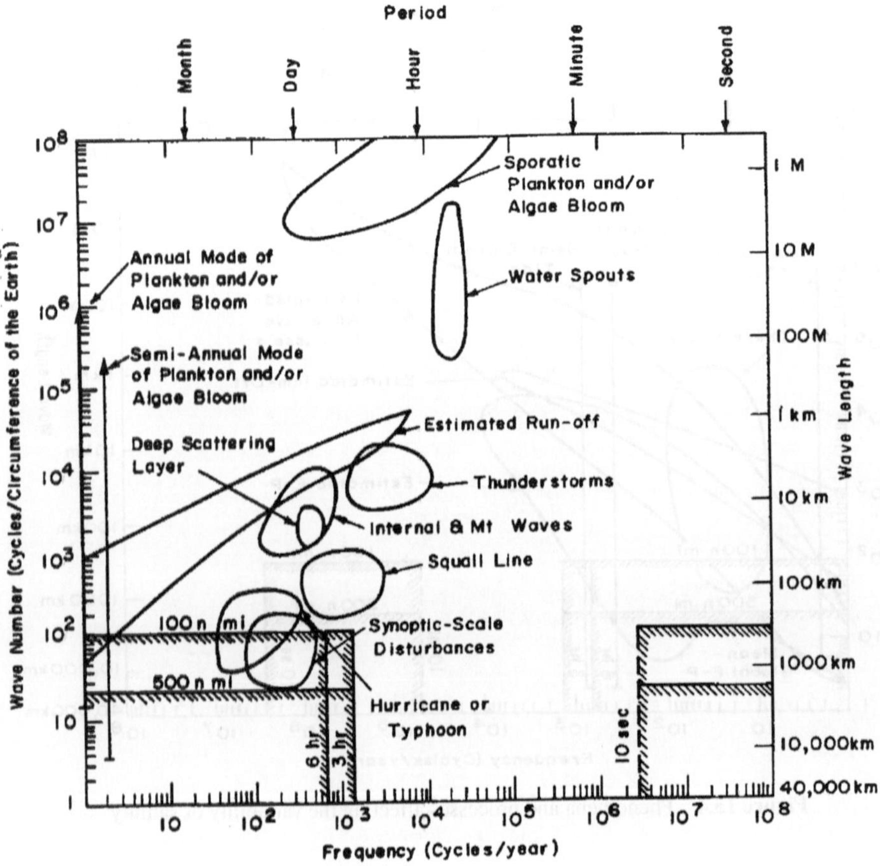

Figure 15.5 Phenomena and processes affecting the variability of transparency

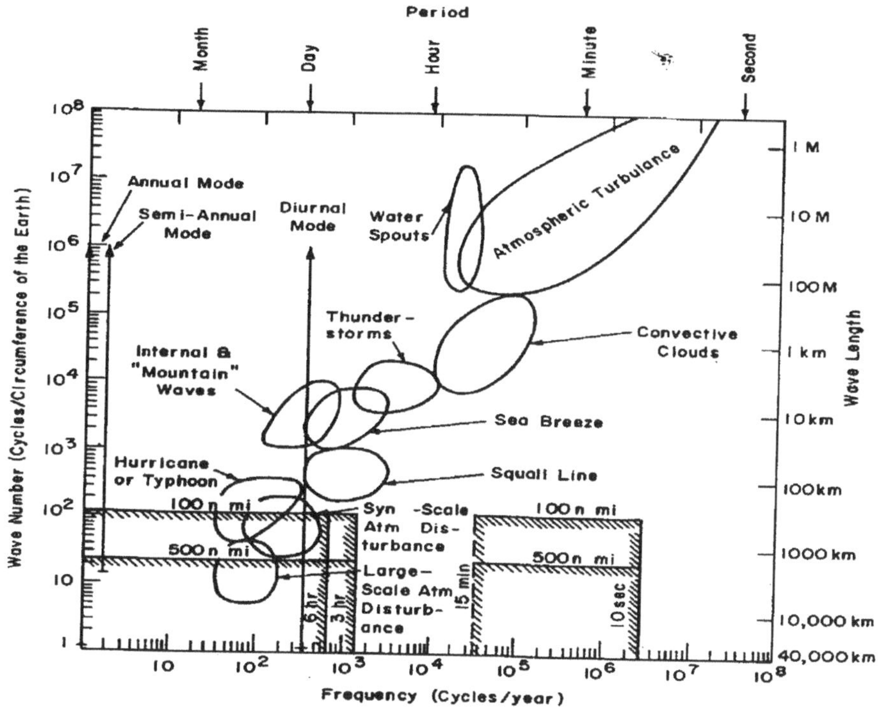

Figure 15.6 Phenomena and processes affecting the variability of wind

References

[1] Urick, R. J., *Principles of Underwater Sound for Engineers*, New York: McGraw-Hill, 1967.

[2] Pryor, C. N., *A Common Methodology for the Calculation of Recognition Differential*, Vol, 1, Appendix G, Space and Naval Warfare Systems Command, Arlington, VA, PMW 182, 15 September 1997.

[3] Jacobs, C. A., Pandolfo, J. P., and Aubert, E. J., " Characteristics of National Data Buoy Systems: Their Impact on Data Use and Measurement of Natural Phenomena," TRC Report 7493-334, December 1968.

16

Modeling Detection and Tactical Decision Aids

Chapter 2, The Sonar Equations, discusses the sonar equation, signal excess, XS or SE, and the concept of figure of merit, FOM. Using that discussion and knowing each parameter of the sonar equation, only simple arithmetic is required to determine the detection range.

16.1 Figure of Merit Range or R50 %

The passive sonar equation is usually written as

$$XS = Ls - Nw - Ln + NDI - Nrd \qquad \text{(Horton [1])}$$

or

$$SE = SL - TL - NL + DI - DT \qquad \text{(Urick [2])} \tag{16.1}$$

The figure of merit, FOM, is defined as the value of transmission loss, TL or Nw, that results in a signal excess of zero:

$$FOM = SL - NL + DI - Nrd \quad \text{Passive}$$

or

$$FOM = SL - NL + AG - DT \quad \text{Active} \tag{16.2}$$

The recognition differential, Nrd, or detection threshold, DT, are normally defined for a probability of detection of 50 %. The range at which this occurs is called the FOM detection range or the R50 % range. Figure 16.1 shows a simple example with the R50 % range of interest being where the FOM line crosses the transmission loss curve. Alternatively, plotting the signal excess versus range, as in Figure 16.2, yields the R50 % range, where the signal excess is zero.

Underwater Acoustics: Analysis, Design and Performance of Sonar Richard P. Hodges

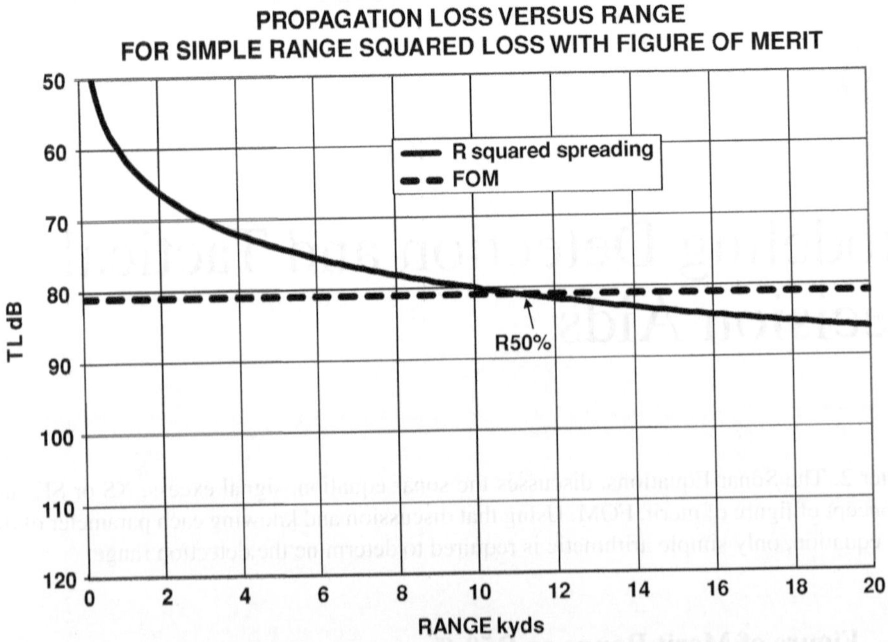

Figure 16.1 Determining the R50 % range using the FOM and transmission loss

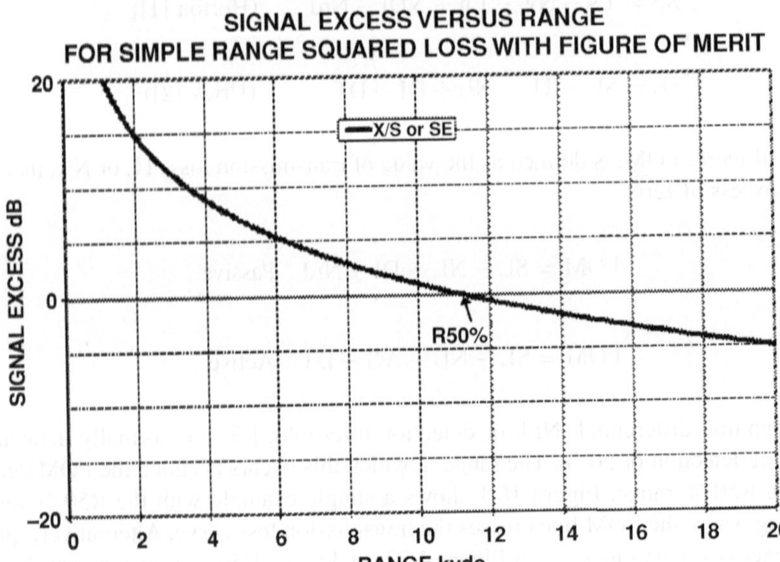

Figure 16.2 Determining the R50 % range using the FOM and signal excess

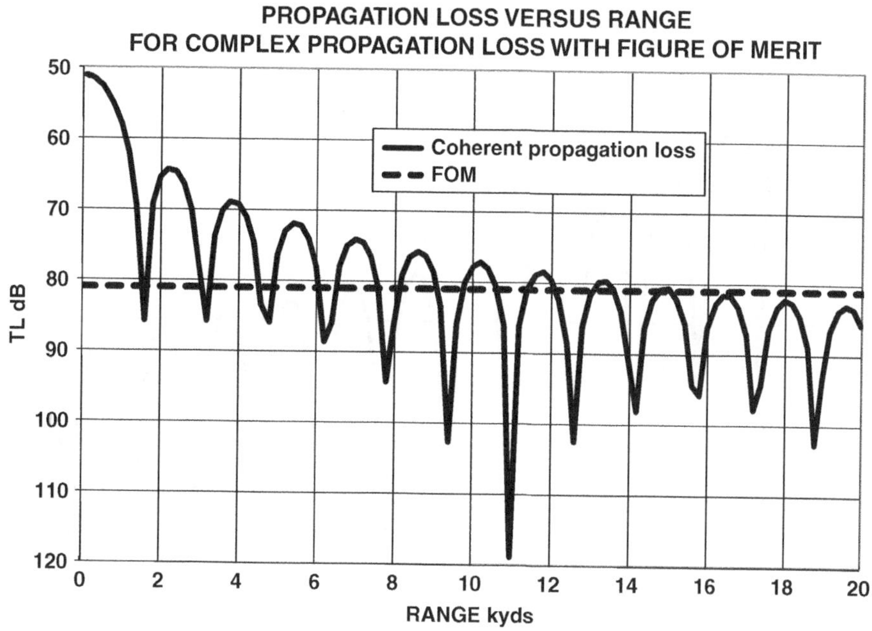

Figure 16.3 Determining the R50 % range using the FOM and complex transmission loss

The process for determining the R50 % range seems to be straightforward, but now consider Figures 16.3 and 16.4. These figures are the same examples as Figures 16.1 and 16.2, except that now a coherent propagation loss curve is used.

A problem arises when picking a single range in Figures 16.3 and 16.4, since the curves have multiple ranges that meet the R50 % range criterion. For cases like this, the maximum range of 14 kyds is used, or the range where contact is expected over 1 kyd range intervals (around 10 kyds) or the range inside which continuous contact is expected (1.4 kyds).

In reality, the problem is even more complex because the target's depth is not known. This complexity has led to the signal excess equivalent of the full field plot for transmission loss, where signal excess (color coded) is plotted versus range and depth (Figure 16.5).

A number of assumptions are being implicitly made in doing such calculations. For example, as derived in Chapter 13, Methodology for Calculation of the Recognition Differential, and Chapter 14, False Alarms, False Contacts, and False Targets, time is associated with the passive detection recognition differential, Nrd or DT. One assumption based on this is that the values of all the terms in the sonar equation are constant over the integration time of the processor. The typical processing times for sonar systems are measured in minutes, so the target could move a significant distance during that time, causing the transmission loss to change. This effect is particularly significant if the detection range is expected to be short. Consider a nondirectional sonobuoy attempting to detect a very quiet submarine; the R50 % range might be 1 kyd. If the assumed processing time is 6 minutes, a 10 kts submarine would move 10 kts $\times$ 6 min $=$ 1 nautical mile or about 2 kyds. Therefore, in an encounter with this sonobuoy, it is unlikely that detection will occur at 1 kyd unless the submarine is circling the sonobuoy at 1 kyd.

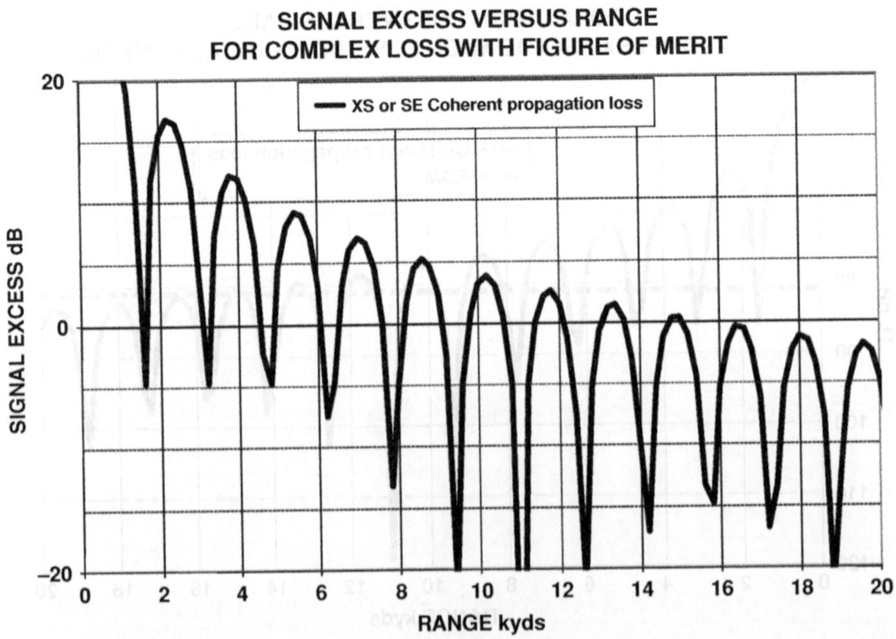

Figure 16.4 Determining the R50 % range using the FOM and signal excess for a complex transmission loss

Each term in the sonar equation is based on a mean or median value, so the result reflects some "global" mean/median/expected value. For any at-sea encounter, the target will probably not be at the class average value, the ambient noise and the sound speed profile may not be what was retrieved from a historical database, the recognition differential may have a distribution, and the propagation loss model might introduce an error. In addition to these

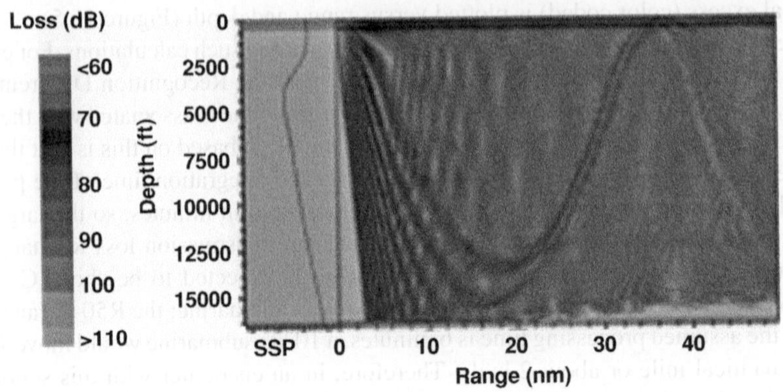

Figure 16.5 Example of a full field plot of signal excess

variables, temporal variability will cause the sound speed profile (SSP), ambient noise, and radiated noise to change with time.

As a result of this variability, the sonar equations should be written as

$$XS_{obs} = XS_{exp} + e \tag{16.3}$$

or

$$SE_{obs} = SE_{exp} + e \tag{16.4}$$

where

SE_{obs} = observed value of SE
SE_{exp} = expected value of SE from the sonar equation
e = an unknown that includes:
temporal variability
the target as a particular ship or a class average
differences between models and the real world
detection performance variability

A random variable is generally used for e in detailed modeling. The parameters of the random variable (distributions, mean, variance, correlation time, and sensor cross-correlation) are discussed in Chapter 15, Variability and Uncertainty. These parameters are a function of whether the modeling is passive or active, and on the state of knowledge for the prediction. For example, a ship at sea may have a noise measurement on each sonar beam and therefore does not need to estimate noise from historical data. Similarly, the ship may have a measured SSP and a recent MODAS (Modular Ocean Data Assimilation System) download, an array of SSPs gridded over an area provided by the MODAS program (see Chapter 5, Transmission Loss). If measured data are available, there is less uncertainty that needs to be accounted for.

The R50 % range is conditional. It is absolutely not the average, median, or mode of detections expected from a typical encounter, as will be discussed in Chapter 18, Tracking, Target Motion Analysis, and Localization. It is best to think of the R50 % range as the range where, if the target is airlifted to that spot and remains there for one ping, or one processing cycle, there is a 0.5 probability the target will be detected.

16.2 Tactical Decision Aids

Tactical decision aids (TDAs) on naval ships support several objectives. First, they help monitor the environment by taking in all available data such as water temperature, sound speed, salinity, bottom depth, sea state, ambient noise, reverberation, SSPs (from expendable bathythermographs launched by the ship or MODAS fields received by communications), etc. This constant monitoring of the environmental conditions allows for the continuous updating of predictions and recommendations.

In addition to being the models used to perform predictions, TDAs also hold a repository of historical databases for all parameters needed for the predictions, such as bathymetry, SSP, bottom loss, ambient noise, target parameters (radiated noise, target strength, sensors), active sensor frequencies and pulse types, and own ship sensor parameters.

Finally, TDAs support the calculations and displays necessary to aid in making tactical decisions. The R50 % range, with all its caveats, is the value most frequently used by TDAs. Despite the uncertainty associated with its calculation, the R50 % range may be the most reasonable metric for making decisions like search depth, how fast to search, and what sonar line-up to use (passive bands, active pulse types, depression/elevation angles, etc.).

Modern TDAs have evolved to support not just a single ship and its sensors, but groups of ships or large fields of sensors working together, by assisting in developing near-optimal arrangements and search patterns. While only sonar has been mentioned here, most TDAs support all weapons, sensors, and communications systems on a platform (RADAR, LIDAR, infrared, low light, electronic counter measures).

There are many TDAs in use in the U.S. Navy. Examples are:

STDA:	Sonar Tactical Decision Aid
SFMPL:	Submarine Fleet Mission Program Library
PC IMAT:	Personal Computer Interactive Multisensor Analysis Training
MEDAL:	Mine-Warfare Environmental Decision Aid Library
ASPECT:	Active Sonar Performance Estimate Computer Tool

Four issues to be explored when discussing the ocean and TDAs are variability and uncertainty, bias errors, relevance of historical metrics, and lack of an ongoing validation effort. Too often, these issues are discussed as though they were all the same; as we will see, they are not.

Detection is a stochastic process. One should not expect to be able to predict the actual observed range for a single case with a TDA. Instead, the expectation should be to determine the right distribution over a large number of events or samples. There is a lot of variability and uncertainty in sonar performance in the ocean. This could be caused by small errors in the knowledge of some parameters, differences in operators or systems, or simply because variables change over time and most measurements tend to be temporally and geographically sparse.

After 30 years of comparing models with at-sea exercises and real world encounters, I believe we have a firm understanding of how to address this part of the problem. Variability and uncertainty should be included in any TDA. Personnel need to be trained in how to manage variability and uncertainty. If the TDA gave the detection range as 6 kyds ($\pm$1 kyd), an operator would want to consider initializing TMA at 6 kyds. If, however, the TDA instead gave 6 kyds (with a range from 20 kyds to 2 kyds), then TMA should probably not be initialized at 6 kyds. Variability is real and using an archival database will confirm that ranges have variation.

Bias errors are usually brought about by wrong inputs. The most frequent sources of bias errors are databases, at-sea measuring systems, or operator errors. Smart operators can usually catch and correct these errors before they cause problems. Two quick anecdotes will illustrate this point:

> A sonar operator managed to prevent shipping level noise from being included when determining the sonar line-up for a search in the Mediterranean. The operator failed to recognize the oversight and followed the recommendation to search for submarines at low frequencies. Given the very high levels of shipping noise at this location, not including the noise in determining the line-up gave the operator misguided recommendations.

> A TDA operator in an exercise off Bermuda was asked the range to the first convergence zone (CZ) and replied "25 kyds." A seasoned chief immediately responded, "25 kyds? You must have the sound speed profile wrong." The chief was right. A decimal point was misplaced for one of the numbers; 4830.3 ft/s was entered as 48303.0 ft/s, which resulted in a severe bending of those sound rays in the model. The expert recognized the CZ stated was a ridiculous answer and sought to correct it.

Bias errors that are small as compared to the variability may not be noticeable, but large uncorrected biases cannot be handled by the variability in the sonar equation. Instead, these errors will result in the wrong distribution of values. For example, if the radiated noise for a target in the database is 14 dB too high, then the predicted ranges will always be much longer than those observed.

It is my opinion that the metrics that have been historically calculated for detection are not what should be presented to the operator for search planning. Instead, we should be calculating and presenting classification information at some level. Requiring an operator to detect a target at *x* kyds, when we have no reason to believe a detection at that range could be recognized as a target, is not useful. Worse, the operator may state, "I never detected at that range," meaning the target was never recognized at the range. I have suggested this many times in my career where the response is usually "it is too hard." So instead, we continue to do the easy thing rather than the most useful thing.

Many TDAs have been developed without a formal ongoing validation effort. This effort should include a comparison of modeled predictions with at-sea results. I am absolutely certain that the naval simulation program I work with, SIM II, would have been out of business years ago if we were not able to point to an extensive and ongoing validation effort as a basis for believing the simulation could estimate the relative performance of systems. The bad news is that a validation program is not cheap or easy. The good news is that big errors and biases are recognized very quickly.

For the first attempt with SIM II, the probability that the exercise results could have come from the SIM II results was about the same size as Plank's constant (6.7×10^{-27}). To validate the model using the exercise results, the SIM II program had to undergo some major modeling overhauls. This early attempt, although not successful, provided direction on what part of the model needed review. The primary problem was that the model had conventional wisdom built in that was simply not true. No amount of verification of the code would have shown us this inaccuracy.

References

[1] Horton, J. W., *Fundamentals of Sonar*, Annapolis, MD: United States Naval Institute, 1957.
[2] Urick, R. J., *Principles of Underwater Sound for Engineers*, New York: McGraw-Hill, 1967.

17

Cumulative Probability of Detection

Unlike the R50 % range discussed in Chapter 16, Modeling Detection and Tactical Decision Aids, the cumulative probability of detection (CPD) attempts to estimate the probability that a target will be detected as it moves relative to the sonar. Typically, CPD is presented as the probability that the target will be detected at least once as a function of range or time. The target and sonar may be assumed to be on a collision or antiparallel course or, more commonly in the real world, both may be following complex tracks. It is important to understand the underlying assumptions when reviewing any curves.

17.1 Why is CPD Important?

The cumulative probability of detection is important because search problems arise regularly in antisubmarine warfare (ASW), mine warfare (MIW), or search and rescue operations. Planners need to know not just at what range a target might be found but also how well the sonar covers an area of interest or sector of their screen.

17.2 Discrete Glimpse and Continuous Looking

In sonar literature, there are two types of developments for computing CPD. The first is the discrete glimpse, which in principle represents active sonars or radars that ping. The second development is continuous looking, which represents passive sensors that look continuously. In practice, when evaluating continuous systems, the total time is divided into smaller time segments making the equations developed identical. Recall from the discussion in Chapter 12, Statistical Detection Theory, that no decision can be made instantaneously, i.e., with zero energy. An important distinction is whether the glimpses are treated as statistically independent or not. If glimpses are treated independently, then the cumulative probability of detection for two sequential time periods is given by

$$\mathrm{CPD} = P_1 + P_2(1 - P_1) \tag{17.1}$$

where

P_1 = probability that the target would be detected if placed there for the glimpse interval (one ping for an active system or the time interval for a continuous system)
P_2 = probability that the target would be detected during the second glimpse interval
CPD = cumulative probability that detection would occur in at least one of the intervals

The cumulative probability for n intervals is one minus the probability that all opportunities fail or

$$\text{CPD} = 1 - \prod_{i=1}^{n} (1 - P_i) \tag{17.2}$$

Unfortunately, statistical independence between small time steps rarely occurs in nature, for, as discussed in Chapter 15, Variability and Uncertainty, passive and active sonar have rather long correlation times.

17.3 Lambda–Sigma Jump Model

The lambda–sigma (λ–σ) jump detection process is a historical method used to introduce correlation by imposing a Poisson process with an exponential correlation function. This methodology was developed in the 1960s when the relatively primitive computers of the time could handle the resulting equations. In this process, a random variable is added to the expected signal excess drawn from a specified distribution. The value stays constant until it reaches a time period drawn from an exponential distribution ($T_{\text{jump}} = -\ln(\text{rand})/\lambda$, where rand is a uniform random number 0–1). The cumulative probability of a single sensor is given by

$$\text{CPD} = \mathrm{e}^{-\lambda\,\Delta t} \max(P_1, P_2) + \left(1 - \mathrm{e}^{-\lambda\,\Delta t}\right)(1 - P_1)\,P_2 \tag{17.3}$$

Equation 17.3 provides for two possibilities: (1) there is no jump in the random process and (2) a jump has occurred, the detection was not made on the first time step, and P_2 is now an independent draw. This process can be applied iteratively to cover additional time steps.

In the intervening years since the development of the λ–σ jump detection process, developing equivalent closed-form equations to cover multiple platforms, with multiple simultaneous sensing channels, has been formidable. It is no longer simple. Computers have developed, thanks to Moore's law, to where simplicity is not the controlling consideration. Studies of sonar recordings have determined that this does not match the behavior of sound in the oceans and predictions based on such a model of performance after detection have proven to be completely in error. One such example, the details of which are classified, was the analysis of a successful at-sea encounter. A λ–σ jump process simulation predicted the probability of success was less than 1×10^{-10}, whereas the equivalent Gauss–Markov process using the same correlation time and variance predicted a 65 % probability of success. The development of the integral equations for the λ–σ jump methodology are available in Wagner *et al.* [1] and will not be reproduced here.

17.4 Nonjump Processes

Virtually all modern programs used to compute CPD use Monte–Carlo methodologies with more smoothly varying random processes. This allows for easy inclusion of correlation effects for signal to signal, sensor to sensor, platform to platform, and time step to time step. This is not to imply that the correct correlation functions and parameters are accurately known.

The importance of applying correlation to the calculation of the probability of detection can be realized by working out a few examples. Consider a target emitting a single signal. If a single hydrophone or sonobuoy is placed in its vicinity, one can calculate the sonar equation and from that estimate the probability of detection. Let us assume that the signal excess came out to be −14 dB (minus 2 sigma, see Chapter 15 Variability and Uncertainty) at 4 kyds, corresponding to about a 2.27 % probability of detection. If 200 sonobuoys were placed on top of each other and assumed to be independent, the probability of at least one detecting the signal would be $1 - (1 - 0.0227)^{200}$ or 98.99 %. Reviewing each term in the sonar equation for these two scenarios reveals that almost nothing is expected to be different between the equations, so the right answer is probably closer to the 2.3 % than to the 99 %. A different answer would be obtained if the 200 hydrophones were beamformed to improve the signal-to-noise ratio.

Consider the radiated noise signature in Figure 10.1. There are 23 lines that could be detected. For simplicity, let us assume that they all have the same signal excess of −7 dB, or minus 1 sigma (a probability of 16% on each). If we assume all 23 noise components are statistically independent, the probability of detection would be $1 - (1 - 0.16)^{23} = 0.98$. In reality, the truth is actually somewhere between the two values. When considering the sonar equation, some of the signals are harmonically related and may be correlated. Some signals are so close in frequency that variations in noise and propagation loss are similar.

Finally, consider two platforms looking at each other. At the same frequency, the propagation loss between them would be the same, but the signals, because they come from different ships, would not be the same and the ambient noise, because of the different look directions, might be different, and the sonars operators, being different people, would be different; again expect partial correlation. All of these differences can be evaluated using a model such as the one described below.

17.5 What Are Appropriate Random Parameters?

Revisiting the introduction to Chapter 15, when the passive or active sonar equations are solved, values for each term are needed. Generally, the values used are estimates of the median or mean values expected, but at any moment at sea the target will not be the class average value, the ambient noise will not be the median for that area and month, the sound speed profile will not exactly match what we retrieve from a database, the propagation loss will come from a model, and the operator performance will not be exactly the median. In other words, each term in the sonar equation is off by an amount that we do not know at any given moment at sea. In addition, temporal variability affects all of the terms with the passage of time. Therefore, the result from a sonar equation calculation is something of a "global mean or median," in other words, an expected value.

The actual observed signal excess will be different by an unknown quantity from the expected value:

$$\mathrm{XS}_{\mathrm{obs}} = \mathrm{XS}_{\mathrm{exp}} + e \tag{17.4}$$

where e includes the temporal variability and all differences between databases, models, and the real world. Models typically use a random process for e. The signal excess (XS) model generally in acceptance is

$$\mathrm{XS_{obs}} = \mathrm{XS_{exp}} + \mathrm{Random} \tag{17.5}$$

where

$\mathrm{XS_{exp}}$ = expected signal excess (i.e., the result of the sonar equation)
$\mathrm{XS_{obs}}$ = observed at-sea signal excess
Random = a sample from a stochastic process that reflects both the temporal variability of the terms in the sonar equation, as well as the platform to platform (operator to operator) variability

The Naval Undersea Warfare Center's (NUWC) SIM II model [2] has been compared extensively to at-sea measurements. The random process used in this model is

$$\mathrm{Random} = R_{\mathrm{corr}} + R_{\mathrm{unc}} + R_{\mathrm{short}} \tag{17.6}$$

where

R_{corr} = long-term correlated random variable, correlated in the sense that all sensors share it
R_{unc} = long-term uncorrelated random variable
R_{short} = short-term uncorrelated random variable

Table 17.1 shows the distributions, parameters, and correlation times for each of these variability terms.

This yields a total standard deviation of about 7 dB [$\sqrt{(4.5^2 + 4.7^2 + 2.7^2)}$] for passive and 12 dB [$\sqrt{(9.0^2 + 6.0^2 + 5.4^2)}$] for active, with correlation times of 5.6 and 6.6 hours respectively.

A Gauss–Markoff process is defined as

$$R_i = \alpha R_{i-1} + \beta r \tag{17.7}$$

where

R_i = ith value of the random series (normal distribution, mean = 0, standard deviation = s, and correlation time = T)
α, β = weighting coefficients, $\alpha = \mathrm{e}^{-\mathrm{d}T/T}, \alpha^2 + \beta^2 = 1$
$\mathrm{d}T$ = time between the ith and $(i-1)$th random variable
r = random variable, normal distribution, mean = 0, standard deviation = s, uncorrelated

Table 17.1 Random variability associated with sonar

Variable	Type	Median	Standard deviation (passive/active)	Correlation time
R_{corr}	Gauss–Markoff	0.0	4.5/9.0	12 hours
R_{unc}	Gauss–Markoff	0.0	4.7/6.0	4 hours
R_{short}	Log Rayleigh	0.0	2.7/5.4	Uncorrelated

The computation of cumulative probability of detection (CPD) along a track requires computing the expected signal excess for each time interval along the track. At each point it is necessary to compute the effect of changes in expected signal excess, as well as the effect of changes in random variables between steps. Although this calculation is mathematically complex, it is easily accomplished with a Monte Carlo approach. Below is an approximate method applicable to a single sensor:

(1) Compute the adjusted signal excess (XS for the time step interval) at each point.
(2) Assume that the "long-term" random variables are constant for the detection period.
(3) Divide the possible values of the long-term random variables into bins. Since it is assumed that the long-term random variable is constant, for each bin, we can accumulate using the independent method described above.
(4) Combine across bins, weighting each with the probability that the value of the long-term random variable would be in that bin.

Figures 17.1 and 17.2 each show a comparison of static or "instantaneous" probability of detection and cumulative probability of detection versus range for different environments. Here, "instantaneous" probability is the probability that XS_{exp} + Random is greater than zero at each point – for this case, a lookup of the probability of XS_{exp} from a Gaussian distribution with a zero mean and a standard deviation of 7 dB. Notice that an approximation is simply applying a maximum function from the long range on the "instantaneous" probability

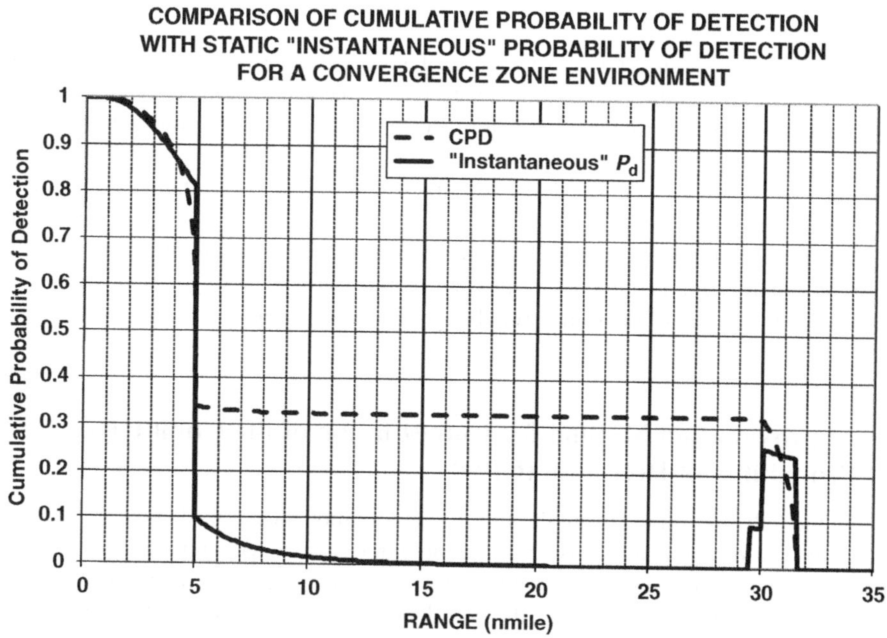

Figure 17.1 Comparison of cumulative probability of detection with static or "instantaneous" probability of detection for a convergence zone environment

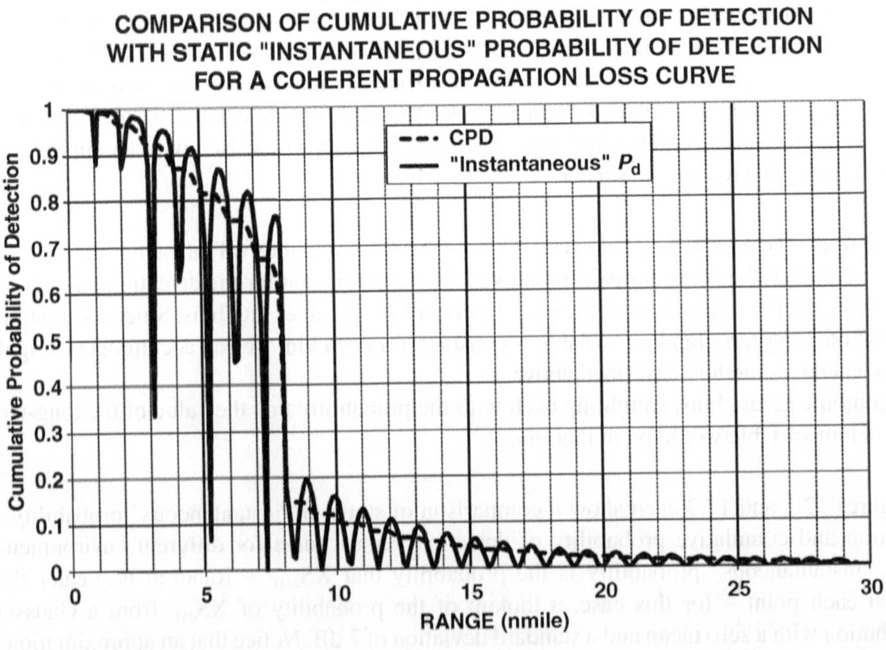

Figure 17.2 Comparison of cumulative probability of detection with static or "instantaneous" probability of detection for a coherent propagation loss curve

curve. Figure 17.1 is a convergence zone type of propagation loss and Figure 17.2 is for a coherent type of propagation loss. The closing rate in these two cases is 10 knots and the assumed integration time is 5 minutes. Therefore, the range changes 0.833 nmile per integration period.

In Figure 17.1, as the CZ is entered, the CPD rises until it equals the static value, at about 5 minutes, and then continues to climb, leveling off after exiting the CZ and again when entering the direct path region (< 5 nmile). The CPD lags for this reason. In Figure 17.2, the fading of the signal is rapid compared to the time spent in any one strong region, so the CPD stays below the maximum of the static probability of detection peaks.

17.6 Approximation Method for Computation of the Cumulative Probability of Detection (CPD)

Evaluating the CPD for a track using the full-up random model described above is best done using a Monte Carlo approach because of the complexities associated with the time variability of the three random processes. I have yet to see a successful attempt at an analytic equation of this methodology. However, because the correlation times for the long-term processes are large (about 6 hours), as compared with the tactical interest detection times, they can be approximated as constant for the time associated with a detection opportunity. This allows a straightforward calculation of CPD.

The first step in calculating CPD is to start at a range beyond which the signal excess is very small or, in the case of an exercise, the starting range or COMEX (commence exercise) range in the Navy. A time step is then selected that is either the time between pings for active sonar or an interval sufficiently short to capture the propagation loss structure or short compared to the integration time for passive sonar. In the SIM II program, typically 0.01 hours is used for the time step (dropping to 0.001 hours for incoming weapons). The procedure used for calculating CPD is a numerical integration start by iterating on the time/range index, i (from 1 to final f), for each choice of long-term random variable, j:

$$\mathrm{cpd}_{i,j} = F(\mathrm{XS}_{i,j})\left(1 - \mathrm{cpd}_{i-1,j}\right) + \mathrm{cpd}_{i-1,j} \tag{17.8}$$

$$\mathrm{cpd}_j(\mathrm{LR}_k) = \text{final cpd for each } j = \mathrm{cpd}_{f,j} \tag{17.9}$$

$$\mathrm{CPD}(\mathrm{LR}_k) = \sum_j p_j \mathrm{cpd}_j(\mathrm{LR}_k) \tag{17.10}$$

where

$\mathrm{CPD}(\mathrm{LR}_k)$ = cumulative probability of detection for the lateral range (LR_k)
I = subscript for time/range step for this lateral range
J = subscript across normal distribution of long-term random variables
$\mathrm{cpd}_{i\text{-}1,j}$ = cumulative probability 1 through $i-1$ in the jth column
= cumulative probability if the long-term random variable is actually the jth value

$$F(\mathrm{XS}_{i,j}) = \left\{1 - \left(\frac{1}{2}\right)^{10^{([XS_{i,j}]/p)}}\right\} \quad if\ \mathrm{XS}_{i,j} > -3 + p\log\left[\frac{\mathrm{d}T}{T(\mathrm{Nrd})}\right]$$
$$F(\mathrm{XS}_{i,j}) = 0 \qquad \text{otherwise} \tag{17.11}$$

where, if $i = 5$, $\mathrm{d}T/T(\mathrm{Nrd}) = 0.125$, then the test is $\mathrm{XS}_{i,j} > -7.5$:
where

$\mathrm{XS}_{\exp(i)}$ = expected signal excess at time/range index, i, the result of the sonar equation adjusted for the passive sonar case to the time step($\mathrm{d}T$) by $p\log[\mathrm{d}T/T(\mathrm{Nrd})]$
$T(\mathrm{Nrd})$ = time associated with the detection recognition differential, usually 5 minutes by convention
p = 5 for passive and 10 for active

$$\mathrm{XS}_{i,j} = \mathrm{XS}_{\exp}(i) + \rho_j^{\mathrm{LTR}(j)} \tag{17.12}$$

where

LTR(j) = long-term random variable, jth interval These are values spanning the range of the distribution. For example, using 0.25 sigma steps from −3 sigma to +3 sigma (with steps depending on the accuracy desired), for the passive case, which has a long-term sigma of 6.5 dB, the series would be −19.5, −17.875, −16.25, ..., 17.875, 19.5.

ρ_j = weighting to be applied to each interval of LTR(j). This is the probability that the long-term random variable will have a value between the LTR(j)− interval and the LTR(j) + interval, where $\Sigma_j \rho_j = 1$. For the example above, the weightings would be 0.0020, 0.0023, 0.0044, 0.0080,. . ., the first value being the cumulative normal from $-\infty$ to -2.875 sigma, the next from -2.875 to -2.75, etc.

The range points (R_i) and associated times (T_i) define the track being evaluated and the speeds/courses of own ship and threat. For a generic search problem, the problem is typically expressed as the relative motion of two nonmaneuvering ships:

$$R_i^2 = \text{CPA}^2 + \sqrt{\left(R_{\text{max}}^2 - \text{CPA}^2 - V_{\text{rel}} \times T_i\right)^{-2}} \tag{17.13}$$

References

[1] Wagner, D. H., Mylander, W. C., and Sanders, T. J., *Naval Operations Analysis*, 3rd edition, Annapolis, MD: Naval Institute Press, 1999.

[2] *SIM II User Reference Manual*, Naval Undersea Warfare Center, Division Newport, Code 60, March 1990, Unclassified.

18

Tracking, Target Motion Analysis, and Localization

Once the target has been detected (classified) the next task is to determine where it is (bearing, range, and depth) and where it is going (course and speed). For passive sonar, tracking normally refers to determining the bearing of the target. Localization includes bearing, range, and maybe depth. Target motion analysis (TMA) includes all parameters (bearing, range, speed, course, and maybe depth). For active sonar, a range estimate is always available. Today, many trackers are multidimensional, i.e., they track several parameters simultaneously. For example, bearing and range or bearing, range, and Doppler.

18.1 Bearing Trackers

Since most sonars form beams, it is possible in principle to determine the bearing by steering the beam looking for a maximum in response. However, for sonar, the received signal to noise has large fluctuations. These are due to many causes, as discussed in Chapter 15, Variability and Uncertainty, including Lloyd mirror fading and short-term fluctuations in noise and propagation loss. Because of this, it is generally impractical to use maximum response beam steering. With active sonar, this is compounded by the intermittent nature of the returns.

The solution to this is called a bearing deviation indicator (BDI), which is designed to overcome the problems associated with peak seeking trackers. There are numerous designs, but they tend to fall into one of two general categories.

18.1.1 Amplitude Difference Method

Two directional hydrophones or beams from an array beamformed in slightly different directions have their outputs subtracted, one from the other. If the signal is nearer the maximum response axis of the first, the output will be positive, if nearer the second, the output will be negative, and if centered, it will be zero. Variations on these, which may include more than two beams and complicated weighting schemes based on the expected shapes of the beams,

Underwater Acoustics: Analysis, Design and Performance of Sonar Richard P. Hodges

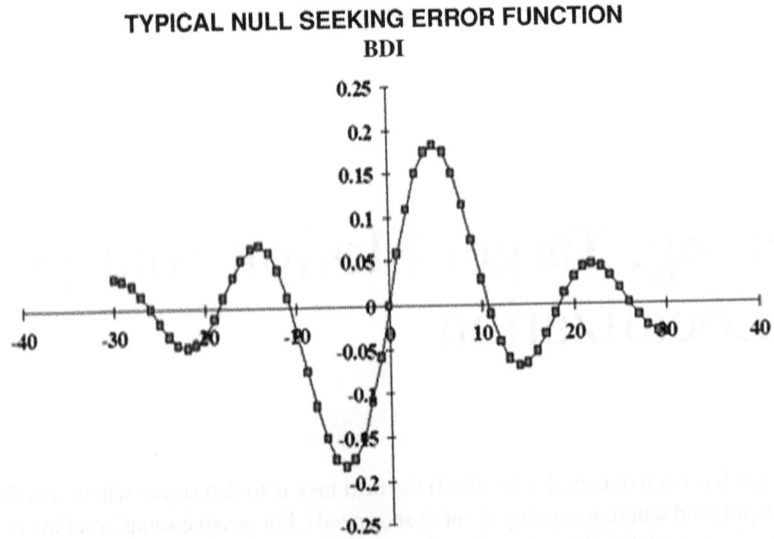

Figure 18.1 Example of a bearing deviation indicator (BDI) curve

are normally referred to as beam interpolation trackers (or BITs). The output response versus angle is referred to a BDI curve, as shown in Figure 18.1. Another advantage of such a response pattern is that the tracker knows which direction to move to correct – for an error, left if the output is positive and right if the output is negative – allowing for the easy design of automatic closed-loop trackers. However, with a peak seeking or maximum response tracker, it must be scanned back and forth to estimate the bearing since left and right errors simply give smaller signals. In practice, a time-average is used to cut down on noise fluctuations.

18.1.2 Phase Difference Method or Cross-Correlation Method

In this case the transducer, or an array aperture, is split into two halves, which are compared to estimate the phase difference or the time delay difference. Figure 18.2 shows the relationship between the bearing deviation and signal phase difference or time delay.

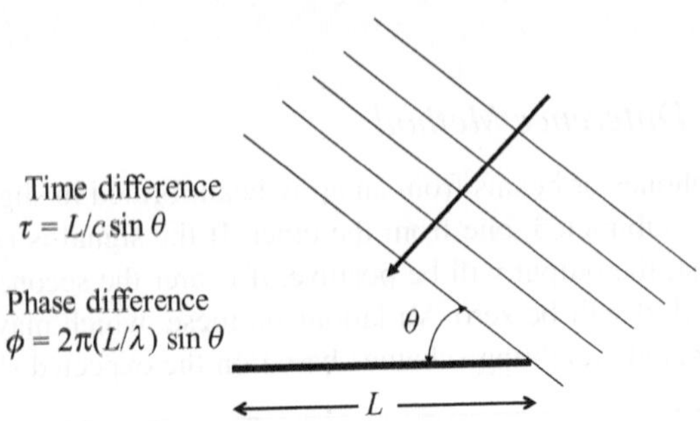

Figure 18.2 Time and phase difference caused by the deviation angle

The advantage of cross-correlation based trackers is that the outputs that are compared each contain the signals that have been subjected to the same sources of fluctuations, thus eliminating some of the problems of variability of the signal with time.

18.2 General Principle of Tracking and Bearing Measurement

For small errors, the time delay error is proportional to the bearing error:

$$\varepsilon\tau = \frac{L}{c}\cos(\theta)\,\varepsilon\theta \tag{18.1}$$

where

τ = time delay
L = baseline for the measurement
c = speed of sound
θ = bearing from the normal direction
ε = error in variable

In terms of standard deviations, this is

$$\sigma(\theta) = c\frac{\sigma(\tau)}{L\cos(\theta)} \tag{18.2}$$

The Cramer–Rao [1 – 4] lower bound methodology can be applied to this problem, yielding

$$\sigma^2(\tau) = \frac{\displaystyle\int f^2\left[H^2(f)N^2(f) + 2H^2(f)S(f)N(f)\right]\mathrm{d}f}{2\,(2\pi)^2\,T_{\mathrm{ave}}\left[\displaystyle\int f^2 H(f)S(f)\,\mathrm{d}f\right]^2} \tag{18.3}$$

where

T_{ave} = tracker averaging time
f = frequency
$N(f)$ = noise power
$S(f)$ = signal power
$H(f)$ = filter response

If we assume that the signal to noise is independent of frequency (the same spectral shape), and $H(f) = S(f)/N^2(f)$ (Eckart filter [5]), then

$$\sigma^2(\tau) = \frac{1}{2\,(2\pi)^2\,T_{\mathrm{ave}}}\left[\frac{1}{\frac{S}{N}^2} + \frac{2}{\frac{S}{N}}\right]\frac{\displaystyle\int f^2\,\mathrm{d}f}{\left(\displaystyle\int f^2\,\mathrm{d}f\right)^2} \tag{18.4}$$

The last term is a constant for the frequency band processed and equals $3/(f_u^3 - f_l^3)$ or assuming $f_u \gg f_l$, then $3/f_u^3$. Converting to the bearing:

$$\sigma^2(\theta) = \frac{(180c)^2}{(\pi L)^2}\sigma^2(\tau) \tag{18.5}$$

or

$$\sigma^2(\theta) = \frac{(180c)^2}{(\pi L)^2}\left\{\frac{1}{2\,(2\pi)^2\,T_{\text{ave}}}\left[\frac{1}{\frac{S}{N}^2} + \frac{2}{\frac{S}{N}}\right]\frac{3}{f_u^3}\right\} \tag{18.6}$$

This is the lower bound for what is called the tracker random bearing error. It is the error resulting after the noise fluctuations have been average for T_{ave}. Note that the error is inversely proportional to the measurement baseline (L). It is the distance between the effective centers of two subarrays and to the frequency to the minus 3/2 power:

$$\sigma(\theta) = k\frac{1}{\sqrt{T_{\text{ave}}}Lf_u^{3/2}}\left[\frac{1}{\frac{S}{N}^2} + \frac{2}{\frac{S}{N}}\right]^{\frac{1}{2}} \tag{18.7}$$

where S/N is now independent of frequency.

Since T_{ave} is a chosen tracker parameter, if we move it to the other side of the equation, then $\sigma(\theta)\sqrt{T_{\text{ave}}}$ is only a function of the aperture baseline, frequency, and signal-to-noise ratio. Thus, if the averaging time is increased the random bearing error will decrease and vice versa. We shall see later in this chapter that $\sigma(\theta)\sqrt{T_{\text{ave}}}$ or $\sigma(\tau)\sqrt{T_{\text{ave}}}$ are the fundamental parameters for ranging accuracy as opposed to bearing error.

Also note that, as shown in Figure 18.3, that $\sigma(\theta)\sqrt{T_{\text{ave}}}$ and $\sigma(\tau)\sqrt{T_{\text{ave}}}$ vary inversely as signal to noise at low signal to noise and as one over the square root of signal to noise at high signal to noise. Typically, narrowband (PNB) trackers can not track at low signal to noise.

The general shape then for the random tracker bearing error has the functional form of

$$\sigma^2(\theta)T_{\text{ave}} = sk^2 + \frac{C_1}{\left(A\frac{S}{N}\right)^2} + \frac{C_2}{A\frac{S}{N}} \tag{18.8}$$

where

$\sigma^2(\theta)$ = variance of the random tracker bearing error
T_{ave} = tracker averaging time
sk = a lower limit usually caused by the details of the beamformer and tracker implementation

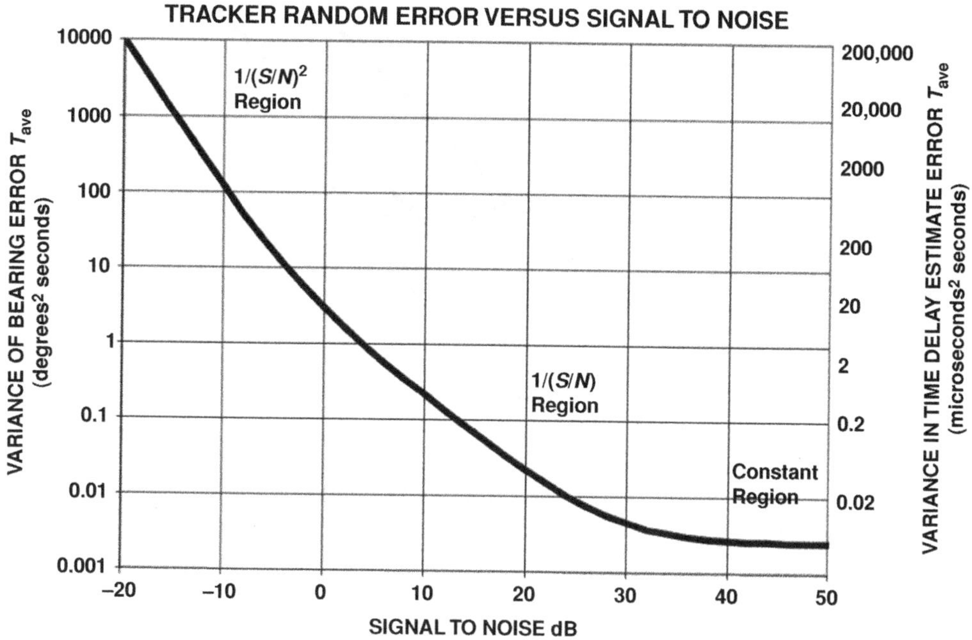

Figure 18.3 Tracker random bearing error versus signal to noise

C_1, C_2 = constants determined by integrating the equation (at least constant if S/N is independent of frequency; otherwise is range dependent)

A = a relative bearing correction, e.g., $\cos(\theta)$ for a linear multispot array or planar array and 1 for a spherical array

For area or volume arrays, normally the aperture is split into four pieces (or four preformed beams are used) for tracking, up/down and left/right, to allow tracking both in azimuth and D/E (depression/elevation).

18.3 Other Sources of Bearing Error for Area Arrays

Frequently a volumetric or area array is located inside a ship or submarine in a free flooding tank. When this is the case, bias errors in bearing occur for several reasons. First, reflections from the structure near the array give rise to multipaths, shown as a solid line in Figure 18.4. Second, shown as a dotted line, is refraction and a differential time delay caused by differences in water temperature inside versus outside the dome. Finally, in gray, as targets approach the baffles the signals will suffer diffraction errors and part of the aperture will be shaded. The net result can look like Figure 18.5 and is frequently called delta bias error, because it will be in one direction on the port side and the opposite on the starboard.

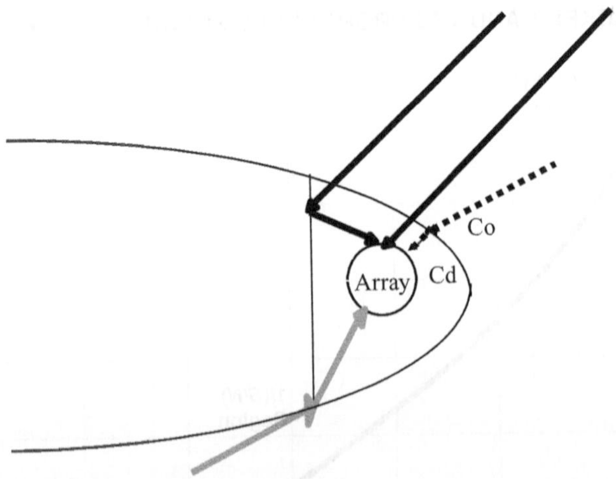

Figure 18.4 Sources of bearing bias error for internal arrays. Multipath is shown in solid line, refraction in dotted line and diffraction (and shading) in gray

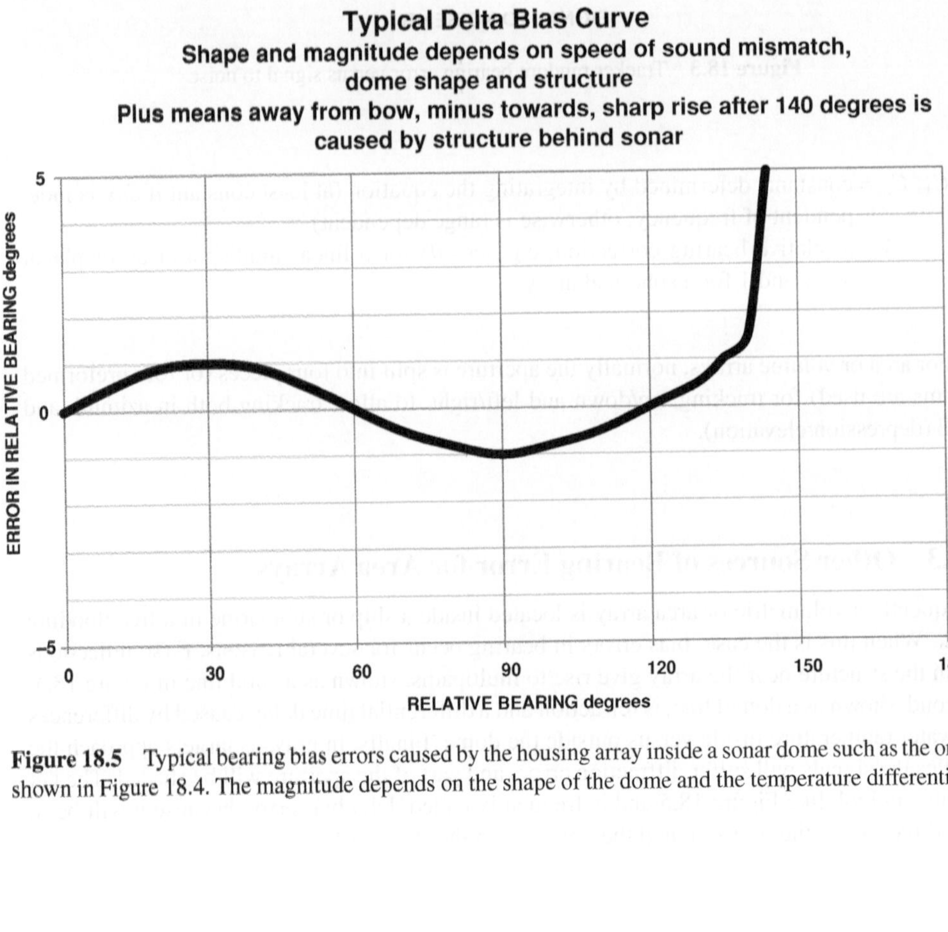

Figure 18.5 Typical bearing bias errors caused by the housing array inside a sonar dome such as the one shown in Figure 18.4. The magnitude depends on the shape of the dome and the temperature differential

18.4 Additional Sources of Errors for Line Arrays

Line arrays have no vertical dimension and therefore any bearing measured is a conical angle; i.e., all directions on a conical surface yield the same response. So the output of a tracker is the same for a signal arriving horizontally at 30° off end-fire and a signal arriving from 30° down (D/E angle) and horizontally at end-fire. The relationship between conical angle (CA), the azimuthal (AZ) angle and D/E angle is:

$$\cos(\mathrm{CA}) = \cos(\mathrm{AZ})\cos(\mathrm{D/E}) \quad (18.9)$$

If the array is not in a baffle, there is a left–right ambiguity (cos(AZ) = cos(−AZ)), meaning that all CAs are between 0 and 180 and there is no way of distinguishing which side, unless a maneuver is performed.

If the line array is a flexible towed array on a flexible cable then additional errors arise. The first of these errors are positional: distance, depth, kiting, cant, and yaw. Figure 18.6 shows the position of a towed array relative to the tow ship. A beamformer will form beams relative to the position and orientation of the array. If the above parameters are not known, then the bearing to the target from the torpedo tube cannot be accurately determined. Cant or droop is caused by an array not having the same density as the water it is in. Cant changes with own ship speed and array depth because of compression. Kiting and yaw are caused by differential currents between the tow ship's depth and the array's depth. The distance behind the own ship depends on the amount of cable used, the systems density, and the own ship speed. Towed arrays and

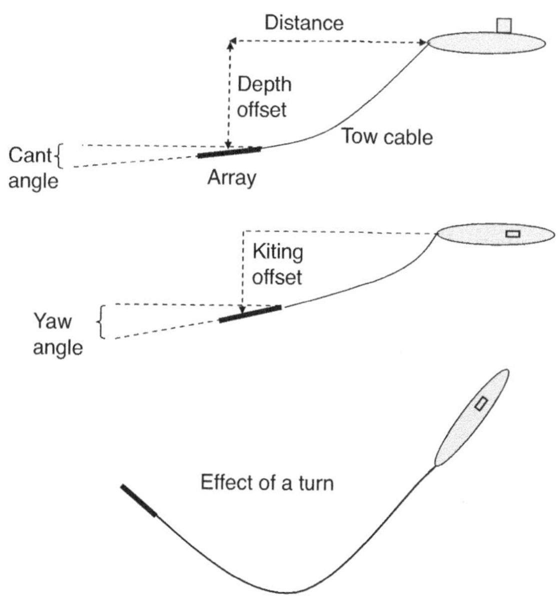

Figure 18.6 Position of a towed array relative to the tow ship

cables are rarely neutrally buoyant and may have lifting surfaces to adjust depth, but usually as tow speed increases the array will end up closer to the tow point depth. The helmsman of the tow ship never steers a perfectly straight course; therefore a damped version of the own ship's track shows up as a shape change or wiggle on the array. If not corrected for, this will result in bearing errors, as well as degradation of the beam shape. When the tow ship makes a course change, the tow cable and array more or less follow it around the corner. For a considerable period of time, the array points in a direction very different from the tow ship's course and it has a large offset position relative to its normal astern position. Arrays may incorporate one or more heading, depth, and tension sensors (because the array changes length with tension), along with suitable software to attempt to correct for these effects. An alternative approach is the placing of transponders on the array to allow the tow ship to determine its position and orientation.

If all of these corrections are applied, there is still a parallax correction necessary to correct the bearing to the tow ship's position, which unfortunately depends on knowing the range to the contact.

18.5 Bottom Bounce

If the path connecting the array and the contact involves bouncing off the bottom of the ocean, additional sources of error occur. The first is that if the bottom has a slope at the reflection point, it will change the trajectory of the signal. The second effect is when the signal penetrates the bottom and returns from multiple layers within the bottom. The final effect is caused by bottom roughness over the region where the reflection takes place, causing decorrelation of the signal. Figure 18.7 shows these effects graphically.

18.6 Manual versus Automatic Tracking

Sonar systems may have an automatic (closed-loop) tracker or may rely on an operator looking at a display and then manually sending a bearing to the tracking party and/or fire control system. The manual process is a very discrete process so the bearings received look quite different. This needs to be understood if the bearings are to be used for determining the range, course,

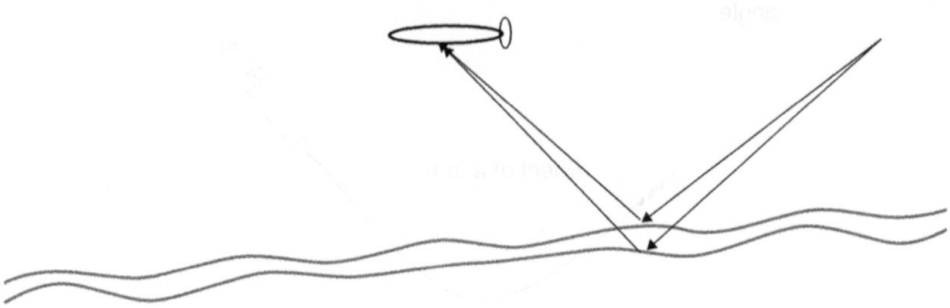

Figure 18.7 Signal being reflected from a sloping, rough, multilayered bottom

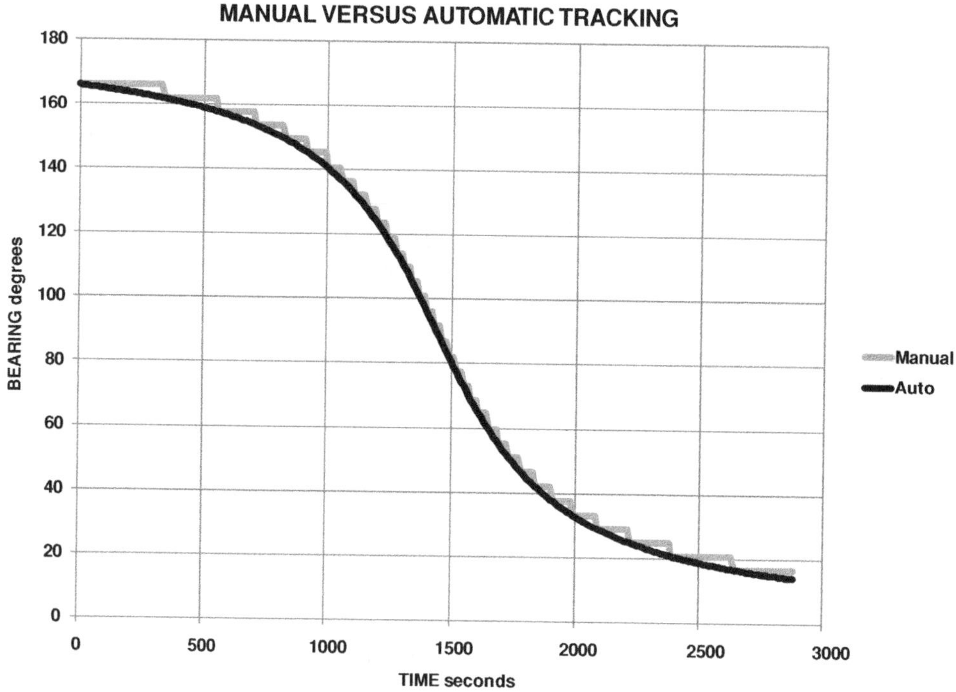

Figure 18.8 Example of bearing from an automatic tracker and manually passed bearing. The jumps are because the operator only periodically changing a bearing

and speed of the contact. Figure 18.8 shows an example of an automatic bearing and manual bearings.

18.7 Localization and Target Motion Analysis

The difference between target motion analysis (TMA) and localization is what is determined. TMA is the estimation of range, bearing, course, speed, and sometimes depth of a contact; localization is less than TMA. For example, localization may estimate range only, range and bearing, speed only, range and depth, etc.

18.7.1 Localization

Active sonar with each received ping gives a range estimate from the travel time between transmission and receipt. If the system is directional, a bearing estimate is also given. The CW range error is proportional to the pulse length and inversely proportional to the SNR:

$$\sigma^2(R) = c^2\left(\frac{\tau^2}{8\left(\frac{S_0}{N_0}\right)^2} + 0.29^2\tau^2 + \alpha^2 R^2\right)$$

$$\sigma^2(\text{SILOS}) = \frac{1}{(0.69 f_0)^2}\frac{1}{4\tau^2\left(\frac{S_0}{N_0}\right)^2} + \frac{0.4^2}{\tau^2} \tag{18.10}$$

where

R = range (ft)
c = speed of sound (ft/s)
τ = pulse length (s)
S_0 = signal level (not in dB)
N_0 = noise level (not in dB)
α = time to range fractional error
SILOS = speed in the line of sight (kts)
f_0 = frequency (kHz)

Assuming no environmental spreading, at a high SNR and with α equal to zero, the ambiguity function for a rectangular CW pulse is $(\sin(x)/x)^2$ on the frequency axis and a triangular distribution ($-\tau$ to $+\tau$) on the range axis.

The spectral power density (Fourier transform) for a rectangular CW pulse is

$$\rho(f) = \left[\frac{\sin(\pi f \tau)}{\pi f \tau}\right]^2 \tag{18.11}$$

This is the frequency distribution about the center frequency.

The Doppler resolution can be defined as either the −3 dB value or as the statistical bandwidth (BW) level. The −3 dB down value occurs at

$$\pi f \tau = 1.39 \tag{18.12}$$

or

$$f(-3\,\text{dB}) = \frac{1.39}{\pi\tau} = \frac{0.44}{\tau} \tag{18.13}$$

The statistical bandwidth of frequency spread is given by

$$W(f) = \frac{3}{4\tau} = \frac{0.75}{\tau} \tag{18.14}$$

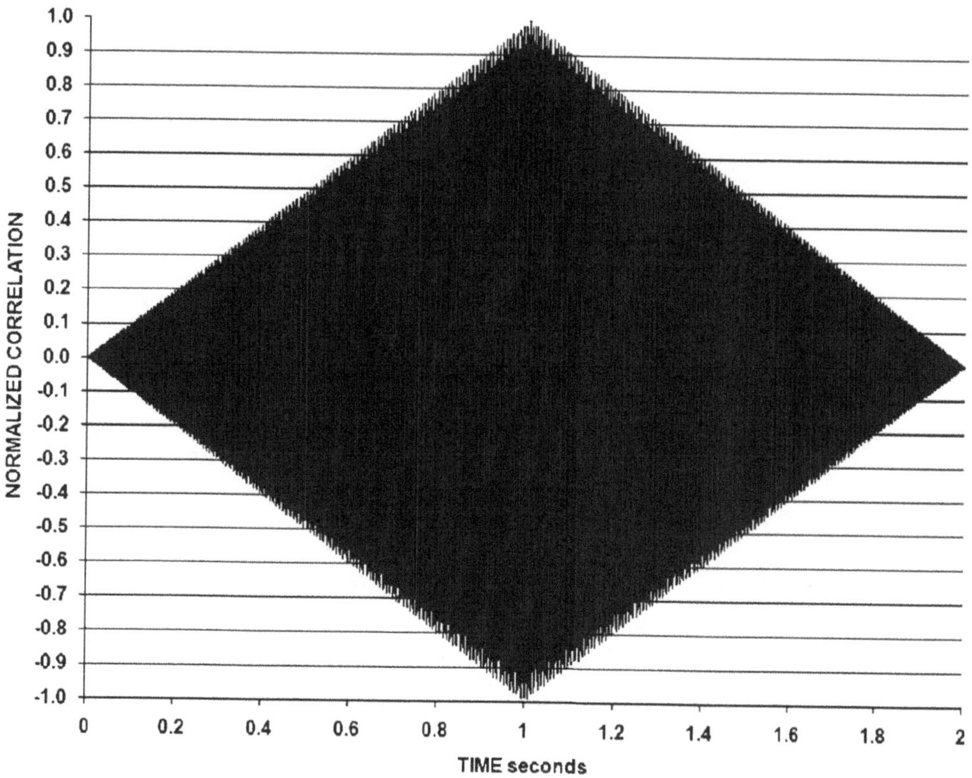

Figure 18.9 Correlation function for a rectangular CW pulse and its replica

$$\text{Doppler resolution}(-3\text{ dB}) = \frac{0.44}{\tau}\frac{1000}{0.69 f_0} = \frac{638}{\tau f_0}$$
$$\text{Doppler resolution}(W) = \frac{0.75}{2\tau}\frac{1000}{0.69 f_0} = \frac{543}{\tau f_0} \tag{18.15}$$

where

τ = pulse length (s)
f_0 = frequency (Hz)
Doppler = limit for high signal to noise (kts)

These results are one sided, i.e., know Doppler ± these values.

The ambiguity function on the range axis for a rectangular CW pulse is a triangular distribution ($-\tau$ to $+\tau$), as seen in Figure 18.9. The two-dimensional ambiguity function is shown in Figure 18.10.

The range resolution can also be defined either as the -3 dB value or as the one standard deviation level. The -3 dB down value occurs at a normalized height of 0.707 or about ± 15 % of the 2τ width:

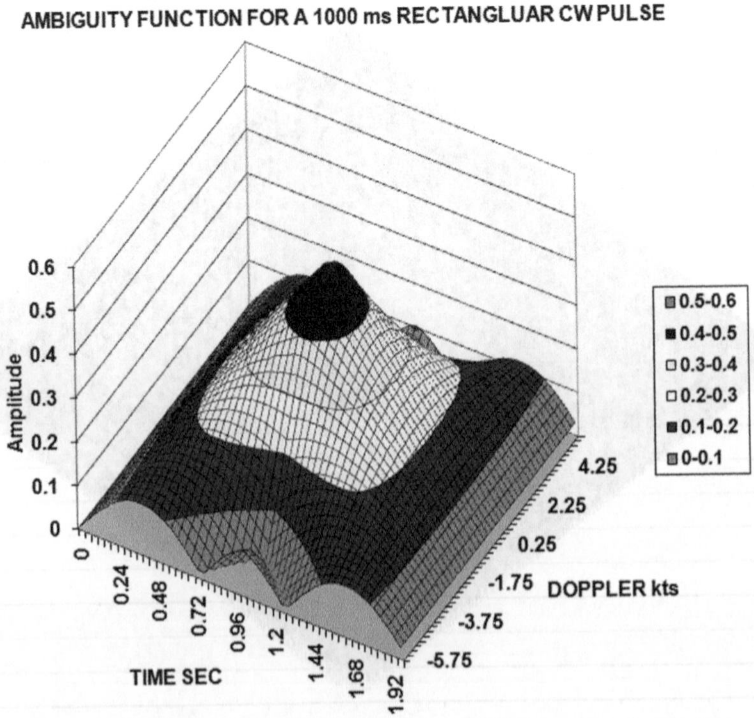

Figure 18.10 Ambiguity function for a 1 second Rectangular CW pulse

$$R(-3\,\text{dB}) = \pm 2c(0.15)\tau \tag{18.16}$$

or

$$R(-3\,\text{dB}) = \pm\frac{4850}{6076}0.3\tau = \pm 0.24\tau \tag{18.17}$$

The standard deviation is

$$\sigma^2(t) = \int_{-\tau}^{0} t^2\left(\frac{t}{\tau^2}+\frac{1}{\tau}\right)\mathrm{d}t + \int_{0}^{\tau} t^2\left(\frac{1}{\tau}-\frac{t}{\tau^2}\right)\mathrm{d}t \tag{18.18}$$

$$\sigma^2(t) = \frac{\tau^2}{6}$$

$$R(\sigma) = \pm\frac{c\tau}{\sqrt{6}} = \frac{4850}{6076}\frac{\tau}{\sqrt{6}} = 0.32\tau$$
$$\text{CW range resolution}(-3\,\text{dB}) = 0.24\tau \tag{18.19}$$
$$\text{CW range resolution}(\sigma) = 0.32\tau$$

where

τ = pulse length (s)
Range resolution = limit for very high signal to noise (nmile)

These results are one sided, i.e., no range $\pm$ these values.

The linear FM (LFM) range error is inversely proportional to the bandwidth and SNR:

$$\sigma^2(R) = c^2 \left[\frac{1}{8W^2 \left(\frac{S_0}{N_0} \right)^2} + (0.4W)^2 + \alpha^2 R^2 \right] \tag{18.20}$$

At a high SNR and $\alpha = 0$, this becomes

$$\sigma(R) = \frac{0.32}{W}$$
$$\sigma(\text{SILOS}) = \frac{5200}{W\tau} \tag{18.21}$$

where the range error is in nmile, the SILOS error is in knots, W is the bandwidth in Hertz, τ is the pulse length in seconds, and α is a speed of sound/path to the horizontal range error. For a linear period FM:

$$\sigma^2(R) = c^2 \left[\frac{1}{8W^2 \left(\frac{S_0}{N_0} \right)^2} + (0.4W)^2 + \alpha^2 R^2 \right] \tag{18.22}$$

At a high SNR this becomes

$$\sigma(R) = \frac{0.35}{W}$$
$$\sigma(\text{SILOS}) = 900\frac{W}{f_0} \tag{18.23}$$

where the range error is in nmile, the SILOS error is in kts, W is the bandwidth in Hertz, f_0 is the frequency in Hertz, and α is a speed of sound/path to the horizontal range error.

For no environmental spreading, a high SNR and α equal to zero, the ambiguity function of a linear (frequency change linearly with time) FM pulse is shown in Figure 18.11. The ambiguity function for a linear FM pulse at high signal to noise is shown in Figure 18.12. Notice the range Doppler uncertainty here.

Other localization methods applicable to passive sonar are wave front curvature ranging (WFCR), multisensor triangulation, multipath or full-field processing, and time delay ranging.

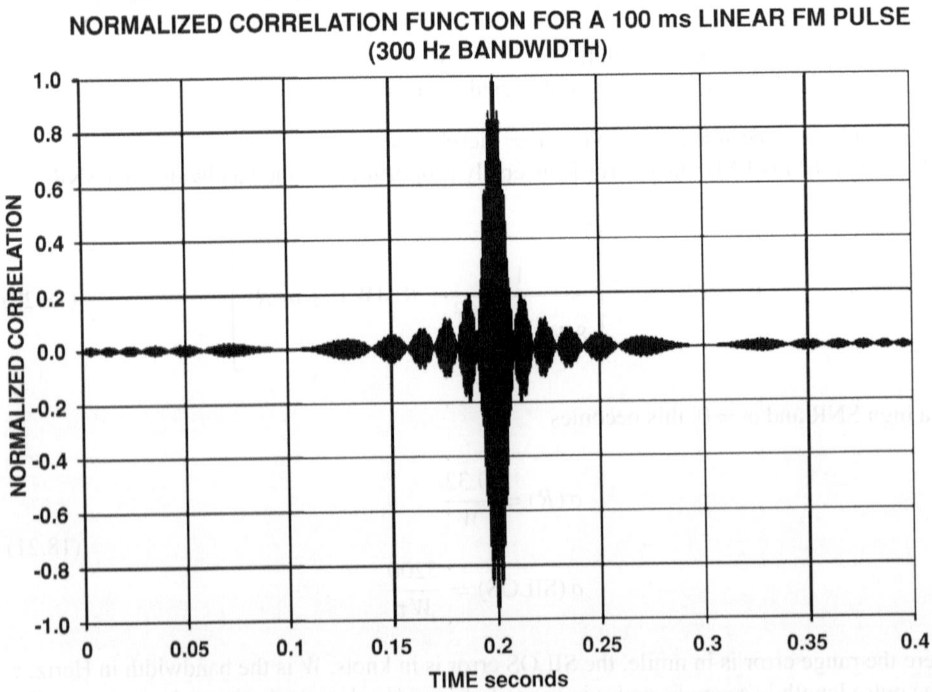

Figure 18.11 Correlation of a linear FM pulse with its replica

18.7.2 Wave Front Curvature Ranging (WFCR)

Wave front curvature ranging uses three arrays or hydrophones. The arrays are processed in pairs with the time differences between the forward–mid arrays being τ_{12} and between the mid–aft arrays being τ_{23}.The equation for the range can be derived as follows. Using the law of cosines with Figure 18.13 yields

$$\begin{aligned}(R + c\tau_{12})^2 &= R^2 + \left(\frac{L}{2}\right)^2 - RL\cos(\theta) \\ (R - c\tau_{23})^2 &= R^2 + \left(\frac{L}{2}\right)^2 + RL\cos(\theta)\end{aligned} \tag{18.24}$$

Adding the two equations and rearranging terms gives

$$R = \frac{L^2 - 2c^2\left(\tau_{12}^2 + \tau_{23}^2\right)}{4c\left(\tau_{23} - \tau_{12}\right)} \tag{18.25}$$

Using the binomial formula on the two equations to find $c\tau_{12}$ and $c\tau_{23}$ yields

$$2c^2\left(\tau_{12}^2 + \tau_{23}^2\right) = L^2\left[\cos(\theta)\right]^2 \tag{18.26}$$

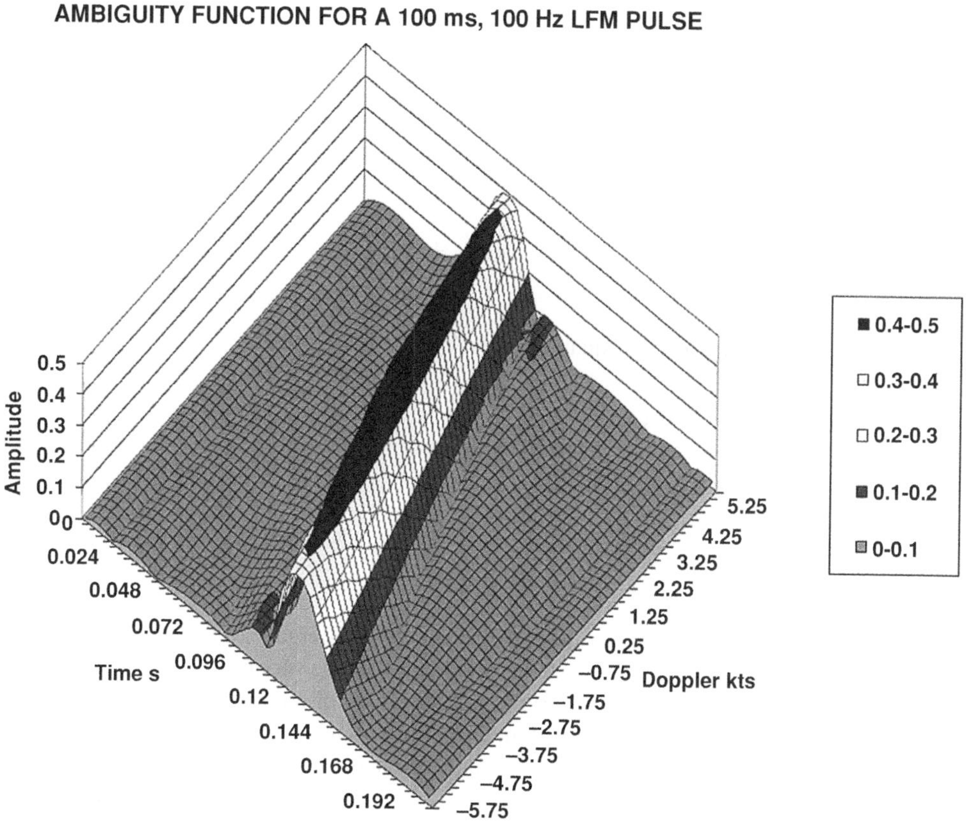

Figure 18.12 Ambiguity function for a 100 ms linear FM pulse at a high SNR

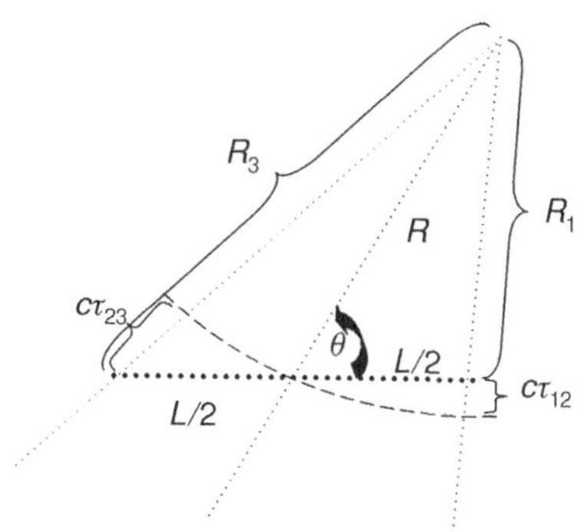

Figure 18.13 Wave front curvature range geometry

Therefore

$$R = \frac{L^2\,[\sin(\theta)]^2}{4c\,(\tau_{23} - \tau_{12})} \tag{18.27}$$

Similarly, if the two equations are subtracted and if we assume that $c^2(\tau_{23}{}^2 - \tau_{12}{}^2)/R$ is very small, then

$$\theta = \cos^{-1}\left[\frac{c\,(\tau_{12} + \tau_{23})}{L}\right] \tag{18.28}$$

An error analysis of the ranging equation gives

$$\sigma^2(R) = \frac{4\,[c\cos(\theta)]^2\,\sigma^2\left(\sum\tau\right)}{[\sin(\theta)]^4}\frac{R^2}{L^2} + \frac{16c^2\sigma^2(\Delta\tau)}{[\sin(\theta)]^4}\frac{R^4}{L^4} \tag{18.29}$$

where

$\Sigma\tau = \tau_{12} + \tau_{23}$
$\Delta\tau = \tau_{23} - \tau_{12}$

and $\sigma(\Sigma\tau)^2$ and $\sigma(\Delta\tau)^2$ are the variance of the sum and difference of the time delays and have variances of twice the individual time delay variances (since variances add), as calculated above for trackers. Notice that the first term is zero at array normal and grows only as range squared. Therefore, for most situations of interest the second term dominates. Under this assumption the ranging equation with error can be written as

$$R + \delta R = \frac{L^2\,[\sin(\theta)]^2}{4c\,(\Delta\tau + \delta\Delta\tau)} \tag{18.30}$$

This form of equation arises for many of the passive techniques. It is important to note that a symmetric error in $\Delta\tau$ gives an asymmetric error in range (see Figure 18.14).

If we look at a single solution, half the time the range solution would be long and half the time it would be short. However, in tactical situations the results are different. Consider a ship that is closing a target and wishes to attack as soon as the target is inside 5 nautical miles; solutions are continuously being generated. Looking at Figure 18.14, only those to the left of 5 nmile represent possible firing solutions; these are all short. Therefore, at-sea data of firing points will underestimate the range to the target. Conversely, if the ship's mission is to follow a target and collision or counterattack are concerns, then the ship may want to stay outside 5 nmile. Figure 18.14 shows that all such solutions are longer than the actual range. At sea, we would expect the estimated range to the target to be long.

18.7.3 Multipath Ranging (MPR)

Multipath ranging is a method of determining range and depth of a target. There are several techniques that are used. Here, the discussion will be limited to simple straight line methodologies, where the assumption will be an isovelocity body of water with a flat bottom so sound rays move in straight lines.

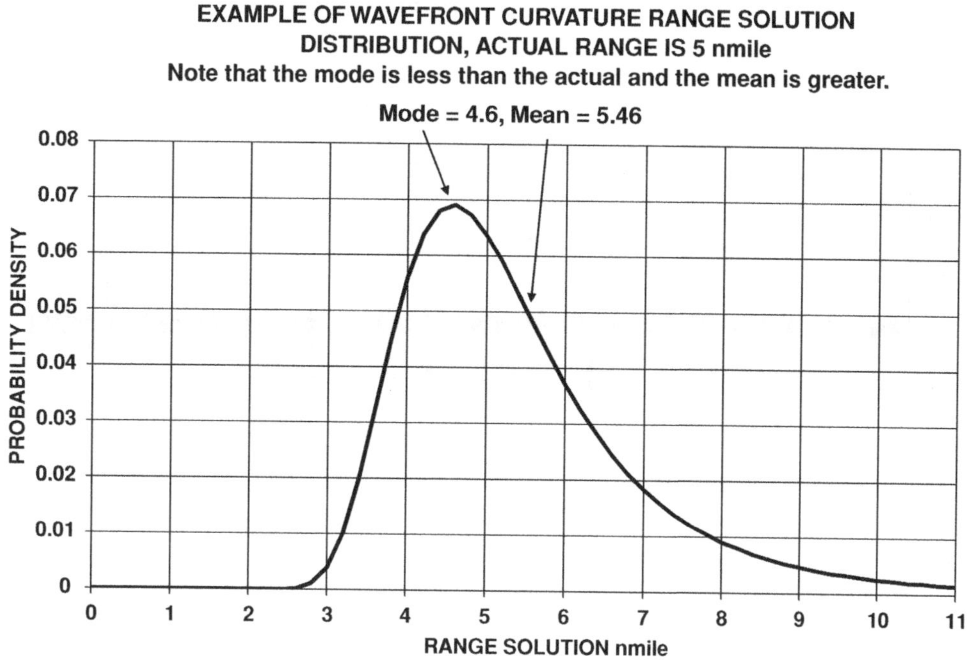

Figure 18.14 An example of distribution of solution ranges from a wave front curvature ranging system, showing the asymmetry

The first method requires a passive broadband sensor that can capture four sets of paths: (1) the bottom bounce, (2) the surface–bottom, (3) the bottom–surface, and the (4) surface–bottom–surface rays, as shown in Figure 18.15.

An autocorrelation process (correlation of the beam output with time delayed copies of itself) is performed on the beam data. The correlation peaks will correspond to the travel time of each path. Figure 18.16 shows an example. In this figure, the peaks at about 5.43 and 8.2 ms are the surface–bottom and the bottom–surface paths, the peak at 13.7 ms is the surface–bottom–surface path, and the peak at about 2.7 ms is the combination of the difference between the surface–bottom and the bottom–surface rays. It should be noted that the order in which the surface–bottom and bottom–surface signals appear depend on the target depth, which is unknown; the solution to this will be discussed later. The four path lengths in question are

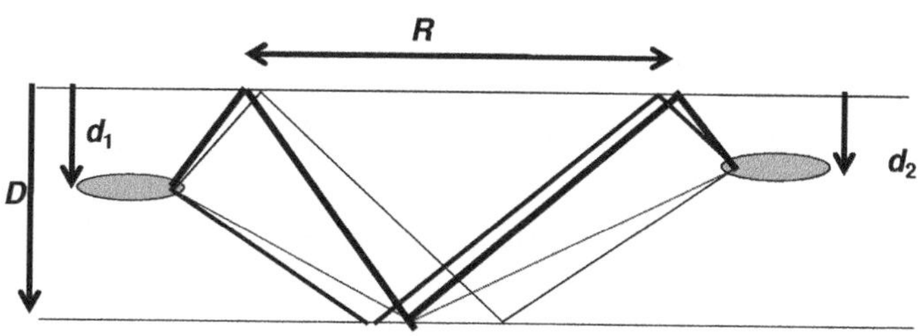

Figure 18.15 Geometry for time delay multipath ranging, using straight line propagation

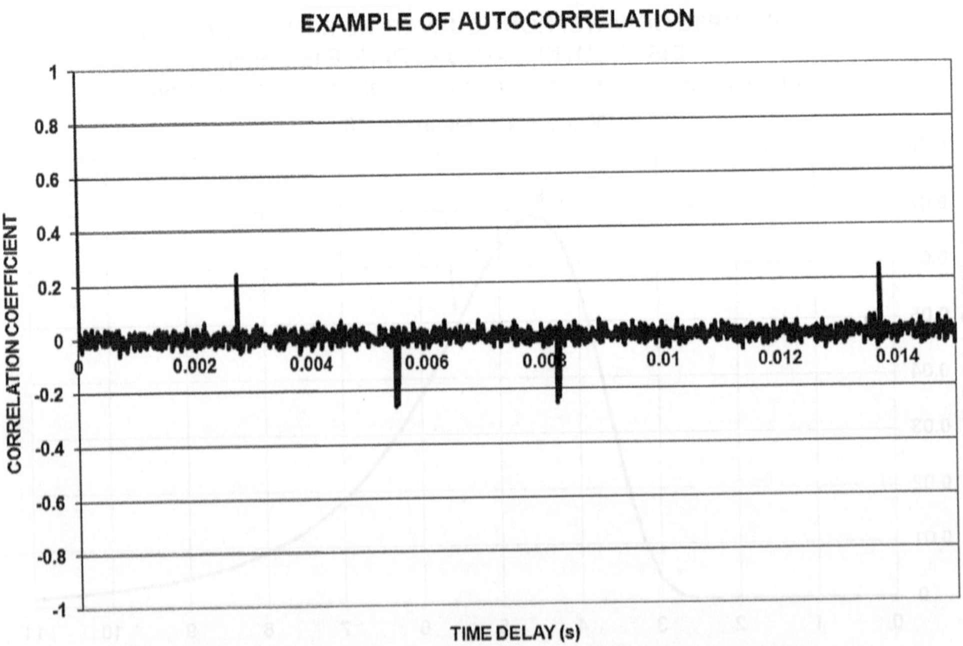

Figure 18.16 Autocorrelation output for time delay multipath ranging

$$
\begin{aligned}
L_{\mathrm{BB}} &= \sqrt{R^2 + (D - d_1 - d_2)^2} \\
L_{\mathrm{BS}} &= \sqrt{R^2 + (D - d_1 + d_2)^2} \\
L_{\mathrm{SB}} &= \sqrt{R^2 + (D + d_1 - d_2)^2} \\
L_{\mathrm{SBS}} &= \sqrt{R^2 + (D + d_1 + d_2)^2}
\end{aligned}
\tag{18.31}
$$

Assuming that the range is much greater than the depths, each of these paths is approximately given by

$$
\begin{aligned}
L_{\mathrm{BB}} &= R + \frac{1}{2R}\left[D^2 - 2D\,(d_1 + d_2) + d_1^2 + 2d_1 d_2 + d_2^2\right] \\
L_{\mathrm{BS}} &= R + \frac{1}{2R}\left[D^2 - 2D\,(d_1 - d_2) + d_1^2 - 2d_1 d_2 + d_2^2\right] \\
L_{\mathrm{SB}} &= R + \frac{1}{2R}\left[D^2 + 2D\,(d_1 - d_2) + d_1^2 - 2d_1 d_2 + d_2^2\right] \\
L_{\mathrm{SBS}} &= R + \frac{1}{2R}\left[D^2 + 2D\,(d_1 + d_2) + d_1^2 + 2d_1 d_2 + d_2^2\right]
\end{aligned}
\tag{18.32}
$$

The differences in path length that will appear in the autocorrelation are

$$L_1 = L_{BS} - L_{BB} = \frac{2}{R}(d_2 D - d_1 d_2)$$
$$L_2 = L_{SB} - L_{BB} = \frac{2}{R}(d_1 D - d_1 d_2) \quad (18.33)$$
$$L_3 = L_{SBS} - L_{BB} = \frac{2}{R}(d_2 - d_1)\, D$$

Each pair of equations above can be solved for range, R, and target depth, d_2, yielding

$$R_{12} = \frac{2d_1 D}{L_2 + \dfrac{d_1}{D - d_1} L_1}, \quad d_2 = \frac{d_1 D}{L_2 \left(\dfrac{D - d_1}{L_1} + \dfrac{d_1}{L_2}\right)}$$
$$R_{23} = \frac{2\left(d_1 D + d_1^2\right)}{L_2 + \dfrac{d_1}{D} L_3}, \quad d_2 = \frac{d_1 D}{\dfrac{\left(\dfrac{1}{L_2} - \dfrac{1}{L_3}\right)}{\left(\dfrac{d_1}{L_2} + \dfrac{D}{L_3}\right)}} \quad (18.34)$$
$$R_{13} = \frac{2d_1 D}{L_3 + \dfrac{D}{D - d_1} L_1}, \quad d_2 = \frac{d_1 D}{\left(\dfrac{D - d_2}{L_2} - \dfrac{D}{L_3}\right)}$$

If all three sets are solved and compared for the two choices for L_1 and L_2, it will be obvious which is most consistent (Figure 18.17). It can be seen that the correct choice solutions cluster more closely and, in some cases, the solution depth may be impossible for a target as 1100 ft or 1200 ft might be well below the maximum depth capability of the expected threat.

18.7.4 Depression/Elevation (D/E) Ranging

If the sensor is capable of measuring the arrival angles of the paths with precision, then the rays can be traced to where they intersect. This would then be the solution.

18.7.5 Triangulation Ranging

Any two sensors that can provide a bearing and are separated by a distance can be used to estimate range (see Figure 18.18).

The figure shows that $\Delta\theta + \theta_1 + \theta_2$ is 180°. The error in the range can be estimated from the error in the two bearing errors:

$$R_s = \frac{L \sin(\theta)}{\Delta\theta + \delta\Delta\theta} \quad (18.35)$$

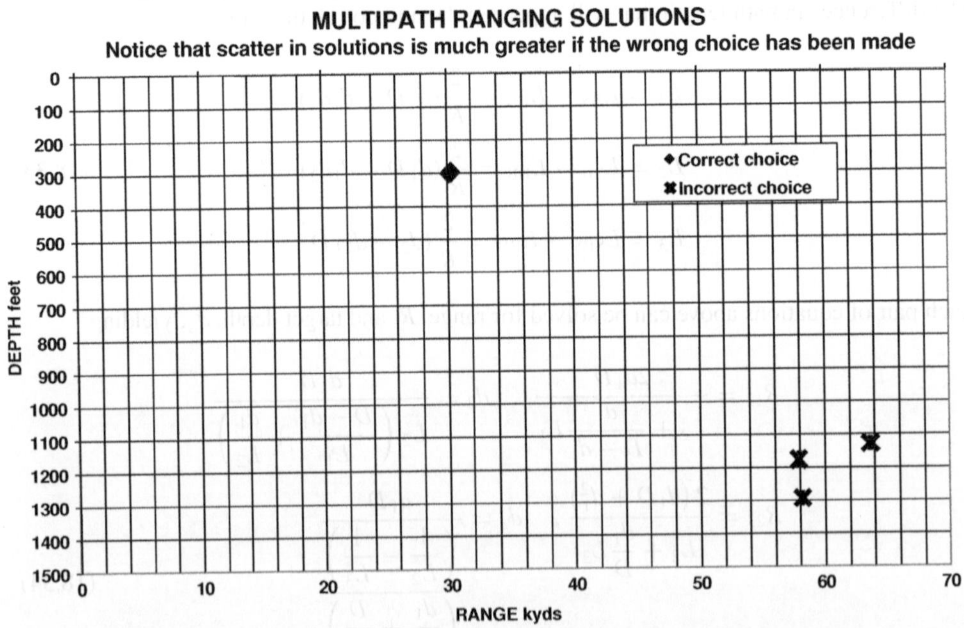

Figure 18.17 Example of multipath arrival time ranging and depth determination

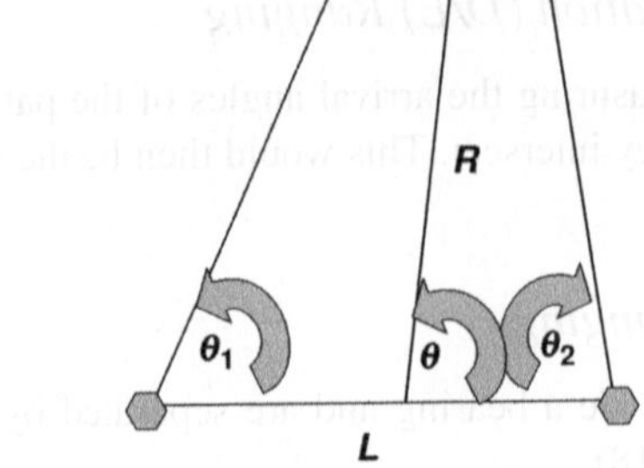

Figure 18.18 Triangulation ranging geometry

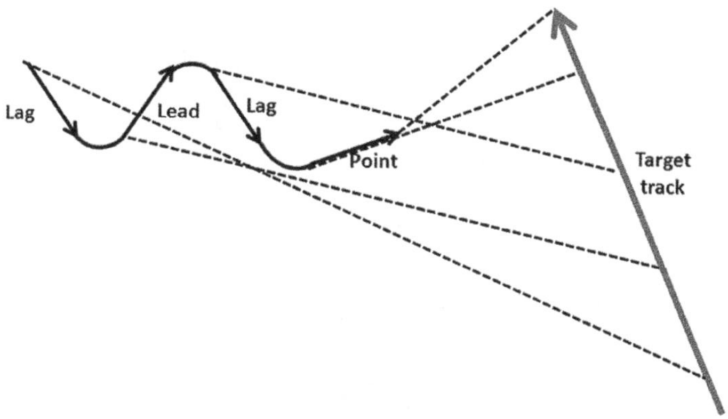

Figure 18.19 TMA jargon

The variance of the uncorrelated random error will combine as the sum of variances:

$$\sigma^2(\Delta\theta) = \sigma^2(\theta_1) + \sigma^2(\theta_2) \tag{18.36}$$

and any bias errors will add.

If one or more of the sensors is a line or towed array, the angles will be conical and the solution range will be the interception between the hyperbole.

18.8 Bearings Only Methodologies

There are certain terms that are associated with passive TMA and tactics. Figure 18.19 illustrates the different types of TMA legs.

These terms can be defined as follows:

Leg: a portion of own ship track (usually assumed to be constant velocity)

Point leg: own ship track aimed directly at a target

Lead leg: own ship track aimed ahead of the target bearing

Lag leg: own ship track aimed behind the target bearing

18.9 Four-Bearing TMA

The classic case of TMA is the four-bearing solution shown in Figure 18.20, which measures or estimates four bearings and solves the linear equations of motion for the target coordinates and velocity components:

$$\tan(\beta_i) = \frac{XT_i - XO_i}{YT_i - YO_i} \tag{18.37}$$

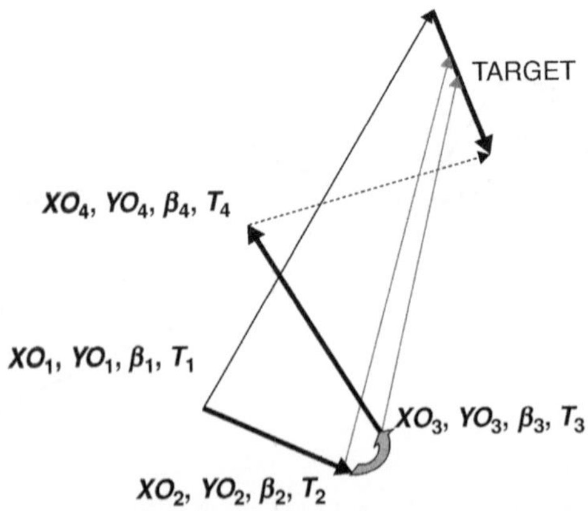

Figure 18.20 Four-bearing TMA geometry

where

β_i = bearing estimate at time i
XT_i = target x coordinate at time i
XO_i = own sensor x coordinate at time i
YT_i = target y coordinate at time i
YO_i = own ship y coordinate at time i

If we assume that the target has constant velocity during the time period in question, then

$$\begin{aligned} XT_i &= XT_1 + VXT\,T_i \\ YT_i &= YT_1 + VYT\,T_i \end{aligned} \tag{18.38}$$

where T_i is the ith time and *VXT* and *VYT* are the x and y components of the target's velocity. This can be rewritten as

$$YT_1\tan(\beta_i) + VYT\,T_i\tan(\beta_i) - XT_1 - VXT\,T_i = YO_i\tan(\beta_i) - XO_i \tag{18.39}$$

Since the β_i, XO_i, and YO_i are all known, we have a linear equation with four unknowns: YT_1, XT_1, *VXT*, and *VYT*. To solve, we need four sets of measurements, at least one of which is linearly independent. This last condition is satisfied if own ship changes velocity during the measurements. In matrix form this is

$$\begin{bmatrix} \tan(\beta_1) & T_1\tan(\beta_1) & -1\text{-}T_1 \\ \tan(\beta_2) & T_2\tan(\beta_2) & -1\text{-}T_2 \\ \tan(\beta_3) & T_3\tan(\beta_3) & -1\text{-}T_3 \\ \tan(\beta_4) & T_4\tan(\beta_4) & -1\text{-}T_4 \end{bmatrix} \begin{bmatrix} YT_1 \\ VYT \\ XT_1 \\ VXT \end{bmatrix} = \begin{bmatrix} YO_1\tan(\beta_1) - XO_1 \\ YO_2\tan(\beta_2) - XO_2 \\ YO_3\tan(\beta_3) - XO_3 \\ YO_4\tan(\beta_4) - XO_4 \end{bmatrix} \tag{18.40}$$

Multiplying left and right by the inverse of the fist matrix gives

$$\begin{bmatrix} YT_1 \\ VYT \\ XT_1 \\ VXT \end{bmatrix} = \text{inv} \begin{bmatrix} \tan(\beta_1)\ T_1 \tan(\beta_1)\ -1\text{-}T_1 \\ \tan(\beta_2)\ T_2 \tan(\beta_2)\ -1\text{-}T_2 \\ \tan(\beta_3)\ T_3 \tan(\beta_3)\ -1\text{-}T_3 \\ \tan(\beta_4)\ T_4 \tan(\beta_4)\ -1\text{-}T_4 \end{bmatrix} \begin{bmatrix} YO_1 \tan(\beta_1) - XO_1 \\ YO_2 \tan(\beta_2) - XO_2 \\ YO_3 \tan(\beta_3) - XO_3 \\ YO_4 \tan(\beta_4) - XO_4 \end{bmatrix} \tag{18.41}$$

The four bearings used need not be the actual measurements at the points shown in Figure 18.20. On each of the TMA legs, a series of measurements can be taken and a least squared fit could be made. From this, the required bearings could be inferred. In general, this would give a better estimate than a single measurement at the beginning and end of each leg. The random error in each would be given by

$$\sigma^2(\beta_m) = \sigma^2(\beta)\frac{2\,(2m-1)}{m\,(m+1)} \tag{18.42}$$

assuming all the individual bearing are equally spaced in time and have equal variances. If 20 bearings were measured on a leg, the bearing error at the end points would be reduced by $\sqrt{}\{2[2(20-1)]/20/(20+1)\}$ or 43 %. This expression and one for the unequal case are derived in Appendix B. The downside of doing this is that the bearings are, in fact, not on a straight line, but rather on a section of a tangent function. Therefore, if the bearing changes are large this approach of fitting to a line will be worse.

18.10 Ekelund Ranging

Ekelund ranging is a localization technique very similar to the four-bearing approach, except that only the bearing rate on each of two legs is used to estimate range. If we assume that the target does not maneuver and that during own ship maneuvers the range does not change significantly, then for each leg we can write

$$\begin{aligned} \text{BDOT}_1\, R &= V \sin(\text{ASP}) - U_1 \sin(\text{AOB}_1) \\ \text{BDOT}_2\, R &= V \sin(\text{ASP}) - U_2 \sin(\text{AOB}_2) \end{aligned} \tag{18.43}$$

where

BDOT = bearing rate (change in bearing with time) on each leg
R = range
V = target speed
U = own speed on each leg
ASP = target aspect (angle between target bow and our line of sight)
AOB = angle off our bow of the contact on each leg (+ to right and – to left by convention)

Subtracting the two equations, the target aspect and speed cancel, leaving

$$R_s = \frac{\text{DSALOS}}{\text{DBDOT}} \tag{18.44}$$

where

DSALOS = change in speed across the line of sight caused by the maneuver
DBDOT = change in bearing rate between maneuvers

An analysis of errors would be as follows:

$$R_s = \frac{\text{DSALOS} + \delta\text{DSALOS}}{\text{DBDOT} + \delta\text{DBDOT}} \tag{18.45}$$

Dividing by DSALOS, this can be rewritten as

$$R_s = \frac{1 + \dfrac{\delta\text{DSALOS}}{\text{DSALOS}}}{\dfrac{1}{R} + \dfrac{\delta\text{DBDOT}}{\text{DBDOT}}} \tag{18.46}$$

The top term contains an error in the own ship speed across the line of sight. The error in the difference in the bearing rate can be estimated as

$$\sigma^2(\text{DBDOT}) = k\frac{\sigma^2(\beta)\,\Delta t}{T^3} \tag{18.47}$$

where

k = a constant
$\sigma^2(\beta)$ = variance in the random bearing error
Δt = time between independent bearings
T = total time on a leg

The derivation of this is in Appendix B and assumes that the bearing measurements are evenly spaced, have the same variance, and that a least square procedure is used to estimate the bearing rates (DBDOT).

Why compute an Ekelund range when the data are available for doing a full bearings only solution? Bearings only methodologies are very sensitive to a variety of variable bias errors (see the delta bias above) that have little or no effect on the Ekelund results. Ekelund ranging can be done quickly in a person's head, whereas bearings only solutions require computers that can invert a matrix. Ekelund ranging does make assumptions that are not true, such as the range does not change during the time involved; even with perfect bearing data the resulting range contains these residual errors.

18.11 Range and Bearing TMA

This approach was developed for active sonar, but can be adapted to any localization method that supplies range and bearing estimates. For example, multipath ranging, or wavefront curvature ranging TMA, is based on a least squares fit to inputs of range and bearing, which yields

$$\sigma^2(R_m) = \frac{2\,(2m-1)}{m\,(m+1)}\sigma^2(R_1)$$

$$\sigma^2(\text{SILOS}) = \frac{3c^2}{m^3-m}\,\frac{\sigma^2(R_1)}{R^2} \tag{18.48}$$

$$\sigma^2(\text{SALOS}) = \frac{3c^2}{m^3-m}\sigma^2(\theta)$$

where

R = actual range between receiver sensor and target platform
M = number of returns based on the ping scale
c = speed of sound in water
SILOS = speed in the line of sight
SALOS = speed across the line of sight
$\sigma^2(R)_m$ = variance of range error for m pings
$\sigma^2(\theta)$ = bearing variance

The methodology for deriving these equations is to do a least squares fit to the range and bearing data, under the assumptions that:

1. There are no target maneuvers.
2. It operates on the target range scale; i.e., the time between pings is $2R/c$.
3. The random components of error are uncorrelated from ping to ping.
4. The variance of the range and bearing errors are constant.

18.12 Other Bearings Only TMA Methodologies

There are many alternative methods of formulating the bearings only problem, each with its strengths and weaknesses. These methods range from brute force to the use of special iterative filters like a Kalman filter. The brute force approach has gained ground with the advent of cheap, fast computers. The process is to assume a solution, calculate the bearing residuals (predicted minus observed), and then vary assumed solution parameters, usually in a parametric way, seeking to find a solution or solution space that minimizes the sum squared of the residuals or the weighted squared residuals (if individual measurements are expected to have different accuracies). This procedure may be manual or automated depending on the fire control system. Included in the iterative methodology may be alternative hypotheses, such as whether all or some of the bearings are coming from a bottom bounce or a direct path, which would effect a conical to azimuthal angle conversion, or whether or not the target has maneuvered (in naval terms, zigged).

18.13 Other TMA and Localization Schemes

All of the above methodologies have assumed that the measured quantities are bearing (horizontal or vertical) or time delays. Other techniques can be built around measuring frequency shift; if the target has a narrow very stable signal, then changes in the frequency as an own ship maneuver or when the target maneuvers can determine the TMA parameter, as well as provide a target zig indicator. The development of these is analogous to the bearings only method except that the observed quantity is Doppler, not bearing, although both can be combined in an algorithm. If, for a narrowband signal, both bearing and frequency are measured simultaneously over a period of time a TMA solution can be obtained without a maneuver.

References

[1] Cramér, H., *Mathematical Methods of Statistics*, Princeton, NJ: Princeton University Press, 1946.
[2] Rao, C., "Information and the Accuracy Attainable in the Estimation of Statistical Parameters," *Bulletin of the Calcutta Mathematical Society*, **37**, 1945, 81–89.
[3] Rao, C., in S. Das Gupta, *Selected Papers of C. R. Rao*, New York: John Wiley & Sons, Inc., 1994.
[4] Kay, S. M., *Fundamentals of Statistical Signal Processing*, Vol. I, *Estimation Theory*, Upper Saddle River, NJ: Prentice Hall Inc., 1993.
[5] Eckart, C., "Optimal Rectifier Systems for Detection of Steady Signals," University of California, Marine Physical Laboratory of Scripps Institute of Oceanography, La Jolla, CA, Ref 52-11, 1952.

19

Design and Evaluation of Sonars

When a program manager of a new antisubmarine warfare (ASW) platform is asked what features they desire in a sonar, some likely responses are: detect at a zillion miles; work well at all speeds, in all environments and in all directions; provide bearing, range, course and speed estimates to a gnat's eye lash; be no bigger than a bread box; and cost almost nothing. Needless to say, some of these requirements are incompatible.

19.1 Choice of Frequency and Size

Long detection ranges require low frequencies because of the attenuation of sound by sea water (see Chapter 5). Unfortunately, in order to get directionality at low frequencies, large arrays are required. As an example, Figure 19.1 has notional data as a function of frequency: radiated noise, ambient noise, and propagation loss at four ranges. Using the inputs from Figure 19.1, Figure 19.2 shows the solution for required directivity versus frequency for passive detection. The grey line shows the minimum required directivity to achieve a zero signal excess at each range. This result was obtained by rearranging the sonar equation to solve for the directivity that would yield zero signal excess. This analysis assumes a broadband detector about one octave wide, centered at each frequency.

In Figure 19.2, the region of minimum directivity as a function of frequency declines as the desired range increases. This is a fairly common result. The minimums here are relatively broad. For example, the 3 kyds curve has a region within 3 dB of the minimum extending from 3 kHz to 20 kHz. Similarly, the 10 kyds curve is within 3 dB of the minimum from 400 Hz to 10 kHz. This example uses a target with very smooth radiated noise and propagation loss. A more realistic choice might yield a sharper minimum or even multiple good regions.

To achieve higher directivity, the product of the size and the frequency must increase (see Table 3.4). For the 10 kyd case above, a directivity of 27 dB at 6 kHz would require a towed array at least 310 feet long or a baffled square array measuring 5.5 ft × 5.5 ft. To achieve the same directivity at 300 Hz would require a 6200 ft towed array or a baffled square array that was 27 ft × 27 ft.

An identical approach can be taken for a noise-limited active sonar. Figure 19.3 shows the required directivity to achieve a zero signal excess at each range versus frequency. These calculations assume an ambient noise-limited monostatic active sonar using the same ambient

Underwater Acoustics: Analysis, Design and Performance of Sonar Richard P. Hodges

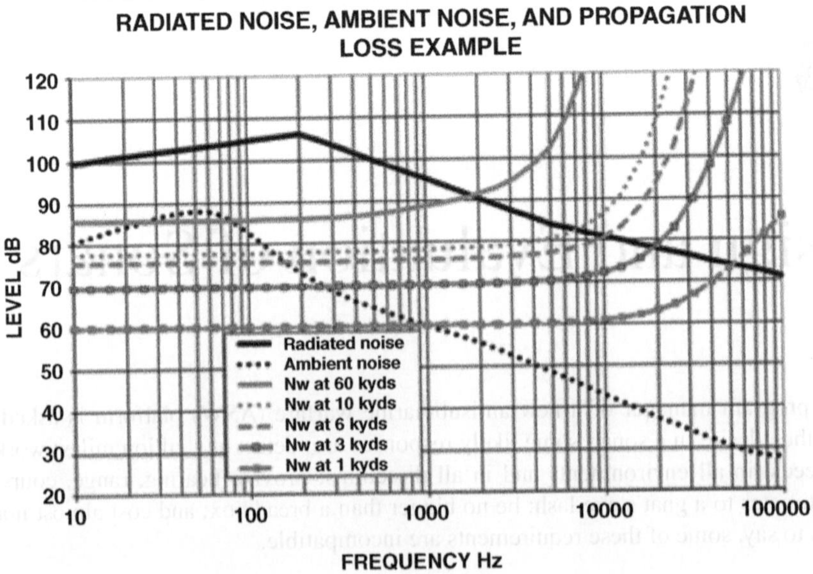

Figure 19.1 Notional examples of passive sonar inputs

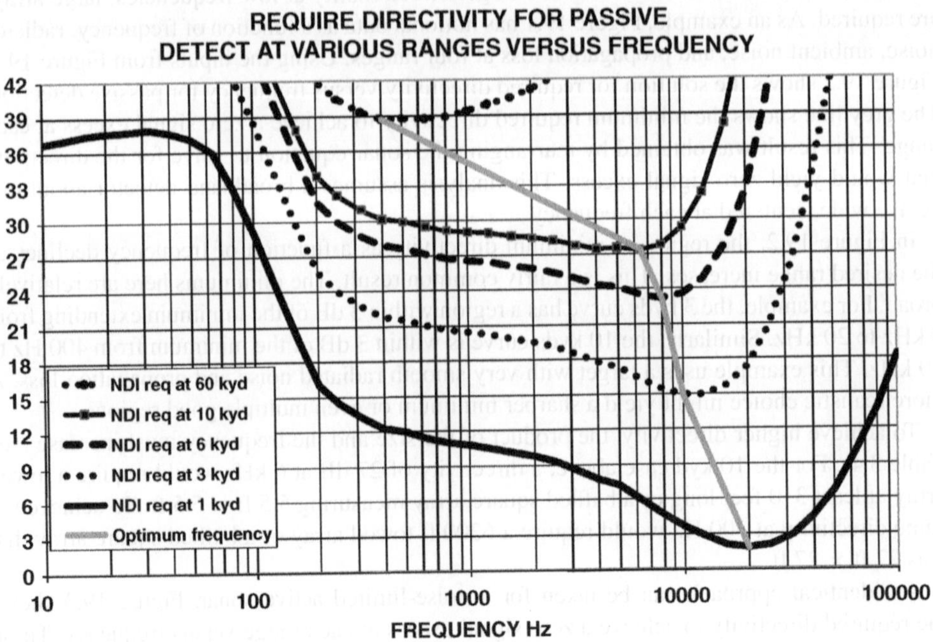

Figure 19.2 Solution for required directivity using the inputs in Figure 19.1

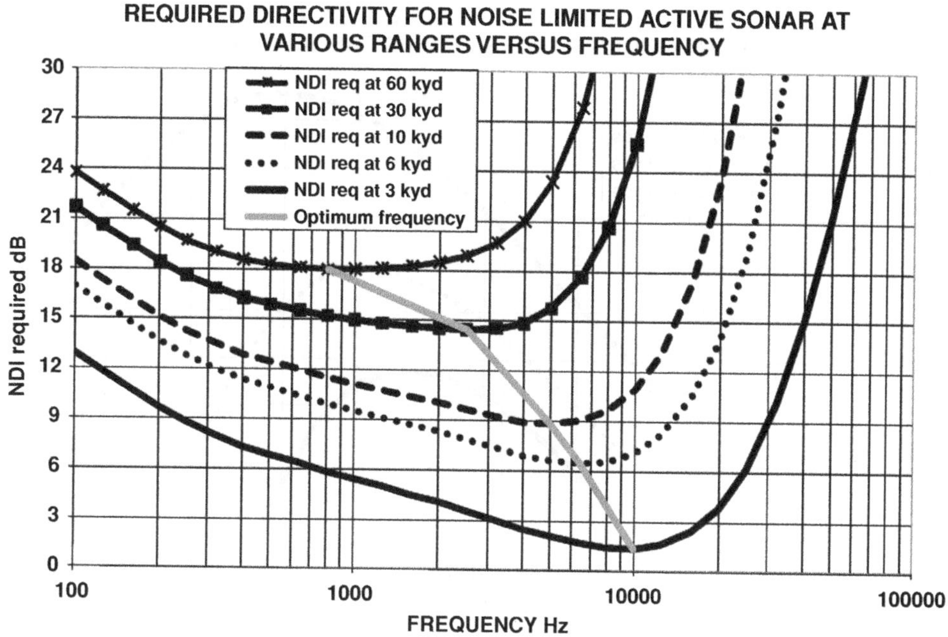

Figure 19.3 Required directivity to achieve a zero signal excess at each range

noise and propagation loss as in the passive case of Figure 19.2, 15 dB active target strength, and a 100 W acoustic output for each hydrophone on transmit. These are notional inputs and should not be interpreted as a general solution.

In Figure 19.3, as the desired range increases the region of minimum directivity versus frequency drops. This is a fairly common result. The minimums here are quite broad, making other criterion such as size and total power (power per hydrophone × number of hydrophones) come into play in the design process. A more realistic choice of inputs might yield a sharper minimum or even multiple good regions.

This type of approach to designing a sonar can also start with reviewing system constraints and then looking for regions that can achieve the desired performance. Figure 19.4 shows the results of such an analysis, where the available transmit power and physical array size were fixed. The performance of three possible ranges of interest is shown. These results are for a small monostatic active system at various ranges versus frequency. The assumptions used are: active target strength is 15 dB, total power transmitted is 250 W, the array is an ambient noise-limited baffled square 6 in × 6 in. These are notional inputs and should not be interpreted as a general solution.

19.2 Computational Requirements

In addition to the choice of frequency and size, the computational requirements of a sonar system need to be considered. For example, a typical towed array used for conventional beamforming, known as a linear multispot array, has the following computational requirements [1]:

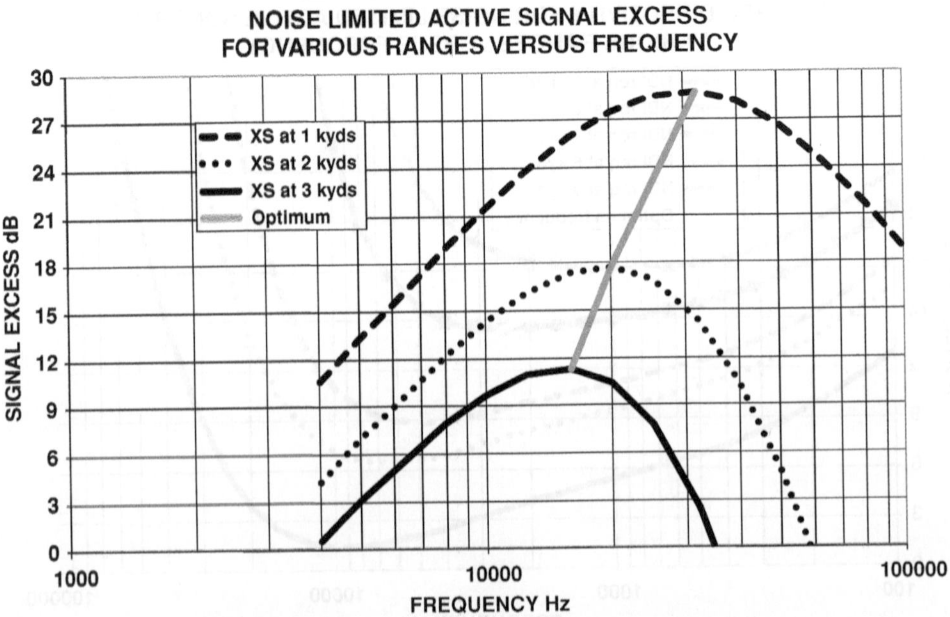

Figure 19.4 Small monostatic active system at various ranges

$$N_e F_s (N_f + N_b) \quad \text{multiplications/s.}$$

$$N_e F_s (N_f + N_b) \quad \text{additions/s.} \tag{19.1}$$

where

N_e = number of elements (hydrophone groups)
N_b = number of beams formed
N_f = number of stages in the FIR filter used for the interpolation process
F_s = sampling rate = $N_s F$ design
N_s = sampling rate multiple (at least 2, usually 6)

While the details will vary depending on the exact implementation, the following equations are adequate for parametric comparisons scaling system performance [1].

19.2.1 Beamforming

For a one-dimensional array (towed or line array), the beamformer computational requirements are given by operations per second (OPS):

$$\text{OPS(1D)} = k(\text{Number of beams}) \text{ (Number of phones/beam) (Sampling rate)} \tag{19.2}$$

Because the hydrophones are separated by a half wavelength at the design frequency, this can be rewritten as

$$\mathrm{OPS(1D)} = k' L^2 F_{\mathrm{d}}^3 \tag{19.3}$$

where

L = acoustic aperture length
F_{d} = design frequency for aperture
k, k' = proportionality constant, which depends on detailed choices, like sampling rate, shading

For a two-dimensional array (area array) like the U.S. Navy's AN/BQQ-5 spherical array, the beamformer computational requirements are given by

$$\mathrm{OPS(2D)} = k''(\text{Number of horizontal beams})\,(\text{Number of vertical beams})\,(\text{Number of phones/beam})\,(\text{sampling rate}) \tag{19.4}$$

Again, because the hydrophones are generally a half wavelength in spacing, this can be rewritten as

$$\mathrm{OPS(2D)} = k''' A^2 F_{\mathrm{d}}^5 \tag{19.5}$$

where

A = array area
F_{d} = design frequency
k'', k''' = proportionality constants, which depend on detailed choices, like sampling rate, angular coverage desired

An important takeaway from these equations is how rapidly the number of computations grows with frequency.

19.3 Signal Processing after Beamformer

19.3.1 Detection

The number of operations per second (OPS) is given by

$$\mathrm{OPS(PNB)} = a(\text{Number of beams})\,(\text{Number of frequency bins})\,(\text{Sampling rate}) \tag{19.6}$$

where

a = proportionality constant, which depends on detailed choices like FFT overlap, NSE implementation

For arrays with multiple apertures, the results for each aperture must be summed.

Similarly, for passive broadband (PBB), the number of operations per second is given by

$$\mathrm{OPS(PBB)} = a'(\text{Number of beams})\,(\text{Sampling rate}) \tag{19.7}$$

where

a' = a proportionality constant, which depends on detailed choices like SSA implementation

For arrays with multiple bands, the results for each band need to be summed.

19.4 Active Pulse Choice

The choice of type and duration of active sonar pulses is a classic example of tradeoffs. There are four basic types of pulses: continuous wave (CW), coded pulses (CP), pseudo random pulses (PRN), and explosive or impulsive pulses. Each type has inherent advantages and disadvantages in both detection and localization. Because of this, many active systems have an assortment of pulses implemented.

There are two basic types of processing: (1) wideband or "incoherent" processing, where a single filter band, wide enough to include the signal and any possible Doppler, is processed, and (2) match filter processing, where the processing attempts to match the expected signals using either multiple frequency bins or replica correlation. Wideband processing may be implemented with a notch around zero Doppler to exclude reverberation and low Doppler signals. This type of processing was used because it was cheap and simple. However, most modern systems do not use this approach because the losses in performance are large relative to match filter processing.

Continuous wave (CW) pulses are very narrow band signals, which may be amplitude shaped or shaded. Long pulses have good Doppler resolution, but poor range resolution. The Doppler resolution provides for discrimination against reverberation and allows for better detection of targets of high speed, in the line of sight when reverberation is high.

Coded pulses (CP) include continuously swept frequency modulated (FM) pulses, which may sweep up or down or linearly in frequency (LFM) or in period, hyperbolic (HFM). These pulses also include frequency stepped signals known as frequency shift keying (FSK). These pulses have good range resolution, but little to no Doppler resolution. These pulses also have high gain against reverberation and are therefore good for detecting low Doppler targets in reverberation.

Impulsive sources, such as air guns, explosives, and sparkers, emit broadband signals because of their short duration. These signals have to be processed incoherently and they have good range resolution and no Doppler resolution. The primary advantage of these sources is their low cost.

Pseudo random pulses (PRN) attempt to exploit all the good qualities of CW and CP by sending a known broadband signal, making good detection, range, and Doppler resolution possible. So why are not all sonars using PRN? First, it is technically difficult to get a transducer that works well in this mode. Second, processing in this mode is difficult because a replica correlator with many range and Doppler bins is needed.

To design and evaluate a sonar system properly, one must consider the sensitivity of performance to system parameters as derived in preceding chapters:

1. Increased source level (Lp or SL) results in:
   a. Increases received signal in direct proportion.
   b. Increases reverberation in direct proportion.

c. Increase in performance only if noise limited as signal to reverberation is independent of source level.
d. The limits of source level are set by cavitation or the transducer power handling or a system's total power capabilities.

2. Increased frequency results in:
   a. Complex environmental effects:
      i. Transmission loss generally increases or degrades.
      ii. Ambient and self noise decreases or improves.
      iii. Surface loss, bottom loss, and backscattering generally increase or degrade.
   b. Target strength generally increases, but may have a complex shape.
   c. If the aperture remains constant and enough hydrophones are provided, then
      i. Source level increases.
      ii. Array gain increases.
      iii. Horizontal beam widths decrease, thereby improving reverberation and tracking accuracy.
      iv. Vertical beam widths decrease, thereby improving reverberation.
   d. For CW signals, improved Doppler resolution, thus giving better performance against nonzero Doppler targets and better target motion analysis (TMA) capability because signal Doppler is proportional to frequency.
   e. The impact on signal processing computations can increase very rapidly with frequency because of the increased sampling rates required (Nyquist) and an increased number of beams formed (see Equation 19.5).
3. Increased pulse length results in:
   a. Recognition differential (Nrd or DT) improvement. However, the net effect of increases in reverberation and processing gain are generally small or a net loss.
   b. The search strategy being impacted negatively by a loss of close-in coverage due to the receive array being used to transmit or the transmitted pulse arriving at the same time as the target signal (called the direct blast). An increase in pulse length may also approach the maximum duty cycle limits of the system; i.e., the system can only operate for 10 % of the time without failing or for so many seconds continuously.
   c. For CW signals:
      i. Improved Doppler resolution thus giving better performance against nonzero Doppler targets and improved TMA capability because the signal Doppler is proportional to frequency and the pulse frequency spread is determined by pulse length, environmental spreading, and source and receiver motion. Until the environmental and motion limits are reached, longer pulses improve performance (see Figure 13.19 and the associated discussion).
      ii. Worse range resolution, which is proportional to pulse length.
      iii. Increased reverberation, which is proportional to pulse length.
4. Increased pulse bandwidth results in:
   a. For CW, the bandwidth is determined by the pulse length.
   b. For CP or PRN pulses:
      i. The noise in the band will increase.
      ii. The recognition differential will improve or decrease. In theory, the recognition differential will decrease to just cancel the noise increase.

iii. Range resolution will improve, thereby improving TMA and improving signal to reverberation.
iv. Energy splitting losses (ESL, see Chapter 13) will increase and eventually limit the gain to be achieved.

5. Increased transmit and receive apertures results in:
   a. Increases in array gain against noise and reverberation.
   b. Increased source level.
   c. Improved tracking accuracy.
   d. At some point, signal decorrelation will limit maximum gain.

19.5 Monostatic, Bistatic, and Multistatic Active Sonars

Most modern, monostatic active sonars in search mode use omni, or wide-sector transmissions, with a highly directional receiver. These receivers have reduced ambient noise, reduced reverberation, and have the ability to provide accurate bearings for tracking. However, there are several disadvantages to using a directional receiver: the requirement for many hydrophones and a multibeam beamformer, as well as very large physical dimensions as compared to a wavelength. Transmitters are usually not highly directional in search mode because this would involve sequential transmissions in many directions before listening for returns, thus resulting in blind spots in coverage. This general difference in the requirements for transmitters and receivers has led to an increase in the use of bistatic active sonars for modern ASW applications. With a bistatic sonar, the transmitter can use a small number of highly tuned efficient projectors and the receiver can use a larger number of untuned, less expensive hydrophones. This will allow the receiver to have good passive performance, as well as serve as the active receiver.

The case for multistatic sonar is more complex. A multistatic sonar is a system where the source and receiver are separated by a considerable distance and are probably not attached to a single platform except by radio link. The question arises, when is a mutistatic system more cost effective than a monostaic or bistatic system? There are two cases to be considered: noise-limited and reverberation-limited conditions. The active sonar equations for a multistatic sonar require multiple solutions of the bistatic sonar equation:

$$\mathrm{XSN} = L_\mathrm{p} + \mathrm{NTS} - N\mathrm{w}_1 - N\mathrm{w}_2 - L_\mathrm{e} - \mathrm{Nrd}_\mathrm{n}$$

$$\begin{aligned}\mathrm{XSR} &= (L_\mathrm{p} + \mathrm{NTS} - N\mathrm{w}_1 - N\mathrm{w}_2) \\ &\quad - \left[L_\mathrm{p} + 10\log\left(\frac{c\tau}{2}\right) + 10\log(\Delta\theta R + m_\mathrm{s} - N\mathrm{w}_3 - N\mathrm{w}_4)\right] - \mathrm{Nrd}_\mathrm{r} \\ &= \mathrm{NTS} - N\mathrm{w}_1 - N\mathrm{w}_2 - 10\log\left(\frac{c\tau}{2}\right) - 10\log(\Delta\theta R) - m_\mathrm{s} + N\mathrm{w}_3 + N\mathrm{w}_4 + \mathrm{Nrd}_\mathrm{r}\end{aligned} \tag{19.8}$$

where

XSN = signal excess with respect to noise
XSR = signal excess with respect to reverberation
L_p = source level

NTS = target strength
Nw_1 = propagation loss from transmitter to target
Nw_2 = propagation loss from target to receiver
Nw_3 = propagation loss from transmitter to reverberation source
Nw_4 = propagation loss from reverberation source to receiver
L_e = level of noise out of beamformer (ambient and flow)
Nrd_n = recognition differential of system with respect to noise
Nrd_r = recognition differential of system with respect to reverberation, given as a function of Doppler
c = speed of sound in water
τ = equivalent pulse length
$\Delta\theta$ = horizontal beam width
R_1 = range from transmitter to target
R_2 = range from reverberation to receiver
m_s = surface or bottom scattering strength

For a monostatic active sonar, $Nw_1 = Nw_2$ and $Nw_3 = Nw_4$, if it is assumed that the propagation losses can be represented in the form:

$$\mathrm{Nw} = k\log(R) + \alpha R \tag{19.9}$$

where

k = 20 for r-squared spreading or 10 for cylindrical spreading
α = attenuation coefficient (dB/kyd)

For low to medium frequencies, where α is small and it is assumed that the range to the source of reverberation is the same as the range to the target, the equations become (for bistatic)

$$\begin{aligned}\mathrm{XSN} &= L_p + \mathrm{NTS} - k\log(R_1) - k\log(R_2) - L_e - \mathrm{Nrd_n}\\ \mathrm{XSR} &= \mathrm{NTS} - 10\log\left(\frac{c\tau}{2}\right) - 10\log(\Delta\theta R_2) - m_s - \mathrm{Nrd_r}\end{aligned} \tag{19.10}$$

For monostatic active sonar where $R_1 = R_2 = R$:

$$\begin{aligned}\mathrm{XSN} &= L_p + \mathrm{NTS} - 2k\log(R) - L_e - \mathrm{Nrd_n}\\ \mathrm{XSR} &= \mathrm{NTS} - 10\log\left(\frac{c\tau}{2}\right) - 10\log(\Delta\theta R) - m_s - \mathrm{Nrd_r}\end{aligned} \tag{19.11}$$

Assuming a noise-limited operation and setting XSN to zero results in

$$\mathrm{FOM} = k\log(R) = L_p + \mathrm{NTS} - L_e - \mathrm{Nrd_n} \tag{19.12}$$

or

$$R^2 = \exp 10\left(\frac{\text{FOM}}{k}\right) \quad \text{for monostatic}$$
$$R_1 R_2 = \exp 10\left(\frac{\text{FOM}}{k}\right) \quad \text{for bistatic} \tag{19.13}$$

If the figure of merits (FOMs) are the same, then $R^2 = R_1 R_2$. A multistatic system could gain coverage by making R_2 small, resulting in a large R_1 or by having several receivers (or one receiver and lots of transmitters) to prevent holes in coverage. For example, if $R = 10$ kyds, then R_2 could be 5 kyds, resulting in a 20 kyds R_1. To get uniform area coverage on a 20 kyds diameter circle, a multistatic system needs 1 transmitter and about 16 receivers as compared to 4 monostatic systems to cover the same area. If buying 1 transmitter and 16 receivers, or vice versa, is cheaper than 4 monostatic systems, then this is a good tradeoff.

The answer is different for the reverberation limited case, where the criterion becomes

$$10\log(\Delta\theta R_2) = 10\log(\Delta\theta R) \tag{19.14}$$

If R_2 is allowed to be smaller than R, the beam width of the receiver can be wider, which is less expensive per receiver.

19.6 Ambiguity Functions

The frequency and time resolution performance of a signal can be expressed as the ambiguity function:

$$A(\delta f, \delta\tau) = \left| \int_{-\infty}^{\infty} s(f, \tau)\overline{s(f + \delta f, \tau + \delta\tau)}\,\mathrm{d}\tau \right| \tag{19.15}$$

where

$s(f,\tau)$ = time domain signal at frequency, f

$s(f + \delta f, \tau + \delta\tau)$ = frequency shifted, time delayed version of the signal

The bar denotes a complex conjugate. Plots for examples of a CW and an FM are shown in Figures 18.9 through 18.12.

These functions show the resolution in frequency and range for any signal designed. Table 19.1 shows the time, range, and frequency resolution for a variety of common pulses.

19.7 Mine Hunting and Bottom Survey Sonars

The frequencies used for these systems are usually high, ranging from 20 kHz to 500 kHz; higher frequencies yield better resolution but smaller ranges. Some mine hunting and bottom survey systems are simply conventional sonars as previously described. However, two special processing techniques are also valuable for this application, side-scan processing and synthetic

Table 19.1 Time, range, frequency, and Doppler resolution for various pulse types

Pulse type	Time resolution (seconds)	Range resolution (kilometers)	Frequency resolution (Hertz)	Doppler resolution (knots)
CW	$0.66T$	$1/T$	$0.88/T$	$1275/(fT)$
CP LPM (linear period)	0.88/BW	0.66/BW	0.6 BW	$900\ \mathrm{BW}/f$
CP LFM (linear frequency)	088/BW	0.66/BW	$3.5f/(\mathrm{BW} \times T)$	$5200/(\mathrm{BW} \times T)$
PRN	0.88/BW	0.66/BW	$0.88/T$	$1275/(fT)$

F = base frequency (Hz), BW = bandwidth (Hz), and T = pulse duration (s)

aperture processing. These two methods of processing are often erroneously discussed as though they were the same. In reality, the processing is very different.

Side-scan sonar processing is used for mapping the sea bed to create an image of the sea floor. It is particularly valuable for identifying objects on the bottom of the ocean, like mines, pipelines, and sunken ships. Side-scan sonars send fan-shaped pulses (similar to the broadside beam of a towed array) toward the sea floor across a wide vertical angle perpendicular to the path of the sensor through the water. The intensity of the acoustic reflections from the sea floor is recorded in a series of cross-track slices stitched together along the direction of motion. These slices form an image of the sea bottom within the swath or coverage width of the beam.

Synthetic aperture sonar (SAS) is analogous to synthetic aperture radar. Synthetic aperture processing combines a series of returns to form an image of much higher resolution than conventional processing. The principle here is to move the sonar along a line and illuminate the same spot on the sea floor with multiple transmissions. This produces a synthetic array size equal to the distance travelled. By coherently processing the data from all the returns, an image is produced whose resolution is set by the track length rather than by the aperture length. This type of processing can improve the resolution of an image by an order of magnitude when compared to conventional sonars.

These types of sonars are used extensively in ocean oil and gas production to inspect underwater platforms, well heads, and pipelines. Not addressed here is oil exploration, where sound sources are used to profile the Earth's structure to locate formations that could contain oil or gas.

19.8 Echo Sounding and Fishing Sonars

Fathometers, echo, or down or depth sounders are some of the oldest uses of sonar. Here, an acoustic pulse is transmitted downward and the sonar listens for the reflected pulse off the bottom. The time delay between transmission and reflection allows for the calculation of the bottom depth. This depth can be used for ocean survey, ship safety from grounding, locating shallow banks preferred by certain fish, or even as a navigation aid.

Fishing is an important global industry, but with world tonnages in decline, there is an increased demand for high-tech sonars to support a variety of fishing needs. Early fish-finders were down sounders, which were capable of detecting large schools of fish, as well as the bottom. The echo returned from a fish is largely due to the air bladder they have for stability (see Section 9.9). Modern fish-finder sonars rival naval ASW sonars in their capabilities. Fishing sonars are being used to maintain the depth of trawl nets and, in some cases, count fish as they enter a net, thus allowing the ship to retrieve the net when it is full.

19.9 Navigation

If accurate bottom maps are available and significant bottom features are present, sonar can be used to establish a position. Active pinging can also be used to determine one's speed over the bottom by tracking bottom features or by measuring the Doppler shift of returns.

19.10 Vehicle Location and At-Sea Rescue

Acoustic beacons are routinely installed on submarines to be activated in the event of an emergency. This is also true of other objects like aircraft, space vehicles that are expected to land in the sea, and possibly drugs that are dropped at sea. These beacons either broadcast continuously or may be command activated, allowing searchers to pinpoint the object. During World War II, rescue rafts on military aircraft carried explosive charges that detonated in the deep sound channel. This sound fixing and ranging (SOFAR) system used multiple receiving stations to determine the location of a downed crew, sometimes from thousands of miles away.

19.11 Intercept Receivers

Intercept receivers are primarily used by the military to detect, classify, and localize active sonar transmissions. These sonars are passive receivers that typically listen over a very wide frequency band and alert the crew when a transmitter is detected. The key design issue with this type of sonar is the large number of transients in the ocean environment due to natural and man-made sources, which are sources of false contacts.

19.12 Communications

In theory, one could speak into a microphone or tap a morse code key and the sound would be transmitted directly into the water, just as amateur radio operators do in the air. In practice, such an approach only works at very short range because of the multipath and fading caused by the environment, which effectively limits the data rate possible without using more sophisticated processing.

The techniques used to overcome these limitations are the same as those used in radio, namely beamforming, diversity, and error correction. Beamforming can limit the effect of the multipath. Diversity, or sending the same information at multiple frequencies or multiple times, limits the effects of fading. Finally, error correction processes are used to recover the message. Another promising approach is "inverse" processing, where a test signal is used to measure the channel characteristics. The results of this method are used to construct a message that compensates for the multipath, assuming a reasonable time constant for the variability.

19.13 Marine Mammals and Active Sonar

Active sonar transmissions may harm people and marine animals if the intensity and duration are sufficiently high and long enough, although the mechanisms and sensitivities are not well understood. The latter concern has been greatly complicated by years of ongoing law suits, with lawyers and judges trying to make decisions in the face of widely varying responses by marine mammals and with a scientific community that is not yet near agreement on the subject. However, it is clear that the number of stranding events of marine mammals associated with the military use of active sonar is tiny compared to the estimated 300 000 marine mammals killed each year by commercial fishing. One outgrowth of this is that the U.S. Navy has published previously classified data on sonars as part of their Environmental Impact Statements. For example,

> The Navy's most powerful surface ship sonar is the SQS-53, which has the nominal source level of 235 dB re 1 squared micropascal-second (μPa2-s) at 1.09 yards (or 1 meter [m]) . . . average source levels of pings varied from a nominal 235 dB SPL (AN/SQS-53C) to 223 dB SPL (AN/SQS-56). The center frequency of pings was 3.3 kHz and 6.8 to 8.2 kHz, respectively [2].

References

[1] Nielsen, R. O., *Sonar Signal Processing*, Boston, MA: Artech House, 1991.

[2] Mariana Islands Range Complex Environmental Impact Statement/Overseas Environmental Impact Statement, Vols 1 and 2, Draft, Mariana Islands Range Complex EIS/OEIS, Naval Facilities Engineering Command, Pacific, January 2009.

Appendix A

Fourier Transforms

The continuous Fourier transform is a specific mathematical form of Fourier analysis. In one dimension, it takes a function of time or position and creates a corresponding function in the frequency domain. The term Fourier transform is used for the frequency domain representation of a function or for the process that transforms one function into the other.

A.1 Definitions

There are several commonly used conventions for defining the Fourier transform. The one most commonly used in signal processing or communications is

$$F(f) = \int_{-\infty}^{\infty} f(t)\mathrm{e}^{-\mathrm{i}2\pi f t}\,\mathrm{d}t \tag{A.1}$$

where

$t =$ time (s) (independent variable)
$f =$ frequency (Hz) (transform variable)

The inverse transform is given by

$$f(t) = \int_{-\infty}^{\infty} F(f)\mathrm{e}^{-\mathrm{i}2\pi f t}\,\mathrm{d}t \tag{A.2}$$

Other frequently encountered notations for $F(f)$ are $F(t)$ or $f(t)$.

In mathematics, the Fourier transform is generally written using angular frequency, $\omega = 2\pi f$, instead of f. When this is done, the units are time in seconds and angular frequency

Underwater Acoustics: Analysis, Design and Performance of Sonar Richard P. Hodges

Table A.1 Summary of common Fourier transform forms

Ordinary frequency (Hz)	Unitary	$F_3(f) = \int_{-\infty}^{\infty} f(t)\mathrm{e}^{-\mathrm{i}2\pi ft}\,\mathrm{d}t = \sqrt{2\pi}\,F_1(2\pi f) = F_2(2\pi f)$ $f(t) = \frac{1}{2\pi}\int_{-\infty}^{\infty} F_3(f)e^{\mathrm{i}2\pi ft}\,\mathrm{d}f$
Angular frequency (radians/s)	Unitary	$F_1(\omega) = \frac{1}{\sqrt{2\pi}}\int_{-\infty}^{\infty} f(t)\mathrm{e}^{-\mathrm{i}\omega t}\,\mathrm{d}t = \frac{1}{\sqrt{2\pi}}F_2(\omega) = \frac{1}{\sqrt{2\pi}}F_3\left(\frac{\omega}{2\pi}\right)$ $f(t) = \frac{1}{\sqrt{2\pi}}\int_{-\infty}^{\infty} F_1(\omega)\mathrm{e}^{\mathrm{i}\omega t}\,\mathrm{d}\omega$
	Nonunitary	$F_2(\omega) = \int_{-\infty}^{\infty} f(t)\mathrm{e}^{-\mathrm{i}\omega t}\,\mathrm{d}t = \sqrt{2\pi}\,F_1(\omega) = F_3\left(\frac{\omega}{2\pi}\right)$ $f(t) = \frac{1}{2\pi}\int_{-\infty}^{\infty} F_2(\omega)\mathrm{e}^{\mathrm{i}\omega t}\,\mathrm{d}\omega$

in radians per second. This convention yields

$$F_2(\omega) = \int_{-\infty}^{\infty} f(t)\mathrm{e}^{-\mathrm{i}\omega t}\,\mathrm{d}t$$
$$f(t) = \frac{1}{2\pi}\int_{-\infty}^{\infty} F_2(\omega)^{\mathrm{i}\omega t}\,\mathrm{d}f \tag{A.3}$$

or the 2π can be distributed evenly in a unity transform:

$$F_1(\omega) = \frac{1}{\sqrt{2\pi}}\int_{-\infty}^{\infty} f(t)\mathrm{e}^{-\mathrm{i}\omega t}\,\mathrm{d}t$$
$$f(t) = \frac{1}{\sqrt{2\pi}}\int_{-\infty}^{\infty} F_1(\omega)\mathrm{e}^{\mathrm{i}\omega t}\,\mathrm{d}f \tag{A.4}$$

Table A.1 shows the common mathematical conventions in use for writing Fourier transforms and their inverses.

A.2 Parseval's Theorem and Plancherel's Theorem

Although both names and equations are often mixed up, these two theorems give equivalent statements. Parseval's theorem is

$$\int f(t)\bar{g}(t)\,\mathrm{d}t = \int F(\omega)\bar{G}(\omega)\,\mathrm{d}\omega \tag{A.5}$$

where the barred functions are complex conjugates.

Plancherel's theorem is

$$\int |f(t)|^2\,\mathrm{d}t = \int |F(\omega)|^2\,\mathrm{d}\omega \tag{A.6}$$

A.3 Properties of Fourier Transforms

Table A.2 Properties of Fourier transforms

Property	Function	Transform
Linearity	$af(t) + bg(t)$	$aF(f) + bG(f)$
Multiplication	$f(t)g(t)$	$F(f) \otimes G(f)$
Convolution	$f(t) \otimes g(t)$	$F(f)G(f)$
Modulation	$f(t)\mathrm{e}^{\mathrm{i}\omega_0 t}$	$F(\omega - \omega_0)$
Scaling	$f(at)$	$\frac{1}{\lvert a\rvert}F\left(\frac{\omega}{a}\right)$
Time shift	$f(t - t_0)$	$\mathrm{e}^{-\mathrm{i}\omega t_0}F(\omega)$
Time reversal	$f(-t)$	$F(-\omega)$
Conjugation	$\bar{f}(t)$	$\bar{F}(-\omega)$

A.4 Localization or Uncertainty Property

In general, the more concentrated $f(t)$ is, the more spread out $F(f)$ is, and vice versa. The scaling property of the Fourier transform may be interpreted as: if a function is "squeezed" in t, its Fourier transform "stretches out" in f. It is not possible to arbitrarily concentrate both a function and its Fourier transform. The tradeoff between the compaction of a function and its Fourier transform can be formalized as an uncertainty principle. For example, for a CW pulse the product of pulse length and the bandwidth is a constant; similarly, for an FM pulse the product of range resolution and the bandwidth is a constant.

Appendix B

Analysis of Errors Associated with a Least Squares Methodology

To obtain a least squares fit to a set of observations, θ_i, from a normal population with a mean, μ, and a standard deviation, σ:

$$\theta = \beta t + \alpha \text{ (line of regression)} \tag{B.1}$$

To obtain a, b, and s^2 estimates of model parameters (α, β, and σ^2) from a set of m experimental data point pairs:

$$P(\theta_i) = \frac{1}{2\pi\sigma^2} \exp\left[\frac{-(\theta_i - \alpha - \beta t_i)^2}{2\sigma^2}\right] = K \exp\left[\frac{-(\theta_i - \alpha - \beta t_i)^2}{2\sigma^2}\right] \tag{B.2}$$

The probability, P, of all observed values occurring is

$$\begin{aligned} P &= P(\theta_1)P(\theta_2)\ldots P(\theta_m) = \prod P(\theta_i) \\ P &= K^m \exp\left\{\frac{-\sum\left[(\theta_i - \alpha - \beta t_i)^2\right]}{2\sigma^2}\right\} \end{aligned} \tag{B.3}$$

The maximum value occurs when α and β are chosen to minimize the sum of the squares in the exponent:

$$\begin{aligned} s &= \sum\left[\frac{-(\theta_i - \alpha - \beta t_i)^2}{2\sigma^2}\right] \\ \frac{\partial s}{\partial \alpha} &= 2\sum(\theta_i - \alpha - \beta t_i) = 0 \\ \frac{\partial s}{\partial \beta} &= 2\sum\left[t_i\,(\theta_i - \alpha - \beta t_i)\right] = 0 \end{aligned} \tag{B.4}$$

Underwater Acoustics: Analysis, Design and Performance of Sonar Richard P. Hodges

Replacing α and β with the estimated a and b gives

$$\begin{aligned} 2\sum(\theta_i - a - bt_i) &= 0 \\ 2\sum t_i(\theta_i - a - bt_i) &= 0 \end{aligned} \tag{B.5}$$

or

$$\begin{aligned} \sum\theta_i &= ma + b\sum t_i \\ \sum\theta_i t_i &= a\sum t_i + b\sum t_i^2 \end{aligned} \tag{B.6}$$

Therefore:

$$\begin{aligned} b &= \frac{m\sum t\theta - \sum t\sum\theta}{m\sum t^2 - \sum t\sum t} = \frac{\text{cov}(t,\theta)}{\text{var}(t)} \\ a &= \frac{\sum\theta t\sum t - \sum t^2\sum\theta}{-m\sum t^2 + \sum t\sum t} \end{aligned} \tag{B.7}$$

The variance about the regression line is estimated by

$$s^2 = \frac{\sum(\theta - a - bt)^2}{m-2} \tag{B.8}$$

where a and b have been estimated from the data, s^2 is based on $(m-2)$ degrees of freedom, and $(m-2)s^2/\sigma^2$ has an χ^2 distribution with $(m-2)$ degrees of freedom. Thus

$$\begin{aligned} \text{ave}(b) &= \frac{\sum(t - \text{ave}_t)[\alpha + \beta(t - \text{ave}_t)]}{\sum(t - \text{ave}_t)^2} = \beta \text{ because } \sum(t - \text{ave}_t) = 0 \\ \text{var}(b) &= \frac{\sum(t - \text{ave}_t)^2\sigma^2(\theta)}{\sum(t - \text{ave}_t)^2} = \frac{\sigma^2(\theta)}{\sum(t - \text{ave}_t)^2} \end{aligned} \tag{B.9}$$

These last two results are independent of the Gaussian assumption for θ. Note that $(b-\beta)/2\sqrt{\sum(t - \text{ave}_t)^2}$ has a Student t distribution with $(m-2)$ degrees of freedom. This may be used either as a test of significance to see if β is likely to have some preconceived value or to establish a confidence interval for β. Thus

$$\begin{aligned} \text{ave}(a) &= \alpha \\ \text{var}(a) &= \sigma^2(\theta)\frac{\sum t^2}{m\sum(t - \text{ave}_t)^2} \end{aligned} \tag{B.10}$$

For example, assume that the time steps are uniform:

$$\begin{aligned} t_i &= i\Delta t \\ \text{ave}_t &= \frac{m+1}{2}\Delta t \end{aligned} \tag{B.11}$$

Then

$$\begin{aligned} \left[\sum (t-\text{ave}_t)\right]^2 &= \sum t^2 - 2\text{ave}_t\, t + \text{ave}_i^2 \\ \left[\sum (t-\text{ave}_t)\right]^2 &= \sum i^2\Delta t^2 - 2\text{ave}_t\, i\Delta t + \text{ave}_i^2 \end{aligned} \tag{B.12}$$

Since

$$\begin{aligned} \sum i &= m\frac{m+1}{2} \\ \sum i^2 &= \frac{2m^3+3m^2+m}{6} \end{aligned} \tag{B.13}$$

then

$$\begin{aligned} \left[\sum (t-\text{ave}_t)\right]^2 &= \Delta t^2\left(\frac{2m^3+3m^2+m}{6}\right) - 2\text{ave}_t\Delta t m\frac{m+1}{2} + \text{ave}_t^2 \\ \left[\sum (t-\text{ave}_t)\right]^2 &= \Delta t^2\left(\frac{2m^3+3m^2+m}{6}\right) - 2\frac{(m+1)}{2}\Delta t\Delta t m\frac{m+1}{2} \\ &\quad + m\left[\frac{(m+1)}{2}\Delta t\right]^2 \\ \left[\sum (t-\text{ave}_t)\right]^2 &= \frac{m^3-m}{12} \end{aligned} \tag{B.14}$$

Under this assumption the variance of b can be written as

$$\text{var}(b) = \sigma^2(b) = \frac{12\sigma^2(\theta)}{\Delta t^2\left(m^3-m\right)} \tag{B.15}$$

and

$$\text{var}(a) = \sigma^2(a) = \sigma^2(\theta)\frac{2}{m}\left(\frac{2m-1}{m+1}\right) \tag{B.16}$$

In Chapter 18, Tracking, Target Motion Analysis, and Localization, these expressions will be used to estimate the improvement of TMA with multiple observations.

The above example assumes that the time steps are uniform and that the variance of the bearings is constant. If this constraint is relaxed, then the variance of b becomes [1]

$$\mathrm{var}(b) = \sigma^2(b) = \frac{\sum_i \frac{1}{[\sigma_i(\theta)]^2}}{\left\{\sum_i \frac{1}{[\sigma_i(\theta)]^2} \sum_i \frac{(\Delta t_{i-1})^2}{[\sigma_i(\theta)]^2} - \left[\sum_i \frac{\Delta t_{i-1}}{[\sigma_i(\theta)]^2}\right]^2\right\}} \tag{B.17}$$

Reference

[1] Hodges, R., *SIM II User Reference Manual*, Appendix A, December 2001.

Index

Printed and bound by CPI Group (UK) Ltd, Croydon, CR0 4YY
16/09/2026
14963385-0004